Metal Oxide–Based Carbon Nanocomposites for Environmental Remediation and Safety

This book focuses on nanotechnology for the preparation of metal oxide–based carbon nanocomposite materials for environmental remediation. It analyses the use of nanomaterials for water, soil, and air solutions, emphasizing the environmental risks of pollution. It further explores how magnetic and activated carbon nanomaterials are being used for a sustainable environmental protection of water and soil, and detection of harmful gases. The status and major challenges of using carbon-based nanomaterials on a large scale are explained, supported by relevant case studies.

Features:

- Exhaustively covers nanotechnology, metal oxide–carbon nanocomposites and their application in soil, water, and air treatments
- Explores pollutants nano-sensing and their remediation towards environmental safety
- Includes economics analysis and environmental aspects of metal oxide materials
- Describes why properties of oxide carbon–based nanomaterials are useful for environmental applications
- Discusses current case studies of remediation technologies

This book is aimed at graduate students and researchers in nanotechnology, environmental technology, and remediation.

Metal Oxide–Based Carbon Nanocomposites for Environmental Remediation and Safety

Edited by
Rayees Ahmad Zargar
and Saleem Ahmad Yatoo

CRC Press
Taylor & Francis Group
Boca Raton London New York

CRC Press is an imprint of the
Taylor & Francis Group, an **informa** business

Designed cover image: © Shutterstock

First edition published 2024
by CRC Press
6000 Broken Sound Parkway NW, Suite 300, Boca Raton, FL 33487–2742

and by CRC Press
4 Park Square, Milton Park, Abingdon, Oxon, OX14 4RN

CRC Press is an imprint of Taylor & Francis Group, LLC

ISBN: 978-1-032-34710-3 (hbk)
ISBN: 978-1-032-34711-0 (pbk)
ISBN: 978-1-003-32346-4 (ebk)

DOI: 10.1201/9781003323464

Typeset in Times
by Apex CoVantage, LLC

Contents

About the Editors

Dr Rayees Ahmad Zargar is working as an assistant professor at Baba Ghulam Shah Badshah University, Rajouri (J&K) in the Department of Physics. Dr Zargar received his PhD on iron-based superconductors in the Physics Department from Jamia Millia Islamia in March 2016. Later, he joined the National Physical Laboratory, New Delhi, as a postdoc researcher. He has wide expertise in the field of superconductors, manganites, and optical properties of semiconductors. During the last years, he has concentrated on diluted magnetic semiconductors, where he has used spin-tronics, sensors, and optoelectronic devices. He is an author and co-author of more than 60 papers with good impact factors, has reviewed many papers and has authored three books and six chapters. He is an editorial board member of international journals and a regular referee of the *Journal of Physics and Chemistry of Solids*, the *Journal of Electronic Materials and Physica C*. He has received a good number of citations according to the Web of Science. He has participated in a number of national and international conferences, seminars, and workshops.

Dr Saleem Ahmad Yatoo has an MSc in environmental science and sustainable development. He has pursued his MPhil and PhD in environmental science and sustainable development from Central University of Gujarat. He is working as a lecturer in the higher education department of Jammu and Kashmir. He has teaching experience at the undergraduate and graduate levels. Moreover, he has published a good number of papers with publishers of international repute like Springer. He has received a good number of citations globally according to Web of Science and other indexes. He has participated in a number of national and international conferences, seminars, workshops, and training programs, as well as outreach courses from Indian Institute of Remote Sensing, Dehradun, India. In addition to his other accomplishments, he has qualified National Eligibility Test and Jammu and Kashmir State Eligibility Test at the national and state levels, respectively. His areas of expertise are remote sensing and GIS, land surface temperature, land use/land cover change, climate change, ecology, biodiversity, environmental nanotechnology, water pollution analysis, wetland diversity, and limnology, among others.

Contributors

Afsar Ali
Department of Chemistry
Indian Institute of Technology Delhi
Hauz Khas, New Delhi, India-110016

Arora, Manju
CSIR-National Physical Laboratory
Dr. K.S. Krishnan Marg
New Delhi, India-110012

Balwinder Kaur
Department of Physics,
Govt. Degree College, R.S. Pura
Jammu, India-181102

Bhat, S. A.
Department of Chemistry
Government Degree College
Women Anantnag
Jammu and Kashmir, 192101

Bhat, Azhar Jameel
Centre for Biodiversity Studies
Baba Ghulabh Shah Badshah
University
Rajouri (J&K), India-185234

Dawane, Vinars
ANCHROM, HPTLC Labs
Mulund, Mumbai
Maharashtra, India-400081
School of Environment and Sustainable
Development
Central University of Gujarat
Gandhinagar, India-382030
Department of Microbiology and
Biotechnology
SardarVallabh Bhai Patel College
Mandleshwar
Madhya Pradesh, India-451221

Dhiman, Tarun Kumar
University School of Basic & Applied
Science
Guru Gobind Singh Indraprastha
University Dwarka
New Delhi, India-110078

Golia, Santosh Singh
CSIR-National Physical Laboratory
Dr. K.S. Krishnan Marg
New Delhi, India-110012

Khan, Saleem
Department of Nanosciences and
Materials
Central University of Jammu
J&K (UT), India-181143

Kujur, Vidya Spriha
Centre for Nanotechnology
Central University of Jharkhand
Brambe, Ranchi, India-835205

Kumar, Bijay Behera
AEBN Division
ICAR-Central Inland Fisheries
Research Institute
West Bengal, India-700120

Kumar, Jayendra Himanshu
School of Life Science, Department
of Biotechnology
Mahatma Gandhi Central
University
Motihari, Bihar, India-8454116

Kumar, Manoj
Department of Chemistry
Maharishi Markendeshwar University
Sadopur Haryana, India-134007

Kumar, Manoj Srinivasan
Department of Biochemistry and
  Biotechnology, Faculty of Science
Annamalai University
Tamilnadu, India-608002

Kumar, Nirantak
Department of Chemistry
IEC University, Solan
Himachal Pradesh, India-174103

Kumar, Pankaj
Integrated Regional Office
Ministry of Environment
Forest & Climate Change (MoEFCC)
Govt. of India Saifabad
Hyderabad, Telangana, India-500004
Department of Environmental Science
Parul Institute of Applied Sciences
Parul University
Vadodara, Gujarat, India-391760

Kumar, Rahul
School of Physical Sciences
Jawaharlal Nehru University
New Delhi, India-110067

Kumar, Sunil Patidar
Department of Chemistry
Shri Neelkantheshwar Govt. Post
  Graduate College
Khandawa, Madhya Pradesh,
  India-450001

Meena, Badri Vishal
Department of Chemical Engineering
Indian Institute of Technology
  Gandhinagar
Gandhinagar, Gujarat, India-382424

Mir, Ab Qayoom
Department of Chemistry
Indian Institute of Technology
Gandhinagar, Gujarat, India-382424

Misra, Vaishali
Department of Nanosciences and
  Materials
Central University of Jammu, Rahya
  Suchani
J&K (UT), India-181143

Munoz, Rodrigo A. Abarza
Federal University of Uberlândia
Institute of Chemistry—NuPE
Santa Mônica, Brazil-38400902

Najar, G. N.
Department of Environmental Sciences
Government Degree College, Kreeri
Jammu and Kashmir, India-193108

Nalini, Namasivayam
Department of Biochemistry and
  Biotechnology, Faculty of Science
Annamalai University
Tamilnadu, India-608002

Pathak, Bhawna
School of Environment and Sustainable
  Development
Central University of Gujarat
Gandhinagar, India-382029

Piplode, Satish
Department of Chemistry
SBS Govt. P.G. College Pipariya,
  Hoshangabad
Madhya Pradesh, India-461775

Poddar, Mrinal
Amity Institute of Nanotechnology
Amity University
Noida, India-201313

Prakash, Man Mohan
Dr B R Ambedkar University of Social
  Sciences
Madhya Pradesh, India-453441

Pratima, Bichandarkoil Jayaram
Department of Biochemistry and
   Biotechnology, Faculty of Science
Annamalai University
Tamil Nadu, India-608002

Premnath, Briska Jifrina
Department of Biochemistry and
   Biotechnology, Faculty of Science
Annamalai University
Tamil Nadu, India-608002

Rani, Anita
Department of Chemistry
IEC University, Solan
Himachal Pradesh, India-174103

Rather, Sajad Ahmad
Department of Economics
Baba Ghulam Shah Badshah University
Rajouri (J&K), India-185234

Ravichandiran, Ragunath
Department of Biochemistry and
   Biotechnology, Faculty of Science
Annamalai University
Tamil Nadu, India-608002

Sahu, Paulami
School of Environment and Sustainable
   Development
Central University of Gujarat
Gandhinagar, Gujarat, India-382030

Sambyal, Sunil
Department of Physics, School of
   Applied Sciences
Shri. Venkateshwara University
Gajraula, Uttar Pradesh, India-244236

Sen, Sonu
Department of Physics
Shri Neelkantheshwar Government Post
   Graduate College
Khandwa, India-450001

Shahabuddin, Farha
Department of Biochemistry
Aligarh Muslim University
Aligarh, Uttar Pradesh, India-202001

Shameem, S. A.
Division of Environmental Sciences
Faculty of Horticulture (FOH)
SKUAST-K
Srinagar, India-190025

Sharma, Archana
Department of Physics
University of Jammu
J&K (UT), India-180006

Singh, Ajay
Department of Physics
GGM Science College (Constituent
   College of Cluster University of
   Jammu)
Jammu, UT-J&K, India-180002

Singh, Avinash Kumar
School of Physical Sciences
Jawaharlal Nehru University
New Delhi, India-11067

Singh, Km Madhuri
School of Environment and Sustainable
   Development
Central University of Gujarat
Gandhinagar, Gujarat, India-382030

Singh, Vishal
Department of Nano Sciences and
   Materials
Central University of Jammu
Jammu, India-181143

Sudan, Sapna
Department of Chemistry
BITS-Pilani KK Birla
   Goa Campus
Goa, India-403726

Swain, Krishna Kumari
Department of Applied Mechanics
IIT Madras, Chennai
Tamilnadu, India-600036

Yatoo, Saleem Ahmad
School of Environment and Sustainable
    Development (SESD)
Central University of Gujarat
Gandhinagar, Gujarat, India-382030

Zargar, Rayees Ahmad
Department of Physics
Baba Ghulam Shah Badshah University
Rajouri (J&K), India-185234

Preface

Environmental nanotechnology is considered to have a great key role in shaping current environmental engineering and science practices. This book, *Metal Oxide–Based Carbon Nanocomposites for Environmental Remediation and Safety*, covers the advanced materials, devices, and system development for use in environmental protection and safety. The development of nanomaterials attaining importance because of the increased environmental challenges due to the impact of modern industrial activities globally. Industrial activity involves the production and use of various toxic organic and inorganic chemicals/by-products that pollute water bodies indirectly and affect aquatic and human life rapidly. Thus, there is a great need to protect the environment through the development of new technologies and by enacting awareness drives for environmental sustainability. The 12 chapters in this volume, all written by subject-matter experts, demonstrate the claim that metal oxide–based nanomaterials have the potential to be the future of industry.

This volume summarizes cutting-edge research on nanomaterial utilization for environmental challenges and protection.

At present, pollution has become a major problem for our environment globally. Many pollutants produced from different sources decompose naturally, but a majority of the anthropogenic pollutants are non-biodegradable and need some treatment. Currently, many technologies have been developed to remediate such pollutants, and nanotechnology is one of the recent technologies that can provide a way to deal with such problems. Heavy metals, toxic gases and dyes are major anthropogenic threats that result in environmental degradation and need to be mitigated using such technology. The removal of these pollutants through environmentally friendly and efficient methods is necessary and thus important to saving our environment. In the recent past, nanomaterials have drawn a lot of attention of researchers, almost from all the existing and new applications in the fields of electronics, health care, energy, optoelectronics, defense, environmental remediation, and others. In addition, because of their special properties, such as increased surface–volume ratio, magnetic nanoparticles are chemically reactive and therefore are good candidates for many environmental remediation applications.

The book *Metal Oxide–Based Carbon Nanocomposites for Environmental Remediation and Safety* highlights the use of nanomaterials for remediating environmental pollution and safety. This book offers both an important volume for academic researchers and a resource for those in industry exploring the applications of nanoparticles in environmental protection and more. The book is divided into three main sections to thoroughly analyze the use of nanomaterials for water, soil and air solutions, with respect to emphasizing environmental risks of pollution. Therefore, the main goal of this book is to collect knowledge and provide information to the public, in general, and researchers, in particular, about how the environmental pollution problem being faced by the world can be overcome/minimized and how to initiate developing low-cost tools for environmental pollution remediation. This book in short explores as well as conveys the concepts to understand the metal oxide–based carbon nanocomposites for environmental remediation and safety.

Key Features:

- Covers the basics of advanced nano-based materials, their synthesis, their development, their characterization and applications and all the updated information related to environmental nanotechnology.
- Closes with a look at the role of nanotechnology for a green and sustainable future.
- Discusses implications of nanomaterials on the environment and applications of nanotechnology to protect the environment.
- Presents up-to-date information on the economic aspect, toxicity and regulations related to nanotechnology in detail.
- Covers the sustainable and environmentally friendly approaches for remediation and cleaning the environment and nanotechnologies available for wastewater treatment and contaminated soil treatment, as well as finding the healthiness of new nanomaterials.
- Discusses major subjects such as nano-sensors, their structural design and properties, carbon nanotube sensors for many applications in industry and those related to the environment and applications of nanocomposites for pollution sensing.
- Presents scanning electron microscopy, transmission electron microscopy (TEM), and high-resolution TEM, giving the scope for their application in environmental protection, environmental remediation, and environmental biosensors for detection, monitoring, and assessment while giving examples of field applications for environmental nanotechnology.
- Covers some important areas of environment nanotechnology that are particularly concerned with the pollution aspect and mitigation measures, techniques and technologies available for use.

1 Synthesis and Characterization of Metal Oxide Nanoparticles for Antimicrobial Application

Santosh Singh Golia, Rayees Ahmad Zargar, and Manju Arora

CONTENTS

1.1 INTRODUCTION

Bacterial, viral and protozoan microorganisms are categorized as microbes. Of these three, bacteria-induced infections are the main contributors to a high mortality rate [1–5]. Antibiotics are used for curing them. The major issue emerges when inadequate, overdose or misuse of the antibiotics for precautionary and corrective purposes have been made without proper medical testing or symptoms indication. Then, bacteria develop a resistance to antibiotics [6–11], and the resistant bacteria develop a super-resistant gene toward almost all antibiotics [6], becoming what is known as super-bacteria. The antibiotics can work on bacterium target either by diminishing their activity or by killing them by the following three ways: (1) attacking the cell wall, (2) disrupt the functioning of RNA and DNA formation parts and (3) the system that involves in production of proteins. Bacteria resistance for both mechanisms is the main reason why continuous research and development for new antimicrobial agents is in demand. Various metals and metal oxide nanoparticles (MO NPs) as alternatives for antibiotics have been tried, and they exhibit good antibacterial activity. It means that nanomaterials as antibacterial agents are complementary to antibiotics with a very promising future because they help in

DOI: 10.1201/9781003323464-1

1

treating bacterial infections when antibiotics are insensitive [12–18]. Nanoparticles as antibacterial agents have the following merits in comparison to antibiotics: (a) They control the resistance processes and restrict biofilm creation, (b) they utilize a multitasking procedure to fight with microbes, and (c) they are good carriers of antibiotics. Biofilm-growing bacteria are the most resistant to antibacterial drugs and the host's immune system. The actual mechanisms behind such a resistance are not clear and have not been fully explored to date [9–11]. Biofilm is basically a structured community of bacteria that is enclosed in a self-generated extracellular matrix of proteins, polysaccharides (carbohydrates), and DNA. The infections produced by such biofilms are very chronic and become very hard to treat, which reflects the seriousness of problems and the immediate need to explore alternative treatment approaches to overcome the problem of increasing resistance of bacteria for recently available antibiotics [6].

1.2   SEVERITY OF THE PROBLEM

Data published in the medical journal *The Lancet* in 2019 [1] based on a survey of 204 countries revealed deaths in all age groups of around 12.7 lakh from antimicrobial drug-resistant bacteria as compared to the deaths caused by HIV/AIDS (8.64 lakh) or malaria (6.43 lakh). By 2050, the deaths by antimicrobial-resistant (AMR) bacteria will tentatively rise to 10 million per annum [1, 3]. This report projected the severity of the problem and encouraged the researchers and medical community, as well as policymakers, to take immediate steps for saving the lives of people and overcoming this worldwide challenge by developing new and more advanced effective AMR materials. The bacteria are found in rod-like (bacilli), spherical (cocci) and helical (spirilla) shapes. They come under two categories: (1) gram-positive bacteria, which have a thick cell wall of peptidoglycan around the cell membrane that is attached to teichoic acids present only in these bacteria, and (2) gram-negative bacteria, which have a very thin cell wall layer of peptidoglycan that is covered by the second lipid outer membrane that prevents the entrance of any foreign material into the bacterial cell [9–11]. The bacterial infection takes place in the human body via a toxic strain of bacteria on the skin or inside the body through touching, inhaling, or consuming stale/contaminated food. Pneumonia, meningitis and food poisoning are illnesses caused by harmful bacteria infecting the chest, bloodstream and stomach, respectively. Only these three illnesses arising from infections developed because of *Escherichia coli*, *Staphylococcus aureus*, *Klebsiella pneumoniae*, *Streptococcus pneumoniae*, *Acinetobacter baumannii*, and *Pseudomonas aeruginosa* are the most harmful antibiotic-resistant bacterial infections, leading to about 79% of the total deaths caused by bacterial infection. The bacteria are becoming more resistant to almost all types of antibiotics used for treatment, meaning that the infection cured previously by antibiotics in a few days now is becoming incurable. There are several factors that lead to the emergence of resistant bacterial strains such as (1) the antibiotics used to attack a single target in the bacteria [19], (2) an over- or improper dosage of the antibiotics being prescribed by the healthcare personnel [20] and (3) in agriculture, the uncontrolled use of antibiotics in feedstock to promote animal growth [21] or spraying over plants to protect them from disease and increase yield [20]. This is really a very grim situation for the health and life of bacteria-infected people. So the present investigations demand pursuing new, advanced strategies for identifying bacterial infections and developing new next-generation drugs or agents by using nanoengineered antimicrobial materials to control and passivate bacterial infections. In the 1980s, methicillin-resistant *Staphylococcus aureus* (MRSA) has been correlated to hospital-associated (HA) infections. But in 2003, a new community-associated-MRSA (CA-MRSA) strain was reported by [22] for infecting healthy people including children also. The key factors that help spread CA- or HA-MRSA bacterial infections are gatherings, physical skin-to-skin contact, and unknowingly sharing infected items like towels contaminated with wound drainage, dirt and unhygienic matter, among others [23]. It means that skin-related bacterial infections can be markedly reduced and prevented by adopting good habits like creating a hygienic atmosphere, avoiding directly touching or exposing skin to the infected part by air and using antibacterial ointments as surface-coating agents.

1.3 MO NPS APPLICATIONS AS ANTIMICROBIAL AGENTS

Several metal oxides are widely used in medicine, agriculture, and environmental protection as antimicrobial/antibacterial agents. Nanocrystalline metal oxides, namely, Ag_2O, CaO, MgO, ZnO, NiO, CoO, CuO, Cu_2O, TiO_2, SiO_2, and Fe_xO_y, have been used as antimicrobial agents owing to their bacteriostatic or/and bactericidal effect. MO NPs' antimicrobial activity path is different from the normal antibiotics used for bacterial/microbial infection treatment because it destroys the bacteria in prokaryotic cells [24] by simultaneously attacking at the biochemical and molecular stages [25], which involves degradation of cell wall, enzyme, lipids, and nucleic acids, among others. The antibacterial activity of promising MO NPs is sensitive to various parameters, which, in turn, affects their physical and chemical properties. These parameters are briefly described in the following:

1. **Structure and crystallinity of MO NPs:** To confirm the formation of the required structure and crystallinity of synthesized MO NPs by a particular synthesis route for antimicrobial application are characterized by the X-ray diffraction technique. These studies give information about the formation of desired phases or mixed phases or the existence of unreacted reactants or impurities, crystalline nature, and preferential orientation along the most intense plane, among others. The values of crystallite size, lattice parameters, and strain in the lattice can be derived from their diffraction pattern by following standard procedures. The crystallinity of MO NPs throws light on their metallic, semiconducting and insulating properties. Metallic and semiconducting MO NPs are produced from the periodic table 3–12 groups metals, and insulating MO NPs are derived from groups 1, 2 and 13–18 metals [26]. Per the electron-excess/cation-excess/anion-deficient or electron-deficient/cation-deficient/anion-excess charge carriers in semiconducting, MO NPs come under the n-type or p-type semiconductor category [27].

The third is intrinsic semiconducting (i-type) MO NPs, which exhibit both insulating and conducting properties. While insulating MO NPs obtained from groups 1, 2 and 13–18 metals of the periodic table show alkaline and electrostatically charged positive natures, respectively. These unique properties of MO NPs are exploited for their interaction with different types of biomolecules present in the bacterial/microbial cell [28]. They decide the effectiveness of the antimicrobial activity for a particular microorganism or for a different variety of microbes.

2. **Chemical composition of MO NPs:** Metal ions' affinity for electrons present in MO NPs play a vital role in their interaction with different atoms present in the multidentate biomolecule ligands in a bacterial cell because some metal ions exist in two or more valence states. The different biomolecules are involved in the formation of protein/nucleus/carbohydrates/lipids/ cell wall of a bacterial cell. They follow hard-soft acid-base (HSAB) theory [29, 30] to form stable complex compounds with metal ions. For example, soft Ag^+, $Hg^{+ \text{ or } 2+}$, and Cd^{2+} ions form complexes with soft bases like sulfur from the thiol (R-SH) functional group present in the proteins of a bacterial cell, while Cu^{2+}, Zn^{2+}, Co^{2+}, Fe^{2+} and Ni^{2+} ions on interaction with hard amine (R-NH$_2$) and soft thiol (R-SH) bases produce stable medium hard–soft-natured complex compounds. The biomolecules with carboxyl (–COOH) and hydroxyl/alcoholic (–OH) functional groups present in bacteria also have a tendency to interact with metal ions to form stable complex salt. The antibacterial activity of the metals ions is tentatively proportional to their affinity for thiol groups.

3. **Particle size, morphological shape, agglomeration into small/medium/ larger-sized ensembles and stable dispersions:** It is a well-known fact that the activity of a nanoparticle depends on its particle size, morphological shape, and agglomeration dependent because on decreasing particle size, the ratio of surface area to volume increases markedly along with the availability of more number of active sites, that is, more number metal ions with unsaturated coordination bonds at the surface. MO NPs acquire different morphological shapes, for example, spherical, nanorod, nanowire, nanosheet, nano/quantum dot, floral/star/diamond/spindle/pyramid, and others, per the chosen synthesis route and parameter optimization for the desired application. The antibacterial activity of MO NPs is also shape-dependent. Since the size of grown nanoparticles is too small, they have a tendency to agglomerate into different-sized ensembles to stabilize themselves by dissipating excessive energy. To overcome this problem, the nanoparticles are stabilized by using hydrophobic or hydrophilic surfactants that help in separating nanoparticles and restrict their aggregation. This helps in forming stable dispersions of MO NPs in aqueous or organic solvents. Hence, MO NPs antimicrobial agents are found to be a better option as compared to their bulk analogues because they facilitate attachments of more bacteria at the surface. Their surface develops a positive, negative, or zero (neutral) charge per the characteristics of metal oxide. The surface charge properties of such nanomaterials are explored from

the zeta-potential measurements. The positively charged surface of MO NPs tends to form strong ionic coordinated bonds with negatively charged functional groups of bacteria-constituting biomolecules and has been found to be very effective antimicrobial agents, while negatively and neutrally charged MO NPs exhibit the least and intermediate antimicrobial activity, respectively.

4. **Antibacterial activity vs. MO NPs concentration:** MO NPs concentration, along with the experimental conditions and the nature and number of bacteria, plays a vital role in destroying bacterial infections. There is no doubt that by increasing the MO NPs' concentration, the antimicrobial activity enhances. But the experiments were carried out to optimize a minimal MO NP concentration to swiftly inhibit the growth of a bacterial infection after incubation. In addition to the concentration of MO NPs, the experimental conditions are also optimized to get the best antibacterial activity for a particular concentration and nature of bacteria.

1.4   SYNTHESIS OF MO NPS

Nanomaterials with a variety of morphological shapes in the form of nanoparticles and thin films are synthesized by various chemical and physical methods through "bottom-up" and "top-down" approaches, respectively. The bottom-up approach involves assembling the initial atomic or molecular dimension constituents for nanoparticle formation. This method is extensively used because it allows their chemical composition, particle size, and shape to be controlled. The versatile bottom-up approach can be employed for the large-scale production of nanoparticles because they work on a self-assembly process. While in the top-down synthesis approach, nanoparticles are obtained by reducing the size of macroscopic/bulk materials to nanometer dimensions. The reduction of the particle size can be achieved through both physical and chemical methods. From the top-down approach, the derived nanomaterials often suffer from imperfections in the structure with lots of impurities. The surface chemistry and physical properties of the nanoparticles are very sensitive to the structure of the material. The obtained nanoparticles have different sizes and shapes with wide particle size distribution. Figure 1.1 summarizes some of these methods in the following schematic.

The economic chemical methods generally require a simple experimental setup, toxic chemical reactants, and solvents. One can optimize nanoparticle size, shape, surface morphology, composition, and structure by varying synthesis parameters and selecting the appropriate synthesis route and reactant concentrations, while physical methods involve using costly instruments, precursors, substrates and trained persons for operation and the maintenance of instruments. In this chapter, different chemical methods are discussed in brief regarding the synthesis of MO NPs for application as antimicrobial agents:

1. **Coprecipitation Method:** The economical and facile coprecipitation method has been widely used for the synthesis of pure and mixed ion ferrite nanoparticles (FNPs) for use as antimicrobial agents. In this method, the first step is MO NP formation in the aqueous or organic liquid phase

Synthesis of Nanomaterials

Chemical Methods   Physical Methods

Precipitation & Co-precipitation

Sol-gel

Microwave-Assisted

Solvothermal/Hydrothermal

Solid State Pyrolysis

Microemulsion

Solution free Mechanochemical

Electrochemical

Sonochemical

Physical Vapor Deposition

Chemical Vapor Deposition

Pulsed Laser Deposition

Ion Implantation

Sputtering

Atomic Layer Deposition

Molecular Beam Epitaxy

Spray Pyrolysis

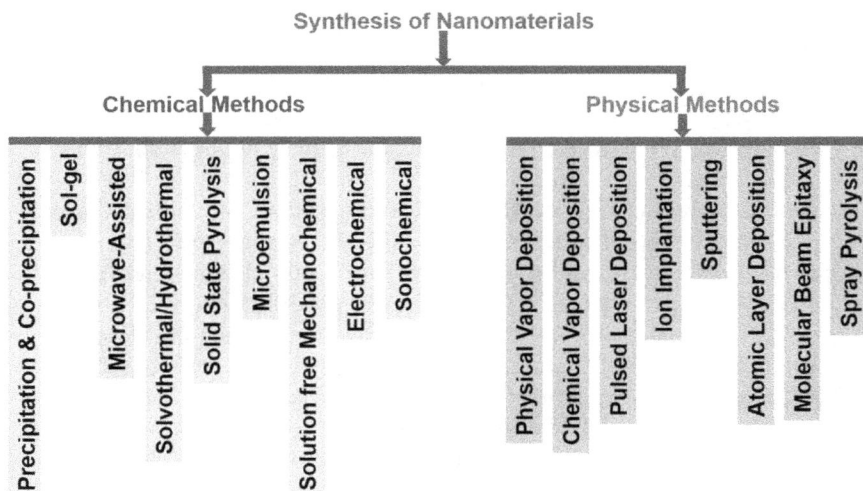

FIGURE 1.1   Schematic presentation of various chemical and physical methods used for the synthesis of nanomaterials.

through oxidation, reduction, and hydrolysis reactions to obtain the desired stoichiometric composition, and the second step helps achieve the required crystal structure and nanoparticle morphology through a thermal treatment. For example, in the synthesis of ferrite nanoparticles by chemical coprecipitation reaction, the aqueous salt solutions of Fe^{2+} and Fe^{3+} chlorides/sulfates/nitrate salts are mixed in a 1:2 stoichiometric molar ratio with a few drops of surfactant to restrict the precipitating nanoparticles' agglomeration. A 25% ammonia solution or a diluted sodium hydroxide solution is generally used to precipitate ferrite oxide hydroxide nanoparticles with continuous stirring in the 8.5–9.0 pH range. The pH of the solution, the stirring rate, the salt concentration, the temperature and the surfactant concentration are the vital parameters that control the phase formation, the chemical composition, the purity, the particle size, the surface area, the shape and the agglomeration of FNPs [31–35]. In the coprecipitation reaction, the nucleation under super-saturation conditions, growth and agglomeration of FNPs via the Ostwald ripening process takes place simultaneously. Fine uniform-sized and -shaped, high-purity, stoichiometric single- and multicomponent FNPs with narrow/wide size distribution under controlled/uncontrolled conditions are obtained on heating precipitated particles at very low temperatures, approximately 80°C. FNPs formation by coprecipitation reaction is represented by the following chemical equation:

$$Fe^{2+} + 2Fe^{3+} + 8OH^- \rightarrow Fe_3O_4 + 4H_2O$$

This method is also used for the synthesis of ZnO nanoparticles. Although it is very difficult to control the particles' size and their distribution due to the kinetic factors, but this problem can be overcome by changing the reactants used for coprecipitation, their ionic strength, the surfactant used for coating nanoparticle surface, the pH and the temperature of precipitating solutions.

2. **Sol-Gel Method:** Initially, in mid-1800s, this sol-gel chemical method was at first used for the synthesis of inorganic ceramic and glass [36, 37]. For MO NP preparation with desired particle size and morphology, this method was successfully used. Metal alkoxide/acetate/nitrate salt as precursors in organic or aqueous solvents with catalysts, stabilizers, and other reactants such as gelling agents are used [36–40]. For example, zinc oxide nanoparticles [40] derived from zinc acetate dihydrate precursor by this method have been described in the following three steps, namely, hydrolysis, hydroxylation, and condensation/polycondensation reactions for sol and gel formation. In Step 1, hydrolysis reaction, a zinc acetate dihydrate aqueous solution forms partially hydrolyzed to basic zinc acetate ($Zn(OH)$ (CH_3COO)), (Zn^{2+} and CH_3COO^-) ions and acetic acid (CH_3COOH). This is sol, a stable colloidal suspension of solid crystalline or amorphous ionic nanoparticles in a solvent. The hydrolysis reaction is affected by environmental conditions like relative humidity and temperature in the formation of basic zinc acetate. In Step 2, hydroxylation reaction, the basic and ionic zinc acetate mixture solution, on heating, forms two molecules of $Zn(OH)_2$. In Step 3, condensation/polycondensation reactions form a chain of Zn-O with terminal OH groups, after completion of these reactions, the more viscous continuous three-dimensional metal oxide–network gel is formed. t gel is aged to separate out the solvent; the Ostwald ripening process helps in forming the solid mass. The obtained hydroxylated ZnO gel is dried at a low temperature, about 80°C, to slowly evaporate the solvent and the moisture. Then this amorphous powder is annealed for 1 hour at 350°C to obtain the desired crystalline stoichiometry of the ZnO nanoparticles. Drying the gel to evaporate the solvent and the annealing temperature, as well as duration, may vary from one metal oxide to another. In this method, one can vary the particle size and shape by varying the precursor salt or concentration, the solvent used for sol formation, the stirring rate, the annealing temperature, and its period parameters to derive a narrow size distribution and uniformly shaped MO NPs. All these merits have attracted researchers to extensively adopt this method for the synthesis of new advanced and existing pure and mixed MO NPs for antimicrobial activity.

3. **Microwave-Assisted Method:** The microwave-assisted method for synthesizing MO NPs has drawn the attention of researchers owing to its high synthesis rate, yield, purity, and uniform, small-sized nanoparticles with a narrow particle size distribution compared to conventional thermal heating, which takes longer time to complete the reaction. This method, in conjunction with precipitation/coprecipitation, sol-gel, microemulsion and hydrothermal, is generally used to markedly increase the reaction rate and drastically reduce the reaction time. Mostly, microwave ovens/reactors operating at a 2.45 GHz frequency with 1 kW maximum output power for

chemical synthesis of materials have been used to avoid the electromagnetic interference effects arising from electronic devices, telecommunication, wireless networks and mobile frequencies. At this operating frequency, with the optimal power for the chemical reaction to proceed, the whole material is uniformly heated without any thermal gradient. The molecules present in the material are vibrating at the rate of 4.9 × 109 times per second at a 2.45 GHz frequency, causing their alignment and realignment with the oscillating field to produce 100°C–200°C temperature, which increases at a rate of 10°C per second. This volumetric heating of material raises the reaction temperature above the boiling point of the solvent, which, in turn, increases the reaction rate by 10–1000 times and duration from minutes to seconds [41]. High-purity, small-sized MO NPs with better physicochemical properties [42] are obtained from this method. One can easily apply this method for the large-scale production of pure and mixed MO NPs because it has overcome the problem of thermal gradient effects [43] for pure and mixed metal oxides, for example, Fe_3O_4 [44], ZnO [45, 46], TiO_2 [47, 48], SnO_2 [49], NiO [50], PtO_2 [51], CeO_2 [52], and CuO [53].

4. **Solvothermal/Hydrothermal Method:** In this method, the precursor metal salts on dissolution in a nonaqueous solvent (solvothermal) or in water (hydrothermal) form complexes that are heated at different temperatures and under high pressure under optimized experimental parameters in sealed vessels in autoclaves for the synthesis of MO NPs. The chemical reaction takes at least 3–48 hours or even more to complete the reaction, which involves nucleation and the growth of nanoparticles. Since, the obtained MO NPs are pure with precisely controlled size, distribution, shape and crystalline phases, no further purification or posttreatment annealing steps are required. Hydrothermal temperature decides that synthesized MO NPs are anhydrous, crystalline or amorphous [54, 55]. The hydrothermal method is one of the most extensively used methods for synthesizing MO NPs. This process does not require costly instrumentation, hazardous chemicals or special conditions to carry out the synthesis. But it still has good control over homogeneity, particle size, chemical composition, phase and morphology of the resultant products.

5. **Solid-State Pyrolysis Method:** The conventional solid-state reaction method has been used to prepare bulk polycrystalline MO NPs by taking their nitrate/carbonate/sulfate/acetate/citrate salts as precursors [56]. Then these bulk materials are ball milled to reduce nanometer particles sized in the range of <100 nm. Ferrite powdered samples from the mixture of metal nitrates of respective elements in stoichiometric ratio. One can also use this method in conjunction with the sol-gel auto-combustion method in which MO NPs are derived by using different ways like (a) metal nitrate and citric acid; (b) polyvinyl vinyl alcohol (PVA)-assisted metal nitrates, ethylene glycol and citric acid; and (c) by using different complexing agents like citric acid, cellulose and cellulose citric acid mixture, among others. The oxidation–reduction reaction takes place during the combustion process in which nitrate ions play the role of the oxidizing agent and the carboxyl

group as the reducing agent. PVA imparts the role of the surfactant, which controls the grain growth and the size of the nanoparticles. The application of different complexing agents helps optimize the desired size nanoparticle, shape, morphology, porosity, microscopic structure and physicochemical properties for efficient antimicrobial agents.

6. **Microemulsion Method:** MO NPs are produced from the thermodynamically stable dispersion of two immiscible liquids microemulsions, for example, water in oil or oil in water, and stabilized by surfactant molecules. It is a well-known fact that stirring two immiscible liquids forms an emulsion in which the lesser amount of liquid has a tendency to form small droplets, a bunch of coagulated droplets or a layer to separate from the higher amount of liquid. The size of droplets in emulsions is in the range from >100 nm to a few millimeters, while microemulsions are transparent and contain droplets of a size in the 1–100-nm range and are stabilized by using ionic or nonionic surfactant and/or cosurfactants. In such microemulsions, metal salts are dissolved and reduced in nanometer-sized monodispersed droplets, which are stabilized by surfactant molecules. In that situation, under some critical concentration, "micelles" or "inverse micelles" are formed, pertaining to the concentrations of water and organic liquid. Micelles are formed with higher amounts of water, and inverse micelles are formed when the concentration of the organic liquid or oil is more. The micelles or reverse micelles on collision results in the formation of controlled size, differently shaped MO NPs [57, 58]. This process can also be easily exploited for the large-scale production of nanoparticles with the advantages of recovering and reusing the surfactant and oil many times for the microemulsion preparation and stabilization of nanoparticles.

7. **Solution-Free Mechanochemical Method:** In this method for preparing MO NPs, when their respective solid salts are ball milled, they then directly absorb the mechanical energy released during the milling process from the power and friction between balls and precursor salt. For nanoparticle formation from its bulk analogue, the size of the particle of salt in the ball mill is reduced due the deformations, fractures and abrasion faced by the salt while grinding within the walls of the balls. In mechanochemical methods, this huge accumulated mechanical energy in the powdered sample is optimum for undergoing chemical reactions to achieve the desired composition of the final product. The basic requirement of this method is high-energy ball mills/shakers/grinders because they are able to develop very high mechanical stress, which on absorption by the precursor compound results in the breaking of the lattice bonds. This creates structural variations in the lattice, and the surface active atomic layers of reactant interfaces are continuously exposed to extreme stress; the absorbed mechanical energy is utilized by the molecules for chemical reactions to form the desired pure and mixed small-sized MO NPs [59–61] in the range of 5–20 nm along with the required crystalline phase, morphology, porosity and specific surface area.

8. **Electrochemical Method:** The very facile and eco-friendly electrochemical method has been used for synthesizing MO NPs. In this process, the anode (negative) electrode is oxidized and released into the surfactant solution; the ions interact in between the electrode–electrolyte interface and synthesize the respective pure and mixed MO NPs [62–65]. The merits of this method are high-purity metal oxides and controlled particle size compared to other synthesis methods. The purity and crystallinity of desired MO NPs size can be optimized by controlling the following parameters: current density, pH, electrolyte concentration, electrode selection and varying the distance between electrodes. At the same time, either by increasing the current density or by decreasing the distance between electrodes, the particle size tends to increase. This is because the closer the electrodes, the faster the reduction–oxidation reaction, which releases more oxidized metal ions into the solution for the rapid formation of MO NPs. The synthesis of ZnO, TiO_2, Fe_3O_4 and NiO nanoparticles by the electrochemical method has been already reported by different groups.

9. **Sonochemical Method:** In this process, the precursor metal salt solution is subjected to different high-powered ultrasound radiations having frequencies in the 20 kHz–10 MHz range for acoustic cavitations to proceed with the chemical reaction to form MO NPs for 30–60 minutes. The ultrasound waves passing through the solution create alternate compression and relaxation. This leads to acoustic cavitation, that is, the formation, growth and implosive collapse of bubbles in the liquid, while the change in pressure develops microscopic bubbles whose collapsing is explosive. This step propagates shock waves within the gas phase of the ceasing bubbles. Hence, the sonochemical reaction is based on the creation, growth, and cessation of bubbles in the solution during the ultrasonication process. In the bubble-growth process, the solute (precursor salt) vapor starts diffusing in the bubble. When the bubble attains its maximum volume, it behaves like a hot spot because of its very high temperature, 5000–25000 K [66], and pressure around 1800 atmosphere. This bubble-cessation process completes instantly, that is, in less than a nanosecond [67, 68], which breaks the chemical bonds of precursor salt followed by a very high cooling rate of about 10^{11} K/s. Such a high cooling rate restricts the organization and crystallization of the product. This is the reason why volatile precursors are selected for this method, in which gas-phase reactions dominate to produce amorphous nanoparticles. The obtained nanoparticles are washed several times and filtered for further calcination processing in the case of MO NPs to get the crystalline with optimal-sized nanoparticles. The sonochemical methods have several advantages, like uniform size distribution, high surface area–to–volume ratio, being fast, purity and the large-scale production of MO NPs. This method has been widely used to synthesize MO NPs [69–72].

10. **Green Chemistry Method:** This method is categorized under the bottom-up approach and found to be better than the chemical methods because it uses the natural nonhazardous plants/algae/fungi/bacteria biological resources as reducing agents. The green synthesis process is designed in

such a manner that it minimizes or entirely removes the use or formation of toxic chemicals or by-products [73–75]. It does not require/or minimize the use of costly chemicals, hazardous solvents/catalysts/surfactants, and the synthesis reaction can be easily executed under mild conditions. This eco-friendly, facile, biocompatible method produces pure MO NPs with the least or without any contamination/impurities. The physicochemical properties of the MO NPs derived by this method are sensitive to the following reaction parameters: precursor chemicals, plant selection, phytochemicals the organic compounds extracts obtained from plant leaves/stem/fruits/flowers/roots/seeds, solvent, temperature, pressure, and the pH of the solution. Plant biodiversity has opened a window, and a wide range of phytochemicals and biomolecules are present in different parts of the plant extracts. The leaves are generally used owing to their abundance and sustainability. The plant extract contains ketones, aldehydes, flavones, proteins, terpenoids, carboxylic/polycarboxylic acids, phenols/polyphenols, polysaccharides, amino acids and vitamins, among others. These compounds have been used for synthesizing and stabilizing MO NPs. The plant derivatives derived MO NPs acquire distinctive morphological shapes like spherical, cubical, cylindrical, needles, stems, prisms and dendrites, among others, with better stability. However, this process takes a long time for MO NP formation owing to the three time-consuming steps: (a) the extraction, (b) isolation and (c) the purification of phytochemicals. MO NPs synthesized by the green chemistry route exhibit enhanced biocompatibility for biomedical applications compared to those obtained by conventional chemical methods. CeO_2, Co_3O_4, CuO, Fe_3O_4, γ-Fe_2O_3, MgO, MnO_2, and ZnO are some examples of MO NPs derived by the green chemistry method for biomedical and antimicrobial applications.

1.5   CHARACTERIZATION OF MO NPS BY VARIOUS ANALYTICAL TECHNIQUES

To confirm the formation, structure, morphology, chemical composition, and particle size of the synthesized MO NPs are characterized by some of the basic techniques, for example, X-ray diffraction (XRD), Fourier transform infrared (FTIR), Raman and ultraviolet–visible light (UV)-Vis spectroscopy, scanning electron microscopy (SEM), transmission electron microscopy (TEM), energy-dispersive X-ray spectroscopy (EDS) and electron paramagnetic resonance (EPR) spectroscopy. The schematic of these characterization techniques is summarized in Figure 1.2.

1. **XRD technique:** XRD patterns of MO NPs are generally recorded by a powder X-ray diffractometer to confirm the formation of the desired composition and structure of the prepared MO NPs. The diffraction pattern consists of various peaks pertaining to Bragg reflections and indexed by Miller indices. This represents the polycrystalline nature with a preferred orientation along particular crystallographic axis/plane and the d-spacing values of the prepared sample are compared with their standard joint

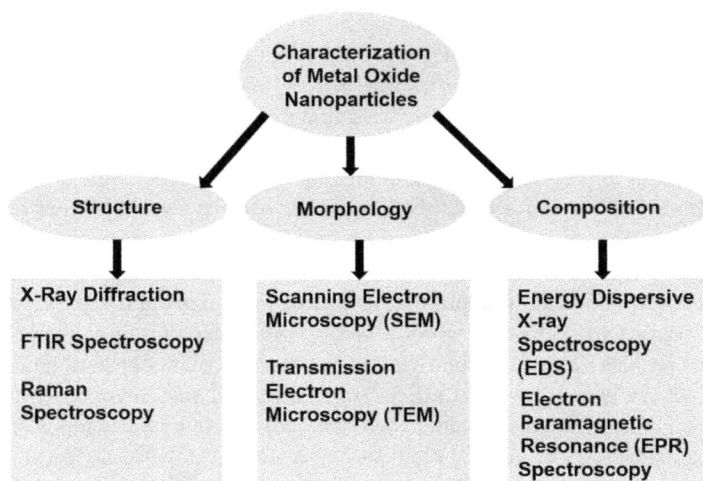

FIGURE 1.2 Schematic of preliminary techniques used for characterization of MO NPs.

committee on powder diffraction standards (JCPDS) card/file. From the diffraction pattern, one can obtain the lattice parameters and particle size. The XRD pattern also reveals the purity of materials and the existence of any other phases of the prepared MO NPs or impurities formed during the synthesis process or unreacted precursor salt. The particle size is calculated using the Scherrer formula.

2. **FTIR and Raman spectroscopy techniques:** The bonding details of MO NPs are inferred by using these two techniques. Their spectra are recorded in 4000–400 cm^{-1} region, the peaks arising from the vibrations of different bonding groups appear as strong, medium and weak intensity peaks. These peaks are assigned to antisymmetric stretching, symmetric stretching and bending modes. Normally the vibrations are active in infrared (IR); then they are forbidden in Raman spectra. But sometimes, some modes are active in both the IR and Raman spectra. This may be due to site symmetry of the bonding group in the lattice or deformations in the lattice induced by interstitial or substitutional atoms/bonding groups and dislocation/missing atoms in lattice. The adsorbed moisture/water stretching and bending modes appear between 3500–3200 cm^{-1} and at about 1600 cm^{-1}, respectively. While the organic functional groups like methyl (–CH$_3$), methylene (–CH$_2$), carboxylic (–COOH), alcohol/phenol (–OH), C-H stretching, amino, nitro and other heteroatom-containing chain/ring structures appear in 3000–800 cm^{-1} region in both the IR transmittance and Raman absorption spectra. M-O stretching modes appear in 600–400 cm^{-1} region. These peaks are the fingerprints of the vibrational modes pertaining to particular functional groups' symmetry and their site symmetry in the lattice structure. This is how these spectra provide information about the structure of MO NPs.

3. **UV-Vis spectroscopy:** The optical absorption spectra of MO NPs are recorded in 200–900-nm region using an easily available UV-Vis spectrometer. These spectra help in distinguishing the monodisperse particles in suspension and agglomeration of nanoparticles in small, medium- or larger sized ensembles. The band-edge absorption of MO NPs has been used for band gap energy and red shifts due to colloidal gel formation.

4. **SEM and TEM:** Both these electron microscopy techniques have been used to investigate the shape and surface morphology of prepared MO NPs. TEM imagery gives more in-depth information about the morphology and precise particle size, while SEM imagery reveals that the prepared MO NPs are in the nano range with particles having spherical/rod/spindle/polygon- or irregularly shaped with different particle sizes. They may acquire dense or porous morphology. However, TEM imagery reveals the formation of fine, nano-range MO NPs as well as their variable sizes, shapes and agglomerate ensembles. The selected area electron diffraction pattern (SAEDP) gives information about the structure of crystalline MO NPs to further confirm the XRD data.

5. **EDS**: The EDS accessory is attached the SEM's main unit. While scanning the MO NPs sample for morphological details, on the simultaneously operating EDS accessory, one can determine the chemical composition qualitatively and quantitatively in terms of the metal oxide–constituting elements and their composition values in atomic % and weight %, respectively. While recording the SEM images, the EDS spectra and data are also recorded to confirm the formation of MO NPs and help detect other metals as impurities.

6. **EPR Spectroscopy:** This sophisticated technology has been used for the qualitative and quantitative analysis of paramagnetic species, for example, oxygen vacancies, deep-level shallow donors in MO NPs (ZnO, TiO_2), free radicals of the surfactant and the nature of magnetic nanoparticles. The oxygen vacancies are generally formed in MO NPs during the synthesis process. The FNPs' (Fe_3O_4) ferromagnetic and superparamagnetic nature is easily characterized by observing their EPR spectra. In ferromagnetic nanoparticles, the broad resonance signal is obtained due to the dominance of dipolar–dipolar interactions among nanoparticles, while superparamagnetic nanoparticles having a size of less than 10 nm exhibit a narrow signal because in this, each nanoparticle behaves as a free magnet. By this method, one can very precisely reveal the nature of paramagnetic species by its g-value and calculate the concentration of oxygen vacancies/superparamagnetic nanoparticles from ppm to ppb level by the comparison method. For the quantitative estimation, 1,1-diphenyl 2-picryl hydrazyl has been used as standard reference material.

The selection of appropriate precursor salt, solvents, surfactant, synthesis route with optimized temperature, pH, time and other parameters and the characterization of MO NPs are very important for assessing their abilities as antimicrobial agents.

In the next section, the work done on the application of MO NPs as antimicrobial agents from 2000 onward is reviewed in brief.

1.6 TENTATIVE MECHANISM FOR ANTIBACTERIAL ACTIVITY OF MO NPS FOR THE DESTRUCTION OF MICROBES/BACTERIA

Extensive work has been reported on the application of MO NPs as antimicrobial agents for the diagnosis and treatment of microorganisms, especially bacteria, owing to their unique physicochemical, electrical, magnetic, optical and biotic properties. The antimicrobial activity mechanism of MO NPs is not fully understood yet, and it is still a challenging problem. The possible mechanisms for drastically reducing the growth or destroying microorganisms/bacteria reported so far are summarized and discussed in lieu of the simultaneous functioning of MO NPs at different constituents of bacterial cells. The process depends on the selection of MO NPs, their nature, their particle size and the chemical reaction. Basically, the MO NPs accumulate at the surface of bacterial cell walls by electrostatic interaction because a negative charge present in the polymers (e.g., proteins carboxylic acid functional group) active groups that attract MO NP metal ions that slowly enter into bacteria by adsorption, dissolution and hydrolysis processes [76] and damages it's cell wall and morphology due to toxic and abrasive nature. Inside the bacterial, MO NPs form superoxide anion (O_2^-), hydroxyl radicals (-OH$^•$), hydrogen peroxide (H_2O_2) and organic hydroperoxides (OHPs) reactive oxygen species (ROS) chemically or by the absorption of light of an appropriate wavelength. The variations in electronic properties and decrease in particle size of MO NPs generate more active groups/ sites at the surface of the nanoparticles which on interaction with oxygen (O_2) and electron donor/acceptor active sites form $O_2^{•-}$, and this species further produces more ROS through Fenton-type reactions [77]. The physicochemical properties of MO NPs, such as surface area, diffusion and electrophilic nature are the parameters, which provide information regarding the quantity of ROS production in the bacteria. $O_2^{•-}$ species obstruct the release of respiratory enzymes by damaging the iron–sulfur (Fe–S) clusters in the electron transport chain and discharge of more ferrous (Fe^{2+}) ions for decreasing adenosine triphosphate (ATP) production. The ferrous ions are oxidized by the Fenton reaction for more OH$^•$ radical production for the destruction of DNA, proteins and lipids present in bacteria cells [78]. H_2O_2, as an oxidant, is fatal to cells; it affects the DNA and proteins in microbes [79]. The intermediate peroxyl free radicals undertake the peroxidation reaction of unsaturated phospholipids present in the membranes for acute damage. This reaction brings conformational changes in the membrane proteins and the membrane fluidity, while the ionic imbalance leads to the migration of more metal ions owing to membrane damage [80]. The presence of malondialdehyde (MDA) has been used as an index for cell membrane damage. Its concentration rises due to lipid peroxidation by ROS [81]. It means the ROS are involved in the disruption of phospholipids present in the cell membranes, lipoproteins and nucleic acids that develop oxidative stress for killing the microbes. Glutathione is a nonenzymatic antioxidant that protects bacteria to overcome oxidative stress. The metal ions released by MO NPs oxidize glutathione for more ROS production. The oxidized glutathione also enhances the lipid peroxidation reaction in the bacterial cell membrane for destroying it. The excessive production of ROS within the bacterial cell disturbs the activity of antioxidant enzymes. This disparity in oxidants and antioxidants in microbes develops oxidative stress, which eventually destroys them. These species simultaneously attack the different parts like the nucleus, lipids, and polysaccharides present in the bacteria/microorganism cell for deactivation or destruction. This fast process restricts the mutation of bacteria/microorganisms to become resistant to the administered drug or ointment, unlike antibiotics, and resulted in the eruption and destruction of the intercellular parts of a microbe/bacterial cell. The penetration of more metal ions into the bacterial/microbial cell interrupts their metabolic activity due to a malfunctioning of enzymes and catalysts [82]. The electrical properties of MO NPs have been captured for interaction with nucleic acids, especially genomic and plasmid DNA [83], which inhibits the cell division process of bacteria/microbes by interrupting chromosomal, as well as the plasmid, DNA replication process. This affects signal transduction in bacteria due to the dephosphorylation of phosphotyrosine. Hence, the restriction of signal transduction in bacteria eventually blocks the bacteria's growth [84, 85]. Metal ions form cross-linkages within or between nuclear material, for example, DNA, which disturbs their helical structured strands. MO NPs' metal ions in the cell are reduced to metal atoms by thiol (–SH) groups present in enzymes and proteins. Hence, this reduction reaction deactivates the proteins, which inhibits the metabolism and respiration activity of bacteria and finally leads to cell death. In addition to this process, the catalytic property of metal ions is exploited, which involves in oxidation of amino acid side chains to protein carbonyls, known as a carbonylation reaction, and acts as a marker to reveal the oxidative protein damage. This reaction decreases the catalytic activity in enzymes which in turn causes protein degradation. In gram-negative bacteria, lipopolysaccharide of the outer lipid bilayer imparts more negative charges than the phospholipid layer [86]. They have more negative charges, that is, stronger strains, compared to gram-positive bacteria [87]. However, for skin bacterial/microbial infections, one can use MO NP photocatalysts in the ointment for external use, which absorb UV/visible light from sunlight/normal light inside the premises for the excitation of electrons from the valence band (VB) to the conduction band (CB). The formed holes in VB and excited electrons in CB initiate the process of ROS formation and follow the same path as described earlier for the destruction of the bacteria cell.

1.7 SUMMARY AND THE FUTURE OF MO NPS AS ANTIMICROBIAL AGENTS

Extensive work has been reported on the use of MO NPs as antimicrobial materials. This chapter has concisely described the introduction and severity of the problem, the synthesis methods for producing MO NPs and their characterization by simple basic characterization techniques. The tentative mechanism is proposed based on available literature. But still, no medically approved standard list of MO NPs for antimicrobial treatment for the healthcare sector is available. An AMR bacterial infection therapy has been tried to reveal the role of MO NPs' action with a tentative mechanism involved in destroying deadly microorganisms. Before being taken to the market as antimicrobial agents, an evaluation and assessment of the risks and adverse effects are required. The risk and severe effects faced by patients during the administration of MO NPs on or into the body have not fully explored and assessed yet. Before marketing them as antimicrobial medicine in place of antibiotics, one should take care of all possible merits and demerits. This process draws special attention to the dosage, patients' immunity, allergies to a particular drug, effectiveness, off-target toxicity of MO NPs and so on. The unique physicochemical properties of MO NPs having a large surface area–to–volume ratio with different morphological shapes have been utilized for antimicrobial activities. The toxicity of MO NPs is one of the major problems, and they are very sensitive to parameters like particle size compared to their bulk analogue, dispersion in solvents, stoichiometric composition, susceptibility and type of the bacterial stains, ROS generation and simultaneous degradation reactions.

REFERENCES

1. C.J.L. Murray, K.S. Ikuta, F. Sharara, L. Swetschinski, G.R. Aguilar, A. Gray, C. Han, C. Bisignano, P. Rao, E. Wool, S.C. Johnson, A.J. Browne, M.G. Chipeta, F. Fell, S. Hackett, G.H. Woodhouse, B.H.K. Hamadani, E.A.P. Kumaran, B. McManigal, R. Agarwal, S. Akech, S. Albertson, J. Amuasi, J. Andrews, A. Aravkin, E. Ashley, F. Bailey, S. Baker, B. Basnyat, A. Bekker, R. Bender, A. Bethou, J. Bielicki, S. Boonkasidecha, J. Bukosia, C. Carvalheiro, C. Castañeda-Orjuela, V. Chansamouth, S. Chaurasia, S. Chiurchiù, F. Chowdhury, A.J. Cook, B. Cooper, T.R. Cressey, E. Criollo-Mora, M. Cunningham, S. Darboe, N.P.J. Day, M.D. Luca, K. Dokova, A. Dramowski, S.J. Dunachie, T. Eckmanns, D. Eibach, A. Emami, N. Feasey, N. Fisher-Pearson, K. Forrest, D. Garrett, P. Gastmeier, A.Z. Giref, R.C. Greer, V. Gupta, S. Haller, A. Haselbeck, S.I. Hay, M. Holm, S. Hopkins, K.C. Iregbu, J. Jacobs, D. Jarovsky, F. Javanmardi, M. Khorana, N. Kissoon, E. Kobeissi, T. Kostyanev, F. Krapp, R. Krumkamp, A. Kumar, H.H. Kyu, C. Lim, D. Limmathurotsakul, M.J. Loftus, M. Lunn, J. Ma, N. Mturi, T. Munera-Huertas, P. Musicha, M.M. Mussi-Pinhata, T. Nakamura, R. Nanavati, S. Nangia, P. Newton, C. Ngoun, A. Novotney, D. Nwakanma, C.W. Obiero, A. Olivas-Martinez, P. Olliaro, E. Ooko, E. Ortiz-Brizuela, A.Y. Peleg, C. Perrone, Ni. Plakkal, A. Ponce-de-Leon, M. Raad, T. Ramdin, A. Riddell, T. Roberts, J. Victoria Robotham, A. Roca, K.E. Rudd, N. Russell, J. Schnall, J.A.G. Scott, M. Shivamallappa, J. Sifuentes-Osornio, N. Steenkeste, A.J. Stewardson, T. Stoeva, N. Tasak, A. Thaiprakong, G. Thwaites, C. Turner, P. Turner, H.R. van Doorn, S. Velaphi, A. Vongpradith, H. Vu, T. Walsh, S. Waner, T. Wangrangsimakul, T. Wozniak, P. Zheng, B. Sartorius, A.D. Lopez, A. Stergachis, C. Moore, C. Dolecek, M. Naghavi, Global burden of bacterial antimicrobial resistance in 2019: a systematic analysis, Lancet. 399 (2022) 629–655.
2. J. O'Neill, *Antimicrobial resistance: tackling a crisis for the health and wealth of nations.* London: Review on Antimicrobial Resistance, 2014.
3. M.E.A. de Kraker, A.J. Stewardson, S. Harbarth, Will 10 million people die a year due to antimicrobial resistance by 2050? PLoS Med. 13 (2016) e1002184.
4. A. Cassini, L.D. Högberg, D. Plachouras, A. Quattrocchi, A. Hoxha, G.S. Simonsen, M. Colomb-Cotinat, M.E. Kretzschmar, B. Devleesschauwer, M. Cecchini, D.A. Ouakrim, Attributable deaths and disability-adjusted life-years caused by infections with antibiotic resistant bacteria in the EU and the European economic area in 2015: a population-level modelling analysis, Lancet Infect. Dis. 19 (2019) 56–66.
5. US Centers for Disease Control and Prevention, *Antibiotic resistance threats in the United States, 2019.* Atlanta, GA: US Department of Health and Human Services, 2019.

6. A.P. Magiorakos, A. Srinivasan, R.B. Carey, Y. Carmeli, M.E. Falagas, C.G. Giske, S. Harbarth, J.F. Hindler, G. Kahlmeter, B. Olsson-Liljequist, D.L. Paterson, L.B. Rice, J. Stelling, M.J. Struelens, A. Vatopoulos, J.T. Weber, D.L. Monnet, Multidrug-resistant, extensively drug-resistant and pandrug-resistant bacteria: an international expert proposal for interim standard definitions for acquired resistance, Clin. Microbiol. Infect. 18 (2012) 268–281.

7. D.A. Hopwood, How do antibiotic-producing bacteria ensure their self-resistance before antibiotic biosynthesis incapacitates them, Mol. Microbiol. 63 (2007) 937–940.

8. D. Mazel, J. Davies, Antibiotic resistance in microbes, Cell. Mol. Life Sci. 56 (1999) 742–754.

9. G. Kapoor, S. Saigal, A. Elongavan, Action and resistance mechanisms of antibiotics: a guide for clinicians, J. Anaesthesiol. Clin. Pharmacol. 33 (2017) 300–305.

10. J.M. Munita, C.A. Arias, Mechanisms of antibiotic resistance, Microbiol. Spect. 4 (2016) 742–754.

11. G.D. Wright, Molecular mechanisms of antibiotic resistance, Chem. Commun. 47 (2011) 4055–4061.

12. L. Wang, C. Hu, L. Shao, The antimicrobial activity of nanoparticles: present situation and prospects for the future, Int. J. Nanomed. 12 (2017) 12–27.

13. R. Singh, M.S. Smitha, S.P. Singh, The role of nanotechnology in combating multidrug resistant bacteria, J. Nanosci. Nanotechnol. 14 (2014) 4745–4756.

14. J.T. Seil, T.J. Webster, Antimicrobial applications of nanotechnology: methods and literature, Int. J. Nanomed. 7 (2012) 2767–2781.

15. R.Y. Pelgrift, A.J. Friedman, Nanotechnology as a therapeutic tool to combat microbial resistance, Adv. Drug Delivery Rev. 65 (2013) 1803–1815.

16. P.K. Stoimenov, R.L. Klinger, G.L. Marchin, K.J. Klabunde, Metal oxide nanoparticles as bactericidal agents, Langmuir. 18 (2002) 6679–6686.

17. A.J. Huh, Y.J. Kwon, "Nanoantibiotics": a new paradigm for treating infectious diseases using nanomaterials in the antibiotics resistant era, J. Control. Release. 156 (2011) 128–145.

18. A. Raghunath, E. Perumal, Metal oxide nanoparticles as antimicrobial agents: a promise for the future, Int. J. Antimicrobial Agents. 49 (2017) 137–152.

19. E. Etebu, I. Arikekpar, Antibiotics: classification and mechanisms of action with emphasis on molecular perspectives, Int. J. Appl. Microbio. Biotech. Res. 4 (2016) 90–101.

20. R.J. Fair, Y. Tor, Perspectives in medicinal chemistry antibiotics and bacterial resistance in the 21st century, Perspect. Medicinal Chem. 6 (2014) 25–64.

21. H.C. Wegener, F.M. Aarestrup, L.B. Jensen, A.M. Hammerum, F. Bager, Use of antimicrobial growth promoters in food animals and enterococcus faecium resistance to therapeutic antimicrobial drugs in Europe, Emerg. Infect. Dis. 5 (1999) 329–335.

22. F. Vandenesch, T. Naimi, M.C. Enright, G. Lina, G.R. Nimmo, H. Heffernan, N. Liassine, M. Bes, T. Greenland, M.E. Reverdy, J. Etienne, Community acquired methicillin- resistant staphylococcus aureus carrying panton–valentine leukocidin genes, Emerg. Infect. Dis. 9 (2003) 978–984.

23. T.E. Zaoutis, P. Toltzis, J. Chu, T. Abrams, M. Dul, J. Kim, K.L. McGowan, S.F. Coffin, Clinical and molecular epidemiology of community-acquired methicillin-resistant staphylococcus aureus infections among children with risk factors for health care-associated infection: 2001–2003, Pediatr. Infect. Dis. J. 25 (2006) 343–348.

24. A. Kumar, A.K. Pandey, S.S. Singh, R. Shanker, A. Dhawan, Cellular response to metal oxide nanoparticles in bacteria, J. Biomed. Nanotechnol. 7 (2011) 102–103.

25. A. Kumar, A.K. Pandey, S.S. Singh, R. Shanker, A. Dhawan, Engineered ZnO and TiO_2 nanoparticles induce oxidative stress and DNA damage leading to reduced viability of Escherichia coli, Free Radic. Bio. Med. 51 (2011) 1872–1881.

26. R.A. Zargar, M. Arora, T. Alshahrani, M. Shkir, Screen printed novel ZnO/MWCNTs nanocomposite thick film, Ceram. Int. 47 (2021) 6084–6093.

27. P.R. Roberge, *Handbook of corrosion engineering*, New York, NY: McGraw-Hill, 2000.

28. R.A. Zargar, ZnCdO thick film: a material for energy conversion devices, Mater. Res. Express. 6 (2019) 095909.

29. R.G. Pearson, Hard and soft acids and bases, J. Am. Chem. Soc. 85 (1963) 35333539.

30. T. Soldatovic, *Correlation between HSAB principle and substitution reactions in bioinorganic reactions*, London: IntechOpen, 2020, http://dx.doi.org/10.5772/intechopen.9168.

31. R.A. Zargar, M. Arora, Study of nanosized copper doped ZnO dilute magnetic semiconductor thick films for spintronic device applications, J. Appl. Phys–A. 124 (2018) 36.

32. A. Bee, R. Massart, S. Neveu, Synthesis of very fine maghemite particles, J. Magn. Magn. Mater. 149 (1995) 6–9.

33. R.A. Zargar, Kundan Kumar, M. Arora, M. Shkir, Structural, optical, photoluminescence, and EPR behaviour of novel Zn0·80Cd0·20O thick films: an effect of different sintering temperatures, J. Lumin. 245 (2022) 118769.

34. K. Petcharoen, A. Sirivat, Synthesis and characterization of magnetite nanoparticles via the chemical co-precipitation method, Mater. Sci. Eng. B. 177 (2012) 421–427.

35. H.C. Roth, S.P. Schwaminger, M. Schindler, F.E. Wagner, S. Berensmeier, Influencing factors in the co-precipitation process of superparamagnetic iron oxide nano particles: a model based study, J. Magn. Magn. Mater. 377 (2015) 81–89.

36. M. Ebelmen, Recherches sur les combinaisons des acidesborique et silicique avec lese thers, Ann. Chimie Phys. 16 (1846) 129.

37. T. Graham, On the properties of silicic acid and other analogue colloidal substances, J. Chem. Soc. 17 (1864) 318.

38. L.L. Hench, J.K. West, The sol-gel process, Chem. Rev. 90 (1990) 33–72.

39. L. Isley, R.L. Penn, Titanium dioxide nanoparticles: effect of Sol–Gel pH on phase composition, particle size, and particle growth mechanism, J. Phys. Chem. C. 112 (2008) 4469–4474.

40. M. Ristic, S. Musi, M. Ivanda, S. Popovi, Sol–gel synthesis and characterization of nanocrystalline ZnO powders, J. Alloys and Compd. 397 (2005) L1–L4.

41. B.A. Roberts, C.R. Strauss, Toward rapid, "green", predictable microwave-assisted synthesis, Acc. Chem. Res. 38 (2005) 653–661.

42. S.C. Motshekga, S.K. Pillai, S.S. Ray, K. Jalama, R.W.M. Krause, Recent trends in the microwave-assisted synthesis of metal oxide nanoparticles supported on carbon nanotubes and their applications, J. Nanomater. (2012) 51.

43. A.B. Panda, G. Glaspell, M.S. El-Shall, Microwave synthesis of highly aligned ultra narrow semiconductor rods and wires, J. Am. Chem. Soc. 128 (2006) 2790–2791.

44. A.A. Al-Ghamdi, F. Al-Hazmi, R.M. Al-Tuwirqi, F. Alnowaiser, O.A. Al-Hartomy, F. El Tantawyd, F. Yakuphanoglu, Synthesis, magnetic and ethanol gas sensing properties of semiconducting magnetite nanoparticles, Solid State Sci. 19 (2013) 111–116.

45. M.G. Ma, Y.J. Zhu, G.F. Cheng, Y.H. Huang, Microwave synthesis and characterization of ZnO with various morphologies, Mater. Lett. 62 (2008) 507–510.

46. M. Yahaya, S.T. Tan, A.A. Umar, C.C. Yap, M.M. Salleh, Synthesis of ZnO nanorod arrays by chemical solution and microwave method for sensor application, Key Engg. Mater. 605 (2014) 585–588.

47. J. Rossignol, P.D. Stuerga, Metal oxide nanoparticles obtained by microwave synthesis and application in gas sensing by microwave transduction, Key Engg. Mater. 605 (2014) 299–302.

48. V. Moghimifar, A. Raisi, A. Aroujalian, N.B. Bandpey, Preparation of nano crystalline titanium dioxide by microwave hydrothermal method, Adv. Mater. Res. 829 (2014) 846–850.

49. A. Cirera, A. Vila, A. Cornet, J.R. Morante, Properties of nanocrystalline SnO_2 obtained by means of a microwave process, Mater. Sci. Engg. C. 15 (2001) 203–205.
50. G.A. Babu, G. Ravi, Arivanandan, M. Navaneethan, Y. Hayakawa, Facile synthesis of nickel oxide nanoparticles and their structural, optical and magnetic properties, Asian J. Chem. 25 (2013) S39–S41.
51. R. Pana, Y. Wu, Q. Wang, Y. Hong, Preparation and catalytic properties of platinum dioxide nanoparticles: a comparison between conventional heating and microwave-assisted method, Chem. Engg. J. 153 (2009) 206–210.
52. C.R. Michela, A.H. Martínez-Preciado, CO sensor based on thick films of 3D hierarchical CeO_2 architectures, Sens. Actuators B. 197 (2014) 177–184.
53. A. Lagashettya, V. Havanoorb, S. Basavaraja, S.D. Balaji, A. Venkataraman, Microwave-assisted route for synthesis of nanosized metal oxides, Sci. Tech. Adv. Mater. 8 (2007) 484–493.
54. S. Li, T. Zhang, R. Tang, H. Qiu, C. Wang, Z. Zhou, Solvothermal synthesis and characterization of monodisperse superparamagnetic iron oxide nanoparticles, J. Magn. Magn. Mater. 379 (2015) 226–231.
55. E. Zhang, Y. Tang, K. Peng, C. Guo, Y.M. Zhang, Synthesis and magnetic properties of core–shell nanoparticles under hydrothermal conditions, Solid State Comm. 148 (2008) 496–500.
56. C. Díaz, M.L. Valenzuela, G. Carriedoc, N. Yutronic, Solid state synthesis of micro and nanostructured metal oxides using organometallic-polymers precursors, J. Chil. Chem. Soc. 59 (2014).
57. T. Koutzarova, S. Kolev, Ch. Ghelel, D. Paneva, I. Nedkov, Microstructural study and size control of iron oxide nanoparticles produced by microemulsion technique, Phys. Stat. Sol. C. 3 (2006) 1302–1307.
58. R.A. Zargar, Kundan Kumar. M. Shkir, Optical characteristics ZnO film: a metlab based computer calculation, under different thickness, Physica B. 63 (2022) 414634.
59. T. Tsuzuki, Mechanochemical synthesis of metal oxide nanoparticles, Commun. Chem. 4 (2021) 143 (11 pp).
60. B. Szczesniak, J. Choma, M. Jaroniec, Recent advances in mechanochemical synthesis of mesoporous metal oxides, Mater. Adv. 2 (2021) 2510–2523.
61. L. Shen, N. Bao, K. Yanagisawa, K. Domen, A. Gupta, C.A. Grimes, Direct synthesis of ZnO nanoparticles by a solution-free mechanochemical reaction, Nanotechnol. 17 (2006) 5117–5123.
62. G. Helen Annal Therese, P. Vishnu Kamath, Electrochemical synthesis of metal oxides and hydroxides, Chem. Mater. 12 (2000) 1195–1204.
63. O. Lebedeva, D. Kultin, L. Kustov, Electrochemical synthesis of unique nanomaterials in inic liquids, Nanomater. 11 (2021) 3270 (32 pp.).
64. S. Costovici, A. Petica, C.-S. Dumitru, A. Cojocaru, L. Anicai, Electrochemical synthesis on ZnO Nanopowder involving choline chloride based ionic liquids, Chem. Eng. Trans. 41 (2014) 343–348.
65. L. Anicai, A. Petica, D. Patroi, V. Marinescu, P. Prioteasa, S. Costovici, Electrochemical synthesis of nanosized TiO2 nanopowder involving choline chloride based ionic liquids, Mater. Sci. Eng. B. 199 (2015) 87–95.
66. K.S. Suslick, S.-B. Choe, A.A. Cichowlas, M.W. Grinstaff, Sonochemical synthesis of amorphous iron, Nature. 353 (1991) 414.
67. R. Hiller, S.J. Putterman, B.P. Barber, Spectrum of synchronous picosecond sonoluminescence, Phys. Rev. Lett. 69 (1992) 1182.
68. B.P. Barber, S.J. Putterman, Observation of synchronous picosecond sonoluminescence, Nature. 352 (1991) 318–320.
69. A. Gendanken, Sonochemistry and its applications in nanochemistry, Curr. Sci. 85 (2003) 1720–1722.

70. Z.X. Tang, Z. Yu, Z.L. Zhang, X.Y. Zhang, Q.Q. Pan, L.E. Shi, Sonication-assisted preparation of CaO nanoparticles for antibacterial agents. Quim. Nova. 36 (2013) 933–936.

71. J.A. Fuentes-García, A.C. Alavarse, A.C.M. Maldonado, A. Toro-Cordova, M.R. Ibarra, G.F. Goya, Simple sonochemical method to optimize the heating efficiency of magnetic nanoparticles for magnetic fluid hyperthermia, ACS Omega. 5 (2020) 26357–26364.

72. A.P. Nagvenkar, I. Perelshtein, Y. Piunno, P. Mantecca, A. Gedanken, Sonochemical one-step synthesis of polymer-capped metal oxide nanocolloids: antibacterial activity and cytotoxicity, ACS Omega. 4 (2019) 13631–13639.

73. P. Aarthye, M. Sureshkumar, Green synthesis of nanomaterials: an overview. Mater. Today Proc. 2021, https://doi.org/10.1016/j.matpr.2021.04.564.

74. I. Bibi, N. Nazar, S. Ata, M. Sultan, A. Ali, A. Abbas, K. Jilani, S. Kamal, F.M. Sarim, M.I. Khan, F. Jalal, M. Iqbal, Green synthesis of iron oxide nanoparticles using pomegranate seeds extract and photocatalytic activity evaluation for the degradation of textile dye, J. Mater. Res. Tech. 8 (2019) 6115–6124.

75. B. Balraj, N. Senthilkumar, C. Siva, R. Krithikadevi, A. Julie, I.V. Potheher, M. Arulmozhi, Synthesis and characterization of zinc oxide nanoparticles using marine streptomyces sp. with its investigations on anticancer and antibacterial activity, Res. Chem. Intermed. 43 (2017) 2367–2376.

76. D. Wang, Z. Lin, T. Wang, Z. Yao, S. Zheng, W. Lu, Where does the toxicity of metal oxide nanoparticles come from: the nanoparticles, the ions, or a combination of both? J. Hazar. Mater. 308 (2016) 328–334.

77. A. Nel, T. Xia, L. Madler, N. Li, Toxic potential of materials at the nanolevel, Sci. 311 (2006) 622–627.

78. M.A. Kohanski, D.J. Dwyer, B. Hayete, C.A. Lawrence, J.J. Collins, A common mechanism of cellular death induced by bactericidal antibiotics, Cell. 130 (2007) 797–810.

79. S.R. Kumar, J.A. Imlay, How escherichia coli tolerates profuse hydrogen peroxide formation by a catabolic pathway, J. Bacteriol. 195 (2013) 4569–4579.

80. T. Saito, T. Iwase, J. Horie, T. Morioka, Mode of photocatalytic bactericidal action of powdered semiconductor TiO2 on mutans streptococci, J. Photochem. Photobiol. B. 14 (1992) 369–379.

81. A. Ayala, M.F. Munoz, S. Arguelles, Lipid peroxidation: production, metabolism, and signalling mechanisms of malondialdehyde and 4-hydroxy-2-nonenal, Oxid. Med. Cell. Longev. 2014 (2014) 360438.

82. A. Gaballa, J.D. Helmann, Identification of a zinc-specific metalloregulatory protein, zur, controlling zinc transport operons in bacillus subtilis, J. Bacteriol. 180 (1998) 5815–5821.

83. K. Giannousi, K. Lafazanis, J. Arvanitidis, A. Pantazaki, C. Dendrinou-Samara, Hydrothermal synthesis of copper based nanoparticles: antimicrobial screening and interaction with DNA, J. Inorg. Biochem. 133 (2014) 24–32.

84. J. Kirstein, K. Turgay, A new tyrosine phosphorylation mechanism involved in signal transduction in bacillus subtilis, J. Mol. Microbiol. Biotechnol. 9 (2005) 182–188.

85. S. Shrivastava, T. Bera, A. Roy, G. Singh, P. Ramachandrarao, D. Dash, Characterization of enhanced antibacterial effects of novel silver nanoparticles, Nanotechnol. 18 (2007) 225103.

86. T.J. Beveridge, Structures of gram-negative cell walls and their derived membrane vesicles, J. Bacteriol. 181 (1999) 4725–4733.

87. Y.C. Chung, Y.P. Su, C.C. Chen, G. Jia, H.L. Wang, J.C. Wu, J.G. Lin, Relationship between antibacterial activity of chitosan and surface characteristics of cell wall, Acta. Pharmacol. Sin. 25 (2004) 932–936.

2 Environmentally Friendly Green Approaches and Applications of Nanoparticles

Anita Rani, Nirantak Kumar, and Manoj Kumar

CONTENTS

DOI: 10.1201/9781003323464-2

2.1 INTRODUCTION

Today, one of the most active and important research areas in modern material science and technology is nanotechnology. There are many different types of nanostructured materials available, such as nanoparticles, nanopores, nanotubes, and so on [1–8]. The nanoparticle is one of the most important components of nanotechnology. The nanoparticles show a broad range of applications not only in electronics, materials science, physics, optics, and chemistry but also in health sciences or biomedical sciences, among others. There are various types of nanoparticles, including gold (Au) [9–12], silver (Ag) [13–16], titanium (Ti) [17–18], zirconium (Zr) [19], strontium (Sr) [20], and others that can be used for various applications. Nanoparticles are particles with a diameter of 1–100 nm; these nanoscale particles possess new, superior, and distinct biological, chemical, and physical properties. Metal nanoparticles are the most studied materials because of their ease of preparation. Furthermore, they can be used as surface-coating agents, detectors, catalysts, antimicrobials, and others. Silver [21, 22], gold [23], platinum [24–27], and palladium are examples of some of the most studied metal-based nanoparticles. The study of these metal-based nanoparticles is particularly useful in the field of health and medicine. Silver has been found to be an effective antimicrobial agent due to the interaction of silver ions with macromolecules present in the cell, like deoxyribonucleic acid, and proteins that has the ability to destroy bacterial cell growth and alter the cell metabolism (DNA). When the Ag ion comes into contact with a living cell, it hinders the synthesis of the protein, reduces the permeability of the membrane, and, finally, is responsible for cell death. Chemically, silver nanoparticles have shown a greater reactive nature than silver in bulk. As a result, silver-based nanoparticles are thought to have better antibacterial properties [28–30].

Several processes, including chemical reduction, can be used to make metal-based nanoparticles. Chemical reduction procedures are frequently utilized because they are more convenient and cost-effective [31]. This approach uses reducing chemicals like sodium citrate or sodium borohydride to reduce metal salts [32]. However, the use of chemicals (such as reducing agents, organic solvents, etc.) in the production of nanoparticles causes the production of hazardous substances that adsorb on the material's surface that cause negative and damaging consequences with its applications [33].

As a result, it is preferable to employ ecologically friendly approaches to synthesize effective and nontoxic nanoparticles. It is possible to synthesize nanoparticles by using only green approaches. By using plants, microorganisms, or natural products, green synthesis methods have been adopted to synthesize metal-based nanoparticles [34].

The content of secondary metabolites as reducing agents has been used in the formation of nanoparticles by using various parts of plants, bacteria, algae, fungi, and yeast, among others, as raw materials [35]. Biological substances are said to serve as stabilizers, reducers, or both in the formation of green nanoparticles [36]. The biological synthesis of many metal nanoparticles was designed by using a variety of plants and their antibacterial activity has been assessed [37]. Nanomaterials have structural characteristics that are halfway between those of atom and bulk materials. While the properties of most micro-structured materials are similar to those of corresponding bulk materials, the properties of materials with nanoscale dimensions of atoms and bulk materials are drastically different. This is due to the small size (nanoscale) of the materials, which results in

1. a high surface energy,
2. a large fraction of atoms present on the surface,
3. reduced imperfections, and
4. spatial confinement.

2.2   TYPES OF NANOPARTICLES

Nanoparticles can be divided into different types on the basis of their composition. On the basis of composition, nanoparticles are classified as organic nanoparticles (e.g., polymers, liposome), inorganic nanoparticles (ceramic, quantum dots, metals) and carbon-based nanoparticles (organic colloids, etc.). However, on the basis of origin, they are classified as natural nanoparticles and anthropogenic nanoparticles (C-containing and inorganic nanoparticles).

Natural and manmade nanoparticles can be distinguished as shown in Table 2.1. The particles can be further divided into carbon-containing and inorganic nanoparticles based on their chemical composition, that is, geogenic, biogenic, pyrogenic and atmospheric.

Fullerenes and carbon nanotubes of biogenic magnetite, pyrogenic or geogenic origin and atmospheric aerosols (organic and inorganic like sea salt and others) are examples of natural nanoparticles. Anthropogenic nanoparticles can be produced accidentally as a by-product, most commonly during combustion, or purposely owing to their unique properties. They're often referred to as engineered or made nanoparticles in the latter situation. Fullerenes and carbon nanotubes (pristine and functionalized), as well as metals and their oxides like silver and titanium, are some examples of designed nanoparticles. Engineered

TABLE 2.1

Type of Nanoparticles on the Basis of Origin

Sr. No.	Mode of origin	Type	Formation	Nature	Examples
1.	Natural nanoparticles	Carbon-containing nanoparticles	Geogenic	Soot	Fullerenes
			Biogenic	Organic colloids organisms	fulvic acids, Humic
			Pyrogenic	Soot	Nanoglobules, CNT Fullerenes, nanospheres onion-shaped
			Atmospheric	Aerosols	Organic acid
		Inorganic nanoparticles	Geogenic	Oxide Clays	
			Atmospheric	Aerosols	Sea salt
			Biogenic	Oxide Clay	Magnetite Au, Ag
2.	Anthropogenic (manufactured engineered) nanoparticles	Carbon-containing	Engineered	Soot	Carbon black Fullerenes Functionalized CNT, fullerenes
			By-products	Combustion by-products Polymeric nanoparticles	Polyethyleneglycol, nanoparticles, CNT, nanoglobules, onion-shaped nanospheres nanoparticles
		Inorganic nanoparticles	Engineered By-products	Oxides Salts Metals Alumino-silicates Combustion by-products	SiO_2, TiO_2, Ag, Fe Metal-phosphates Zeolites, ceramics, clays and metals of Platinum group

nanoparticles are the focus of contemporary environmental research, but many of them are also found naturally, such as fullerenes and inorganic oxides, among others. The various sorts of natural and man-made materials are discussed in the following sections.

2.2.1 Natural and Unintentionally Fabricated Carbon Nanotubes and Fullerenes

Fullerenes and carbon nanotubes are regarded as manmade nanoparticles; they are also considered as natural particles or have near environmental cousins. While some of these fullerenes may have come to earth via comets or asteroids [38], the vast majority are thought to have been generated from algal matter while metamorphosis at a temperatures range between 300 to 500°C in the presence of elemental sulfur [39] or during the natural processes of combustion.

2.2.2 Natural and Unintentionally Generated Inorganic Nanoparticles

Natural inorganic nanoparticles can come from the atmosphere, geology or biological agencies. Inorganic nanoparticles are found everywhere in geological systems [40, 41]. Nanoparticles are also known as common aerosols in the environment, and they are pioneers in the creation of bigger particles that are known to have a significant impact on atmospheric chemistry, global climate, visibility and pollution transport on a regional and global scale [42]. Soil dust and sea salt are examples of primary atmospheric nanoparticles, although coarse particles make up the majority of the mass fraction. Based on mass, the average particle size of mineral dust (airborne) is in a range between 2 to 5 mm. The average particle size depending on the number of particles is roughly 100 nm, with a significant percentage falling below this figure [43]. Rhodium (Rh)-and platinum (Pt)-containing particles created by car catalytic converters are a unique type of inadvertently produced nanoparticles. Although the majority of Rh and Pt are linked with coarser particles, roughly 17% has been discovered to be related with tiniest aerosol fraction, that is, 0.433 mm [44].

2.2.3 Engineered Fullerenes and Carbon Nanotubes

Buckminsterfullerene C_{60} is by far the best studied among the large family of fullerenes. Fullerenes are primarily proposed for use in composites of fullerene polymer, electro-based optical devices, thin films and numerous biological applications [45, 46]. Due to the limited water solubility of fullerenes, much effort has been devoted to functionalization and a plethora of C60 derivatives have been synthesized, each with their own set of characteristics and properties [47]. Carbon nanotubes are the trendiest topic in the area of physics right now [48]. A range of distinct carbon nanotubes with highly varying properties are created, which depend on the synthesis method, the methodology employed for the separation from amorphous by-products, following cleaning stages and finally different functionalization [49]. The potential of altering the characteristics of carbon nanotubes is being investigated in biological and medical applications [50].

2.2.4 Soot

Anthropogenic and natural combustion processes, which occur in fixed and mobile sources, produce a diverse range of particles, so-called ultra-fine particles that fit the traditional definition of nanoparticles. The scope of this study is limited to the "soot" component of the black carbon combustion continuum and refers to nanosized black carbon as soot. Soot is discharged into the atmosphere because of recondensation after incomplete combustion of renewable and fossil fuels [51], where it spreads throughout the hemisphere and settles on soils and aquatic bodies. Carbon black is a type of soot that is used in industry for a variety of purposes, including filler in rubber compounds, notably in automotive tires, and others. Carbon black particles are partially nanometer-sized with approximate values ranging from 20 to 300 nm for various nanoparticles [52, 53].

2.2.5 Organic Colloids

In natural waters, colloidal matter is described as molecular assemblies, particles and macromolecules with a size of about 1 nm and hence fall into the size range of nanoparticles to some extent. Inorganic colloids and big biopolymers, such as polysaccharides and peptidoglycans, are examples of environmental colloids. Colloids must be regarded as an important component of the nonbiological medium that supports life [54]. However, our understanding of natural colloid structure and environmental impact has improved greatly in recent years; hence, their composition and precise function remain unknown.

2.2.6 Aerosol

Recently, the methods of synthesis based on aerosol have gained popularity because of their ease, versatility, low cost, small particle shape and size and the capability of synthesizing particles with negligible toxic environmental effects. The techniques based on aerosol provide nanoparticles with very high purity. In comparison to traditional wet-chemical methods, this method does not involve a large number of steps [55]. This system of nanoparticles preparation is being most widely acquired on a commercialized scale as it permits the collection of nanoparticles in a single step and results in little waste production [56].

2.2.7 Engineered Polymeric Nanoparticles

In medicine, nanoparticles made from organic polymers have piqued the attention of researchers as medication carriers. Polymers are particularly used to design nanoparticles with customized properties due to their ability to manipulate morphology, size, composition surface and charge. These nanoparticles are ultimately taken up by various living cells, and their capability to traverse the blood-brain barrier is being investigated [57]. For soil and groundwater remediation, several forms of polymeric nanoparticles have been produced and proposed [58, 59]. It has been found that micelles, such as amphiphilic polyurethane particles with exterior side (hydrophilic) and inner core (hydrophobic), are well adapted for withdrawing hydrophobic contaminants from the soils. The nanoparticles remove polycyclic aromatic hydrocarbons in the same way that micelles do, but unlike micelles the polymeric nanoparticles, do not bind to soil particles. Dendrimers, which act as water-soluble chelates, are another polymeric nanoscale substance [60].

2.2.8 ENGINEERED INORGANIC NANOPARTICLES

Engineered inorganic nanoparticles are made up of a variety of materials, including metal, metal salts and oxides of metals. Element silver (Ag) is employed as a bactericide in numerous conditions [61], whereas elemental gold is being studied for its catalytic activity [62]. The utilization of zero-valent iron (nanoscale) for remediating groundwater is the most studied nanotechnological application based on the environment [63].

2.2.9 METAL OXIDES

Nanoparticles obtained from oxides of metals are one of the most commonly used nanoparticles [64]. For many years, bulk materials of iron oxide, titanium oxide, aluminum oxide and silicon oxides have been manufactured. However, they have lately been synthesized in a nanoscale form and have already made their way into the consumer market, such as ZnO in sunscreens [65]. Pigments, photocatalysis and cosmetic additives are just a few of the applications for TiO_2 nanoparticles.

2.3 EFFECT OF CHEMICALLY SYNTHESIZED NANOPARTICLES ON ENVIRONMENT AND LIVING ORGANISMS

Environmental impacts of nanoparticles require detailed knowledge of characteristics, such as physicochemical properties, identification, method of emission and toxicity level for the environment [66]. The toxicity impact depends on the number of and frequency that nanoparticles reach the environment. The quantitative analysis and evaluation of nanoparticles released into the environment need deep research right from the raw materials required to produce them to their emission into the environment [67].

2.3.1 ON SOIL

A large number of nanoparticles is found to penetrate the environment and the soil. They may be accidently passed into the soil through various agencies, like wind and in excretory material from the laboratory during synthesis and so on, and show their potential toxicity, hence disturbing the natural behavior of soil [68]. The environment's soil can be examined by measuring the activity of the enzymes present in soil. Along with soil, nanoparticles affect the organisms found in soil, like earthworms, insects and others. The burrowing activity of earthworms is very important and is found to be crucial for increasing the fertility power of soil, increasing water filtration and minimizing the erosion effect. That's why earthworms were used to observe the effects of nanoparticles, and some experiments were designed to detect their effect on them. One of the most important methods used to find the biological changes in earthworms is targeting the apoptotic process. Apoptotic cells are found in intestinal and cuticle epithelium. These parts were directly exposed to nanoparticles, and it has been observed that they affect the mucus, antibacterial molecules, nutrient absorption and immune protection supported by chloragogenous tissue. *Eisenia fetida* (species of earthworm) was exposed to 4 nm Co nanoparticles over seven days, and evidence shows that this organism retained the nanoparticles for eight weeks and that only 20% of the ingested nanoparticles were discharged from the body. These nanoparticles were found in the cocoons and blood along with in spermatogenic cells [69].

Similarly, nanoparticles of titanium oxide enter several layers of soil, but they are found to be inert. TiO_2 nanoparticles show toxic effects in the presence of sunlight and water as they form reactive species, that is, free radicals. These nanoparticles damage the DNA of humans and animals in the presence or absence of sunlight [70, 71].

2.3.2 On Terrestrial Animals

Terrestrial animals include all the animals living on earth, especially human beings. Due to their large surface area and very small size, nanoparticles can enter the body through various modes, such as inhalation, transcutaneous mode and others, but in all modes, the small size of nanoparticles enables endocytosis to infiltrate the cell and then transcytosis to enter other cells. So upon inhalation, they reach the nervous cells of the olfactive epithelium, then the axons, and then the olfactive bulbs in the brain, and here, they affect the neurons. Experiments with mice and rats confirm the actual harmful effects of nanoparticles on brain. At about 1.5 µg/ml concentration, Cu nanoparticles produce the expansion of endothelial cells in brain capillaries. In another experiment, Ag nanoparticles affected the blood-brain barrier after an exposure of 24 hours [72]. It has been found that the size of nanoparticles plays an important role in the spread of disease. For example, in adult rats, Ag nanoparticles with a particle size of 25–40 nm induce more cytotoxic effects compared to nanoparticles with 80 nm and result in the reduction of its locomotor activity [73]. Various nanoparticles also have the ability to reach in lungs, liver, bone marrow, spleen, lymph, heart and more. In these body parts, they are responsible to induce oxidative stress, antioxidant activity, prooxidant activities and the like [74].

2.3.3 On Amphibia

Amphibians or semiaquatic animals are those species that spend their larval life in water and breathe with gills. When they become terrestrial, they develop lungs and their gills disappear. Hence, their circulatory system also gets modified. Apoptotic cells have also been found in tadpoles [75]. As metamorphosis is under the control of the pituitary and thyroid glands, so amphibians represent a unique model for showing the toxic effects of nanoparticles. An experiment was performed on a species of tadpole, that is, *Lithobates catesbeianus*, to observe the effect of TiO_2 nanoparticles. The results were evaluated after exposure of 20 nm of nanoparticles through genes transcription encoding of the receptors of thyroid hormones (thrb and thra) implicated in the metamorphosis, catalase (cat), superoxide dismutase (sod), for rlkI (larval keratine type I) and hsp30 (stress proteins) [76]. Different nanoparticles of TiO_2, CuO and ZnO did not result in the death of the embryo but affected the intestine in concentrations of more than 50 mg/L. In particular, nanoparticles of ZnO that activate the adverse effects on the intestinal barrier, and hence, allow nanoparticles to reach the connective tissue [77]. However, some ZnO-based nanoparticles can also improve the different visual functions in the body. A study of electroretinograms identified that nanoparticles enhance the wavelength amplitude in tadpoles adapted to anonymity on exposure to sunlight. Also, they improve visual sensitivity and decrease the time of light-sensitive pigment (rhodopsin) regeneration [78].

The identification of gold nanoparticles has also been done in other organs of amphibians such as the spleen, liver, intestine, kidney, muscle stomach and so on. They have shown the presence of about 0.09% on direct ingested of nanoparticles and 0.12% when bullfrogs were fed earthworms. These results clearly reveal the possibility of a trophic transfer of gold nanoparticles [79].

2.3.4   ON AQUATIC INVERTEBRATES AND VERTEBRATES

Some freshwater species such as *Thamnocephalus paatyurus* and *Daphnia magna* have been used to observe the toxic effects of nanoparticles. Results have shown that exposure to TiO_2, CuO and ZnO nanoparticles leads to the accumulation of these particles in the gastrointestinal tract and other parts of the body [80]. Similarly, chronic effects, such as effects on reproduction and growth, have also been found [81]. The effect of exposure to CuO nanoparticles has also been reported in the form of cellular toxicity and cell death through biochemical pathway of cell structure, transcription regulation, cytoskeleton, oxidative stress, proteolysis and apoptosis [82]. The zebrafish, *Danio rerio*, is a fish that is accepted by regulatory agencies for investigating nanoparticle toxicity because of its well-known biology. During the zebrafish's early development and larval development stages, many nanoparticles, such as Cu nanoparticles, have shown toxic effects, such as increased tissue damage and decreased hatching and survival rate [83–85]. Metal ions liberated from some metal-based nanoparticles, that is, TiO_2, ZnO and Ag, are also responsible for their toxicity in water [86–88].

2.3.5   ON ALGAE AND PLANTS

Various studies have been done to observe the effect of nanoparticles on algae and plants. The harmful effect is due to the dissolution, release and finally uptake of nanoparticles. It has been observed that nanoparticles are responsible for retarded growth of marine phytoplankton species like *Duanliella teriolecta*, *Skeletonema marioni* and

Thalassiosira pseudonana [89–91]. In plants, exposure to aluminum nanoparticles has been shown to affect the growth of roots [92]. However, high concentrations of ZnO nanoparticles slow the process of seed germination in ryegrass and result in toxicity from both particle dissolution to Zn ions and their particle-dependent effects [93].

2.3.6 On Microorganisms

A broad range of nanoparticles are found to show antimicrobial activities and hence contribute as effective antifungal and antibacterial agents [94, 95]; for example, Nano-Ag, Nano-Cu, Nano-Ni and Zn have shown the ability to inhibit the growth of *Staphylococcus aureus* [96, 97]. *Escherichia coli* has also been found to show the toxic effects of nanoparticles such as ZnO, TiO_2 and Ag, among others. Along with it, other microorganisms like protozoa and yeast are also affected by nanoparticles of ZnO, TiO_2, CuO and others [98, 99], and studies have shown that some nanoparticles of silver do not show their harmful effects not only on bacteria but also on mammalian (liver, brain, etc.) cells, fish, algae, fungi, plants, bacteria, crustaceans, soli-forming chemolithotrophic bacteria and nitrogen-fixing bacteria, among others [100–104].

2.3.7 On Human Beings

The optical, chemical and electronic properties of nanoparticles are responsible for the discovery and development of novel nanoparticles that have a wide range of applications. However, this development and expansion of different nanoparticles result in serious risks to the health of living things, especially humans [105]. Nanoparticles are known to affect the skin, gastrointestinal tract, lungs and other body parts in humans as shown in Table 2.1. Hence, "nanotoxicology" represents the concept that nanoparticles are a particular and unique class of toxins that differ from other pathogenic particles [106–108]. The mechanism for nanoparticle exposure to living things and the environment is different. The prime exposure can occur to workers during research-scale synthesis, at the production site of nanoparticle-based materials and during handling of raw materials, packaging, characterization and transportation, among others. Even the introduction of nanoparticles into the human body can also take place through different mechanisms:

• Penetration of nanoparticles through skin
• Insertion by respiratory system through inhalation
• Intake by digestive system through ingestion

2.3.7.1 Nanoparticle-Induced ROS Generation and Oxidative Stress

Antioxidant defenses include low-molecular-weight agents like proteins, enzymes and others that play the role of scavenger for ROS. Excessive ROS may result in the generation of diseases in the body, which is regarded as the condition of oxidative stress. So oxidative stress is due to an imbalance between oxidants and antioxidants that results in the overall enhancement in the ROS level [109]. Evidence indicates that oxidative stress has been caused by different nanoparticles and hence negatively affects the cells of the body. Metal-based nanoparticles such as nickel (nano-Ni), silver (nano-Ag), cobalt (nano-Co) and zinc (nano-Zn), among others, are known to induce an enhancement in intracellular ROS by using a probe, that is, H_2DCF-DA [110–113]. H_2DCF-DA is administrated in the form of diacetate, which is further deacetylated in living cells and converted to less permeable H_2DCF. The H_2DCF gets transformed into DCF (fluorescent dichlorofluorescein) by the action of ROS. However, DCF is capable of photo amplification for eliminating nonspecific reactions. HE (hydroethidine) has also been used to evaluate the production of intracellular ROS with submission to different metal nanoparticles [114, 115].

2.3.7.2 Systemic Toxicity

Due to their physical resemblance to physiological molecules such as proteins, nanoparticles show tendency to revolutionize medical imaging, therapeutics, diagnostics and various biological processes. It has been found that the nanoparticles have the ability to translocate into the blood from the lungs, which results in the systemic revelation of internal organs. However, neuronal uptake can be a translocation route from airways [116]. The extent of toxicity varies with the site of deposition and the mode of administration of nanoparticles.

2.3.7.3 Genotoxic Effects of Nanoparticles

Evidence indicates that the toxicity of some nanoparticles increases with a decrease in particle size [134, 135]. Nanoparticles of arsenic (As), chromium (Cr), cadmium (Cd) and nickel (Ni), among others, are found to be carcinogenic in both humans and

TABLE 2.2
Harmful Effects of Nanoparticles of Living Things

S. No.	Nano Particle	Target	Concentration (time/size)/ Route of Administration	Cellular Target	Animal Target	Major Effects	Ref
1.	Silica-based nanoparticles	Lungs	Approximately 10 to 100 g/ml for 24 h, 48 h and 72 h	A549 lung cancer cells of human		The Oxidative stress indicated the mode of Cytotoxicity.	[117]
2.	Silica-based nanoparticles	Lungs	Approximately 0–185 g/ml for 24 h.	endothelial cells (EAHY926) J774, human lung cancer cells monocyte (A549) macrophages		Enhancement in dose-dependent cytotoxicity.	[118]

(Continued)

TABLE 2.2 *(Continued)*
Harmful Effects of Nanoparticles of Living Things

S. No.	Nano Particle	Target	Concentration (time/size)/ Route of Administration	Cellular Target	Animal Target	Major Effects	Ref
3.	Silica-based nanoparticles	Lungs	Approximately 20 mg for 1 or 2 months through Intratracheal administration		Wistar rats	Lower pulmonary fibrosis induced by nanosized silica particles as compared to particles with micro-size due to the transfer of ultrafine nanosilica from the lung parenchyma.	[119]
4.	Silver-based nanoparticles	Dermal	Concentration 0.76–50 g/ml for 24 h	A431 (skin carcinoma in human)		Rise in lipid peroxidation, GSH and SOD at concentrations between 6.25–50 g/ml. DNA fragmentation results in cell death due to apoptosis.	[120]
5.	Silver-based nanoparticles	Dermal	dressing of wound with silver-coated Acticoat™ for 7 days		burns patient (human)	Discoloration of area of skin under treatment (argyria-like), elevated concentration of silver in urine and plasma with increasing amount of liver enzymes.	[121]
6.	Silica-based nanoparticles	Dermal	Approximately 70, 300 and 1000 size in nanometers	Cells of murine Langerhans (XS52)		Enhancement in toxicity with faster cellular uptake due to smaller particles and concomitant toxicity.	[122]

S. No.	Nano Particle	Target	Concentration (time/size)/ Route of Administration	Cellular Target	Animal Target	Major Effects	Ref
7.	Silver-based nanoparticles	Dermal	Approximately 50 and 100 g/m treatment for 24 h	Mouse fibroblasts (NIH3T3)		ROS-associated cellular apoptosis in mitochondria.	[123]
8.	TiO_2-based nanoparticles	Dermal	Applied approximately 2 mg/cm² as sunscreen on volar forearm for 1st 3 days		Volunteers (human)	Less than 1% of total amount of sunscreen applied could be identified within pilosebaceous orifices.	[124]
9.	Gold-based nanoparticles	Dermal	Approximately (13 nm) 95, 142 and 190 g/m and (45 nm) 13, 20 and 26 g/ml for 3 to 6 days	Fibroblasts Cells of Human dermal (CF-31)		Cytotoxicity was found to depend on dose and size. Particles with 45 nm are responsible for greater toxicity at 10 g/ml while particles with 13 nm exhibited cytotoxicity at 75 g/m concentration.	[125]
10.	Gold-based nanoparticles	Dermal	Concentrations of 0 to 0.8 g/ml with 14 nm in size treated 2, 4 or 6 days.	Fibroblasts Human dermal cells		Reduction in the proliferation of cell that depend on a dose of nanoparticle.	[126]
11.	Gold-based nanoparticles	Liver	Approximately 0.14 to 2.2 mg/kg with 13.5 nm in size applied for 14–28 days through intravenous, oral or intraperitoneal route.			Administration via tail vein show less toxicity as compared to oral administration.	[127]

(Continued)

TABLE 2.2 *(Continued)*

Harmful Effects of Nanoparticles of Living Things

S. No.	Nano Particle	Target	Concentration (time/size)/ Route of Administration	Cellular Target	Animal Target	Major Effects	Ref
12.	Silver-based nanoparticles	Liver	A concentration of 23.8, 26.4 or 27.6 g/ml as single or repeated doses of 20, 80 and 110 nm, respectively, once in a day for 5 days through Intravenous route of administration.		Wistar rats	Repeated administration of nanoparticles show implications for toxicity in tissues.	[128]
13.	Silica-based nanoparticles	Liver	Approximately 0.001 g/ml for 1, 3, 7, 15 and 30 days Intravenously.		ICR mice	Liver, lungs and spleen were found with the accumulation of nanoparticles. Hepatocyte necrosis and hepatic portal area were affected.	[129]
14.	Gold-based nanoparticles	Brain	40, 200 and 400 g/kg for 8 days with 12.5 nm size administered intraperitoneal.		C57 or BL6 mice	Nanoparticles with small concentrations were found to cross the BBB but evident neurotoxicity could not be induced.	[130]
15.	Silver-based nanoparticles	Brain	Approximately 10, 25 or 50 g/ml for 1 h in a day.	Homogenates and Wistar tissue of rat		Oxidative stress and cell damage occur by decreasing the respiratory chain complexes I, II, III, and IV in mitochondria.	[131]

S. No.	Nano Particle	Target	Concentration (time/size)/ Route of Administration	Cellular Target	Animal Target	Major Effects	Ref
16.	Silver-based nanoparticles	Brain	With 60 nm size, 30, 300 or 1000 mg/kg for 28 days orally.		Sprague-Dawley (rats)	Oral introduction results in the accumulation of nanoparticles in the brain and other parts of the body. High dose may also result in the enhancement in cholesterol level that indicate hepatotoxicity.	[132]
17.	CdSe QD–based nanoparticles	Brain	50 nmol cadmium with concentration of 0.68 mg containing for six hour and 13.5 nm size intraperitoneally.		ICR mice	High concentration of Cd ions accumulates in brain tissue but no signs of parenchymal damage and inflammation were observed.	[133]

animals [136–138]. Different routes have been suggested to explain nanoparticle-induced carcinogenesis as follows:

- Effects of redox signaling and ROS on the normal functioning of cells and on the mutation of genomic DNA
- Decreasing the defenses of the antioxidant
- Activation of various factors of nuclear transcription, such as NFAT, NF-κB and AP-1
- Apoptosis induced by metals
- Metal effects on the growth regulation of cell

Metal-induced tumor suppressor and oncogene activation such as p53 suppression show the genotoxic effects in living organisms [139–141]. Metals like cadmium, nickel, arsenic and chromium disturb the methylation level of DNA and gene-specific histone-tail post-translational modification marks. These nanoparticles have the ability to perforate cells by using various methods such as receptor-mediated endocytosis, clathrin and passive diffusion, as well as diffusion within the nuclear membrane and nuclear pore, or may persist in nucleus.

The presence of nanoparticles inside the nucleus results in a direct interaction between DNA or DNA-related proteins and nanoparticles, which ultimately leads to physical and chemical changes in genetic material [142–144].

2.3.7.4 Inflammation and Nanoparticles

Inflammation due to injury is an ordinary condition, but an excess of inflammation can cause inflammatory disorders, including inflammatory bowel disease, arthritis and dermatitis, among others [145–148]. Excessive immune responses activate the function of immunomodulatory cells, like macrophages, neutrophils and others, and are responsible for the release of various proinflammatory mediators, like ROS and cytokines. Persistent inflammation is a major reason for various diseases that result in the generation of many diseases of heart and lungs. [149–151]. Few *in vivo* and *in vitro* results have shown that nanoparticles enhance the rate of inflammation in the body. Some nanoparticles are found to damage the mitochondria without any inflammation and ultimately cause cell death. [152]. The administration of some nanoparticles synthesized using metals like Ag, silica, aluminum and others into the body causes the activation of inflammation in macrophages that leads to the excretion of the proinflammatory cytokines IL-18 and IL-1β, among others, which is mediated by ROS production and lysosomal disruption [153–155]. Several studies have revealed that exposure to silica, asbestos and aluminum salts increases the secretion of IL-1β by activating NLRP3 [156, 157]. Exposure to some nanoparticles, like nano-TiO$_2$ and nano-C, induces pulmonary inflammation due to the excretion of IL-1β [158]. Thus, it is observable that IL-1β plays a crucial role in the toxic effects generated by nanoparticles.

2.4 PRINCIPLES OF GREEN CHEMISTRY

To reduce the toxic and harmful effects of chemically synthesized nanoparticles, a remedial solution for a green approach to nanoparticle synthesis has been adopted. This is found to be an environment-friendly approach and was developed to conserve and use natural resources. A green approach aims to synthesize nanoparticles with minimal energy consumption and negligible waste production while using fewer natural raw materials [159]. There are about twelve principles of green chemistry which are shown in Figure 2.1.

- **Prevention:** This principle emphasizes prevention rather than production. It means that measures should be taken to prevent the generation of waste rather than to treat it.
- **Atom economy:** This principle refers to the synthetic procedures that must be designed to amplify the involvement of whole raw materials in the final product.
- **Less hazardous chemical syntheses:** This principle refers to the substances used and generated should not possess toxicity to the environment and animals.
- **Designing safer chemicals:** This principle refers to designing such chemical products that should give maximum efficiency and minimum toxicity.
- **Safer auxiliaries and solvents:** This principle refers to avoiding, if possible, auxiliary materials like separating agents, solvents, intermediate and others that can be toxic.

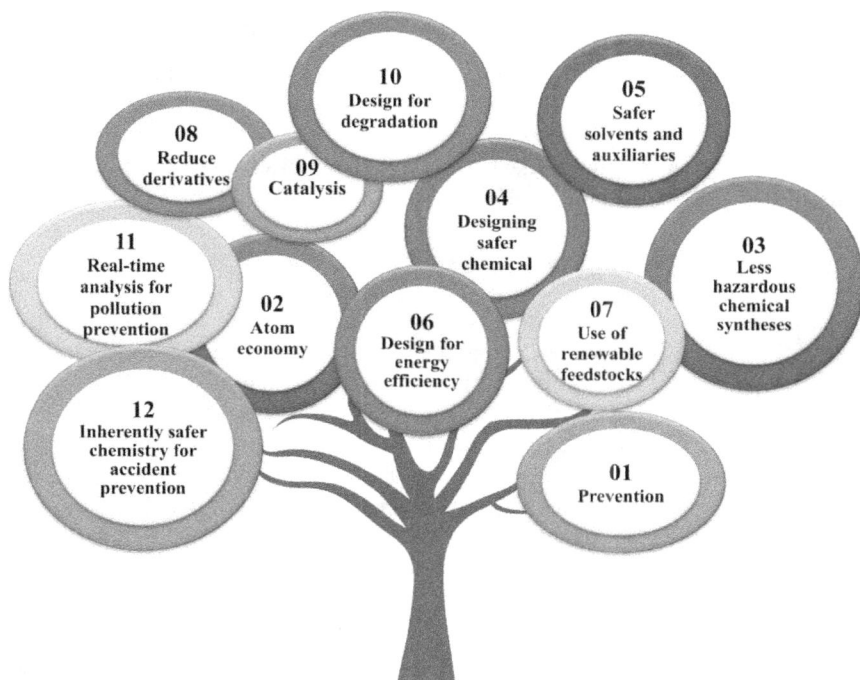

FIGURE 2.1 Principles of green chemistry.

- **Design for energy efficiency:** Arrangements should be done to do the synthesis at ambient pressure and temperature to minimize the energy requirements.
- **Use of renewable feedstocks:** Renewable starting (raw) materials must be used rather than exhaustible materials to conserve the environment.
- **Reduce derivatives:** The use of protection/deprotection, blocking groups, chemical and physical processes, among others, should be minimized to avoid using hazardous reagents.
- **Catalysis:** Selective catalytic reagents should be used in place of stoichiometric reagents.
- **Design for the degradation:** Synthetic products must be designed in a manner so that on degradation their by-products should not preserve in the environment.
- **Real-time analysis for prevention of pollution:** Analytical methods should be used right from starting through the final preparation of the product so that knowledge of hazardous substances can be taken in before reaching the final step.
- **Inherently safer chemistry for accident prevention:** Raw materials and intermediates that have the capabilities of minimizing chemical accidents, fire and explosions should be chosen.

2.5 METHODS OF SYNTHESIS OF GREEN NANOPARTICLES

Basically, there are three methods of nanoparticle synthesis, chemical, physical and biological, as well as the green method of synthesis. But in this chapter, we discuss only the green method of synthesis of nanoparticles. Synthesis can be done in one step using biological organisms like bacteria, actinomycetes, yeasts, algae, fungi and plants, among others, as shown in Figure 2.2.

2.5.1 PLANTS

The advantage of utilizing plants to synthesize nanoparticles is widely applicable because they are easily available, economical, safe to handle and show a broad variety of metabolites. A large number of plant parts have been used to synthesize metal-based nanoparticles. Nanoparticles of nickel, copper, silver, cobalt, gold, zinc and others have been synthesized using parts of *Helianthus annus* (sunflower), *Medicago sativa* (alfalfa) and *Brassica juncea* (Indian mustard), among others. Many plants are recognized for accumulating large concentrations of metals compared to others, and these are termed as hyperaccumulators.

For example, *Brassica juncea* has excellent metal-accumulating power and later assimilating the same metals as nanoparticles [160]. The role of phytochemicals is found to be very important in the preparation of nanoparticles. The main phytochemicals which are responsible have been recognized as ketones, aldehydes, terpenoids, carboxylic acids, flavones and amides, among others. Every part of the plant such as stem, root, flower, leave, seed, bark and so on is used in the preparation.

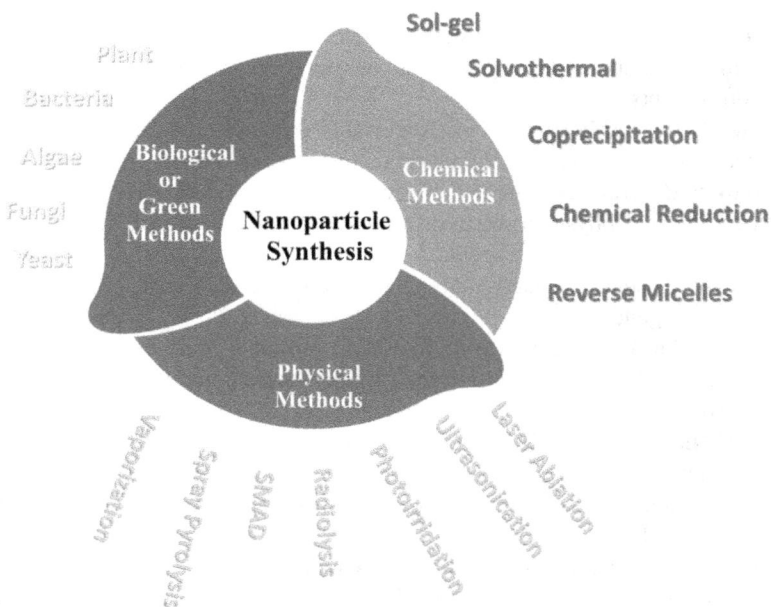

FIGURE 2.2 Methods of synthesis of nanoparticles.

Liquid extracts of plant parts and metal solutions are prepared in appropriate solvents under suitable conditions and then subjected to synthesized nanoparticles using various processes. Gold nanoparticles have been prepared using the liquid extracts of *Diospyros kaki* and *Magnolia kobus* leaves. After identifying the nanoparticles, it was reported that particles of polydisperse with an approximate size of 5–300 nm were recovered at lower temperatures while higher temperatures assisted in the formation of spherical and smaller nanoparticles [161].

2.5.2 BACTERIA

Bacteria are the greatest choice for nanoparticle manufacturing among all other methods because of their exceptional ability to reduce the number of heavy metal ions. Some species of bacteria have developed the ability to use specialized defense mechanisms to fight stressors like heavy metals or metal toxicity. Some of them (*Pseudomonas stutzeri* and *Pseudomonas aeruginosa*) have been found to remain and proliferate in high metal ion concentrations [162, 163]. Furthermore, it has been observed that when spread on elemental sulfur as energy source, *Sulfolobus acidocaldarius*, *Thiobacillus ferroxidans* and *Thiobacillus thiooxidans* were able to transform ferric ions to ferrous ions. *T. thiooxidans* was able to aerobically reduce iron in a low pH medium. Ferrous iron could not be oxidized by *T. thiooxidans*, while the biological reduction of ferric iron by *T. ferrooxidans* was not found to be aerobic due to the quick reoxidation of ferrous iron when oxygen was present [164]. Other phenomena of biomineralization have been reported, including the development of tellurium in *E. coli* [165]. The direct reduction of Tc(VII) by resting cells of *Shewanella* has also been reported. *Enterobactercloacae*, *Desulfovibrio desulfuricans* and *Rhodospiriillum rubrum* reduce selenite into selenium [166]. Researchers have shown the ability of *P. Aeruginosa*, *Bacillus cereus*, *E. coli* and *B. subtilis* to remove La^{3+}, Ag^+, Cu^{2+} and Cd^{2+} from the solution. They discovered that the cells of bacteria may hold significant amounts of metallic cations of nanoparticles. Furthermore, some of these bacteria, such as Magnetotactic, may produce inorganic nanoparticles like intracellular magnetite nanoparticles [167]. The majority of the bacteria involved in the production of nanoparticles are depicted in Table 2.3.

TABLE 2.3
Green Biosynthesis of Nanoparticles Using Bacteria

Bacteria	Nanoparticle	Morphology	Size (nm)	References
Bacillus cereus	Silver	Spherical	20–40	[168]
Aeromonas	Silver	-	6.4	[169]
Bacillus megatherium D01	Gold	spherical	1.9 ± 0.8	[170]
Bacillus subtilis	Silver	Triangular and Spherical	5–50	[171]
Bacillus subtilis 168	Gold	Octahedral	5–25	[172, 173]
Clostridium thermoaceticum	Cadmium sulfide	Amorphous	--	[174]

(Continued)

TABLE 2.3 *(Continued)*
Green Biosynthesis of Nanoparticles Using Bacteria

Bacteria	Nanoparticle	Morphology	Size (nm)	References
Desulfobacteraceae	Zinc sulfide	Spherical	2–5	[175]
Bacillus subtilis	Silver	Triangular and Spherical	5–50	[176]
Corynebacterium	Silver	-	10–15	[177]
Escherichia coli	Cadmium sulfide	Wurtzite crystal	2–5	[178]
Escherichia coli MC4100	Gold	Rod-shaped, triangular, spherical and hexagonal	Less than 10 to 50	[179]
Escherichia coli	Silver	Spherical	8–9	[180]
Geobacillus sp.	Gold	Quasi-hexagonal	5–50	[181]
Lactobacillus strains	Gold	Crystalline, hexagonal, triangular and cluster	20–50 and above 100	[182]
Lactobacillus casei	Silver	Spherical	25–50	[183]
Rhodobacter sphaeroides	Zinc sulfide	Spherical	8 (Average diameter)	[184]
Rhodopseudomonas capsulata	Gold	Spherical and nanoplate	10–20	[185, 186]

2.5.3 ALGAE

Algae are found to accumulate heavy metal ions and possess an extraordinary ability to remodel them into more malleable states [187]. Because of these unique attributes, algae act as model organisms for preparing bionanomaterials. The extracts of algal consist of polyunsaturated fatty acids, carbohydrates, minerals, proteins, oil, fats and many bioactive compounds like antioxidants (tocopherols, polyphenols), and pigments such as carotenoids (xanthophylls, carotene), phycobilins (phycocyanin, phycoerythrin) and chlorophylls [188]. All these active compounds have been illuminating as reducing and stabilizing agents. Research reveals that to synthesize nanoparticles using yeast, the following are involved in the preparation:

a. an algal extract,
b. a metal precursor solution and
c. algal extract incubation with the metal precursor solution.

The initial step is mixing the liquid algal extract with the metal precursor solution. The change in color of the reaction mixture confirms the formation of nanoparticles [189, 190]. In the extract, the bioactive component supports the nanoparticle synthesis in which temperature, time, pH and concentration are controlling factors.

Initially, the intracellular synthesis of algal nanoparticles was known to form, but later, the extracellular mode of synthesis was also recorded [191, 192].

2.5.4 FUNGI

Fungi are known as eukaryotic organisms that live in a variety of ordinary environments and usually form decomposer organisms. Only roughly 70,000 different species of fungi have been identified out of which only 1.5 million species are estimated on the planet. According to more current research, roughly 5.1 million fungal species can be discovered using high-throughput sequencing approaches [193]. It's worth noting that these organisms are capable of digesting extracellular food and releasing specific enzymes to hydrolyze the complex or bigger components into simpler or small molecules that can be absorbed and used as sources of energy. The importance of investigating the role of fungi in nanobiotechnology is emphasized. Due to their toleration and bioaccumulation ability with metals, fungi have received greater attention in the research on the biological synthesis of metallic nanoparticles [194]. The ease with which fungi can be scaled up is a well-defined advantage of using them in manufacturing nanoparticles because of various advantages as shown in Figure 2.3. Because fungi are efficient secretors of extracellular enzymes, large-scale synthesis of different enzymes is possible [195]. Another advantage of using a green approach mediated by fungus is to produce metallic nanoparticles with cost-effectiveness and using biomass proficiently. Furthermore, because a lot of species develop quickly, cultivating and maintaining them in the laboratory is quite simple [196]. Most fungi have very high wall-binding and the ability to absorb intracellular metal [197]. Fungi can make metal nanostructures and nanoparticles (meso) by using a specific reducing enzyme, whether intracellularly or extracellularly or by using a bio-mimetic mineralization method.

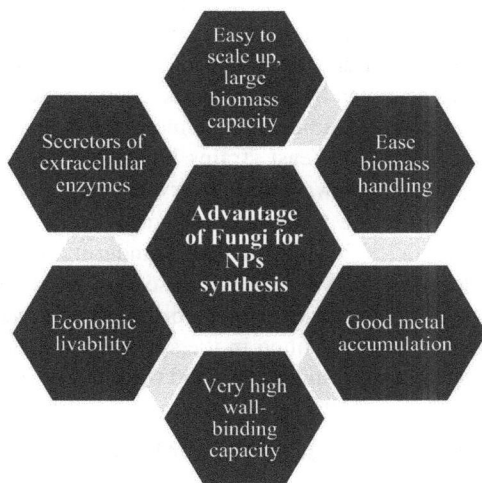

FIGURE 2.3 Advantages of fungi as bio-factories for the production of nanoparticles.

The study of fungal species is relatively fresh in terms of nanoparticle processes in nanotechnology. The filamentous fungus *Verticillium sp.* synthesizes silver nanoparticles extracellularly, according to one of the first studies on the manufacture of metallic nanoparticles by using fungi [198]. Among all the fungal species found for synthesizing nanoparticles, the (filamentous) fungus *Fusarium oxysporum* has been mostly used for this objective. The creation of extracellular nanoparticles has been documented as well as the creation of MoS_2, PbS, CuS and ZnS nanoparticles by the first used fungus. The identification of proteins in an aqueous solution indicates that a potential sulfate-depleting enzyme-based nanoparticle was created using the same fungus, but they appeared separately. They can also appear as a clump with a rapidly shifting shape and in a size range of 5 to 50 nm. However, the result other research confirms that spherical silver nanoparticles in the size of range 20 of 50 nm could be produced using *F. oxysporum*; by comparing the results of the two different studies [199, 200], the differences between the size and morphology of nanoparticles could be attributed to variations in temperature, even though the size of nanoparticles did not appear to depend on a variation in time [201]. Although quasi-spherical nanoparticles are the most common, other morphologies can be obtained depending on the metallic ion solution and incubation conditions. The synthesis of nanoparticles with various metals has been carried out using *F. oxysporum*. Extracellular production is described as having a wide range of sizes as well as distinct forms in all circumstances. The reduction in metal ions caused by this fungus has been linked to NADH-based reductases and an extracellular shuttle quinone process [202].

2.5.5 YEAST

The preparation of nanoparticles using yeast strains has additional advantages over bacteria because of mass production, as well as their ease of control under laboratory conditions, the synthesis of multiple enzymes and quick growth with by using some basic nutrients [203]. Manufacturing metallic nanoparticles using yeast has been the subject of some research. However, one of the key approaches in applying biological materials was achieved by utilizing eukaryotic systems, notably *S. pombe* and *Candida glabrata*. A few studies have demonstrated the potential usefulness of yeast-produced nanoparticles. Yeast strains have also been used to synthesize silver and gold nanoparticles. It was also observed that silver nanoparticles were created extracellularly using the silver-tolerant yeast strains, that is, MKY_3, with hexagonal silver nanoparticles 2–5 nm in size. Standardizing and documenting of the silver nanoparticles were also done [204] based on the varied temperatures of the situations. A changed concentration of chloroauric acid was used to incubate *Yarrowia lipolytica* cells, which resulted in the synthesis of gold nanoparticles that were affected by the salt concentrations used and the number of cells [205].

2.6   CHARACTERIZATION OF NANOPARTICLES

The identification of green nanoparticles has been done by observing changes in the color of the solution after adding the appropriate constituents, that is, biological

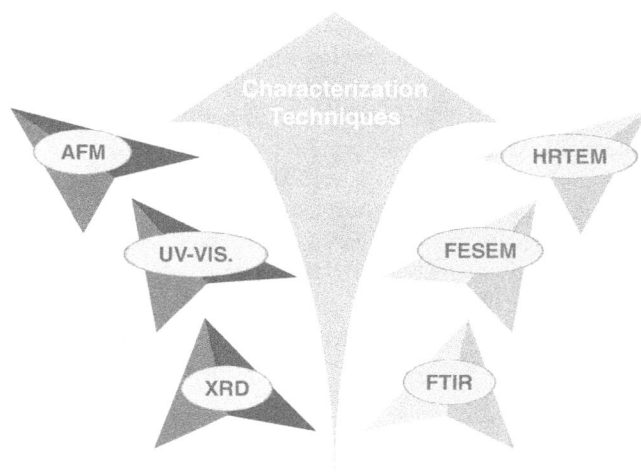

FIGURE 2.4 Different characterization techniques used to identify nanoparticles.

extract, metal solution and solvent, among others. Further confirmation has been done by using various spectral and analytical techniques as given in Figure 2.4.

2.6.1 OPTICAL SPECTROSCOPY

Optical spectroscopy constitutes the measurement of nanoparticles using different methods of spectroscopy, that is, atomic absorption spectroscopy, photoluminescence, Fourier transform infrared spectroscopy, Raman spectroscopy, dynamic light scattering, fluorescence correlation spectroscopy, zeta potential and UV-visible spectroscopy, among others. The technique of atomic absorption spectroscopy is used to determine the concentration of mass in a solid or liquid sample. The principle is based on the excitation of electrons from the ground to excited state by absorbing particular wavelengths. As the amount of absorbed energy is related to total the number of atoms in the path of light, the mass sample of concentration can be quantified by comparing the signal to calibration standards at known concentrations. Photoluminescence at low temperatures gives spectral peaks giving information regarding host material and impurities. The surface chemistry of nanomaterials has been analyzed by using infrared spectroscopy. This technique helps identify different ligands adhere to the surface of nanoparticles and information regarding transformation of functional groups has also been recorded [206]. Raman spectroscopy is also used for characterization of nanoparticles. It gives information about the size distribution; the chemical, structural and electronic properties; protein-metallic nanoparticle conjugates; and others. It records tissue abnormalities in nanoparticles. Advanced techniques in Raman spectroscopy, like increased spatial resolution (SERS), enhanced RS signal (SERS) and topological information of nanomaterials (SERS, TERS) etc., are further used to characterize the nanoparticles more efficiently

[207]. Similarly, dynamic light scattering analyzes the intensity of light as a function of time, which gives information regarding the size of the particles. Fluorescence spectroscopy gives information about the binding kinetics and hydrodynamic dimensions of nanoparticles as this technique studies concentration effects, chemical kinetics, molecular diffusion, conformation dynamics and more. Electrokinetic potential or zeta potential measures the effective electric charge and quantifies the stability of the charge on the surface of nanoparticles. The oppositely charged ions move with the nanoparticle, and the layer of surface charges and oppositely charged ions is known as an electrical double layer. In the same way, UV-visible spectroscopy is used to assess the light absorbed and scattered by the sample. The obtained measurements are then compared at each wavelength. The data are obtained in the form of spectra plotted as an extinction as a function of wavelength.

2.6.2 Electron Spectroscopy

Electron spectroscopy comprises characterization techniques such as X-ray absorption spectroscopy. It includes EXAFS (extended X-ray absorption fine structure) and XANES or NEXAFS (X-ray absorption near edge structure). This technique evaluates the coefficient of X-ray absorption as a function of energy. Every element in a nanoparticle consists of absorption edges, which correspond to various binding energies of the electrons. In very low concentrations, EXAFS is regarded as a convenient and simple way to confirm the chemical state of the targeted species [208].

2.6.3 X-Ray Diffraction Analysis

X-ray diffraction analysis (XRD) analysis is performed to determine the parameters of the lattice, orientation, purity, size and crystallinity of nanoparticles. The positions of the peaks are then compared with the ICPDS (International Center of Diffraction Data card) and confirmed for various phases in the fabricated nanoparticles. The size of crystalline nanoparticles was determined by Debye-Scherrer's Eq. [209]. On the XRD spectrum, the appearance of a broad peak confirms an amorphous nature of the nanoparticle while a narrow peak indicates a crystalline nature. The size obtained from crystalline nanoparticles shows that XRD-derived size is bigger compared to the magnetic size due to the existence of small domains in nanoparticles aligned in same direction.

2.6.4 Electron Microscopy

Electron microscopy includes various important techniques of characterization, that is, SEM (scanning electron microscopy), FESEM (field emission scanning electron microscope), TEM (transmission electron microscopy), atomic force microscopy (AFM), scanning tunneling microscopy (STM), near-field scanning optical microscopy and others. SEM is the most widely used technique to identify nanoparticles. It gives information about the chemical composition and morphology of the compound. This characterization technique not only provides topographical information but also gives an idea of the chemical composition near the surface; hence, it is very useful in determining the chemical composition of the tested material [210]. However, FESEM provides less electrostatically distorted and clearer images with a resolution of 1.5 nm and is found to be 3 to 6 times better than SEM. An important tool used for the direct imaging of nanoparticles is TEM. It is used to find the grain size, morphology and size distribution in nanoparticles. TEM images are formed by using a beam of transmitted electrons. The pattern of electron diffraction gives information about the atomic arrangement, orientation and structure of narrow regions in the nanoparticles. However, AFM provides data regarding sorption, aggregation, size distribution, dispersion, structure, surface properties and surface mapping, among others, while direct spatial resolution at the atomic scale is provided by the STM technique.

2.7 APPLICATIONS OF GREEN NANOPARTICLES

A broad range of applications in different areas has been shown by green nanoparticles as shown in Figure 2.5.

2.7.1 MEDICINAL AND THERAPEUTIC APPLICATIONS

The medicines derived by using nanoparticles are termed as nanomedicine. Nanomedicines obtained by using a green synthesis approach offer an alternative to conventional ways of disease therapy, diagnosis and understanding the biological system at the cellular level [211].

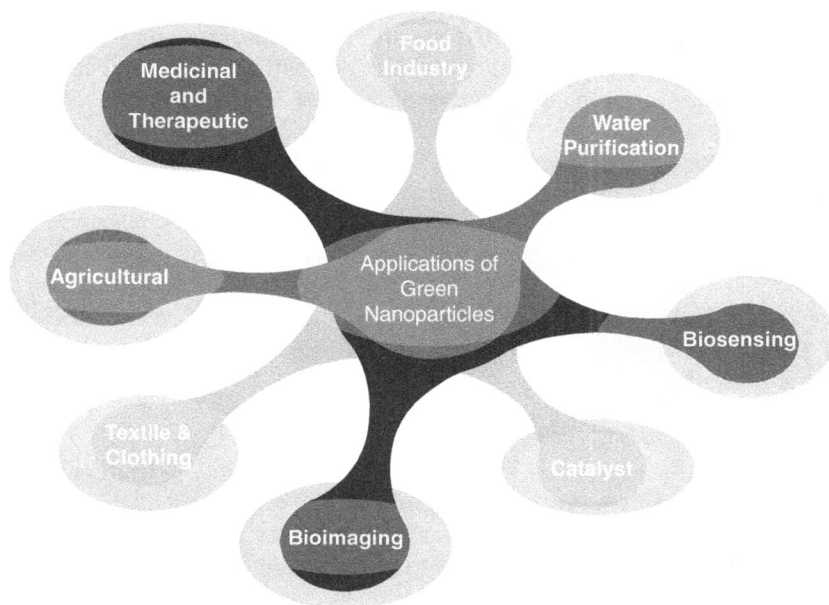

FIGURE 2.5 Applications of green nanoparticles.

2.7.1.1　Antidiabetic Therapy

The biosynthesis of silver nanoparticles has been done by using leaf extract of *Lonicera japonica*. These nanoparticles were investigated for antidiabetic activity, and results have shown the effective inhibition against enzymes like α-glucosidase and α-amylase with IC50 values 37.86 and 54.56 µg ml^{-1}, respectively [212]. Similarly, silver nanoparticles of the aqueous leaf extract of *Pouteria sapota* have been studied for in vivo and in vitro antidiabetic activity in streptozotocin-induced rats, and a significant reduction in the level of blood sugar was found [213]. Green nanoparticles obtained from gold have been synthesized from the plant *Fritillaria cirrhosa*. The application of these nanoparticles on streptozotocin-induced rats shows that gold nanoparticles activate the islet cells of the pancreas, hence indicating its antidiabetic activity [214].

2.7.1.2　Antimicrobial Activity

Chemically synthesized antibiotics have limited use due to their side effects. Due to their nontoxic nature, green nanomedicines give a promising and safe solution that has the ability to replace toxic and existing antibiotics. Silver nanoparticles have been found to be very effective against various pathogens. Leaf extract of *Pisidium guvajava* [215], *Swertia chirayita* [216] and *Rhizophora lamarckii* (mangrove plant) [217] has been used to prepare nanoparticles of magnesium oxide, and they all are found to show excellent antibacterial activities against various strains of pathogens.

2.7.1.3　Anticancer Properties

Silver nanoparticles of *Bacillus funiculus* were found to exhibit cytotoxic effects on MDA-MB-231 (breast cancer cells). Various concentrations of Ag nanoparticles, that is, 5 to 25 µg/ml, were subjected to cure the breast cancer on treatment for 24 h. Nanoparticles were observed to inhibit the growth of cancer cells in a dose-dependent mode by activating ROS generation, caspase-3, lactate dehydrogenase and caspase-3 [218]. An aqueous extract of *Alternanthera sessilis* was used to prepare silver nanoparticles, and their anticancer activity against prostate cells was observed. These nanomedicines were found to decrease the cancer cells and increase antiprolif-eration activity. Changes in the morphology of the cancer cells like oxidative stress, biochemical responses, cell shrinkage and coiling were identified, which ultimately led to apoptosis. For prostate cancer cells, a 95% apoptosis was found at a concentration of 25 µl/ml, which shows 100% inhibition on the growth of breast cancer cells [219]. *Origanum vulgare* has been used to treat indigestion, urinary tract disorders, respiratory disorders, rheumatoid arthritis and dental issues, among others [220].

2.7.1.4　Wound Healing Ability

A large number of nanoparticles were found with the capability of healing wounds without causing any serious toxic effects. The fruit peel of *Lansium domesticum* was used to obtain silver nanoparticles and was investigated as a cure for burned-related wounds [221]. These nanoparticles were introduced into Pluronic F127 gels to examine their healing capacity. It was found that upon injection of 0.1% w/w of silver nanoparticles, the extent of healing activity was increased, and this efficiency was further increased by using hydroxyproline. Plant-extracted Ag nanoparticles from

Indigofera aspalathoides have also been used as hydrogel dressing for new burns, cuts and dry wounds without using preservatives [222]. Gold and silver nanoparticles obtained from actinobacterial metabolites also show excellent wound healing ability. A comparison of standard ointments with a concentration of 10% silver nanoparticles and silver or gold nanoparticles heal 100% of the excision wound after treatment for 19 to 21 days. Along with this, they are also applied for scarless wound healing when clinical practice is occurring [223].

2.7.1.5 Neuroregenerative Therapy

Spinal cord disease, Parkinson's disease, brain injury, stroke and hampered neuronal integrity, among others, come under the category of neurodegenerative diseases. Regeneration therapy is used to slow these degenerations [224]. Silver-based nanoparticles 30 nm in size were biosynthesized from bacterial *E. coli* to cure SH-SY5Y (human neuroblastoma cells). Silver nanoparticles showed their activity by increasing the length of neurite and the divergent neuronal markers such as Map2 expression, neurogenin-1, Drd-2, Gap-43, β-tubulin III and synaptophysin [225].

2.7.2 Bioimaging Applications

Leaf extract from *Olax scandens* has been used to synthesize colloidal green Ag nanoparticles, and it has been reported that these nanoparticles can exhibit red fluorescence in $B_{16}F_{10}$ cells, which can be used in the field of bioimaging [226]. The proteins and fluorescent phytochemicals found in the methanolic leaf extract of *O. scandens* link themselves to the silver nanoparticles, showing strong fluorescent properties inside living cells [227].

2.7.3 Biosensing Applications

Biosensors are measuring or sensing devices that can estimate a material on the basis of a biological interaction. Scrutinizing and analyzing the biological interactions are unique in the field of environmental issues, medical diagnosis and screening for food quality. Silver nanoparticle–based room temperature ammonia sensors were obtained from the polysaccharide *Cyamopsis tetragonoloba* were prepared and used for detecting the level of ammonia in biological fluids, such as sweat, plasma and saliva, among others [228]. Similarly, gold nanoparticles were obtained from *C. tetragonaloba* to prepare an optical ammonia sensor that shows an efficient reproducibility of results.

2.7.4 Water Purification

Water obtained from natural resources is not fit for drinking or consumption because of large amounts of inorganic, organic, biological and radiological contaminants like surfactants, dyes, pesticides, bacteria, algae, viruses and ions like arsenic, fluoride, mercury copper, plutonium, cesium, uranium and others [229, 230]. Catalysts and absorbents based on nanoparticles are very good alternatives to harmful chemicals and are eco-friendly solutions for treating and purifying water [231]. Green nanoparticles by using silver with biopolymer xylan have been used to detect Hg^{2+} in a sample of impure water [232]. The purification of water using silver nanoparticles has been approved by the World Health Organization [233]. Nanoparticles prepared from grapheme, ferric oxide (Fe3O4), manganese oxide, titanium oxide (TiO_2), magnesium oxide and zinc oxide are commercially used as adsorbents to treat water containing azo dyes, heavy metals and other pollutants [234, 235]. Zinc oxide nanoparticles obtained from cassava starch and aloe vera are used as adsorbents for copper ions and show a high removal efficiency as the concentration of adsorbents increases [236]. The incorporation of metal-based green nanoparticles like alumina, titanium, silica and zeolite into a polymeric membrane enhances the membrane's permeability and hydrophobicity [237–239]. Photocatalytic activity of green-synthesized nanoparticles by using Pt, Ag, Pd and Au metals are being used for the degradation of various dyes [240–242].

2.7.5 AGRICULTURAL APPLICATIONS

Agriculture is a common source of income of more than 50% of the total population of world. Using sustainable natural resources has led to some major problems in recent times such as urbanization, the accumulation of fertilizers and pesticides, changes in climate and so on [243]. These problems have been solved by green nanoparticles due to their detecting and controlling ability of disease in plants [244]. The reason behind using green nanoparticles in agriculture is to decrease the high cost of fertilizers and harmful environmental inputs because the green nanoparticles are prepared by using greener principles of synthesis that result in a reduction of the harmful exposure and emission of methane, carbon dioxide and nitrous oxide in the environment, thereby reducing the health risks to farmers and increasing crop productivity [245]. Due to their large surface area and small size, nanoparticles can penetrate the interior of a seed and activate phytohormones, which are essential for seed germination and growth [246]. The treatment of *Spinacia oleracea L.* seeds with 0.25% TiO_2 nanoparticles results in an improved rate of photosynthesis and assimilation of nitrogen [247, 248]. However, the treatment of *Citrullus lanatus* seeds with Fe_2O_3 nanoparticles enhances the rate of germination and improves fruiting behavior and the development of plant [249]. Utilizing green nanoparticles in the field of agriculture is more advantageous because of their feasibility, nontoxicity and safety [250]. Nanoparticle-based fertilizers and plant production products include nanoemulsions, nanoparticles, viral capsids and so on that are not only used to control pest in plants but also used as excellent delivery systems of active ingredients for various diseases. Nanofertilizers contain nanoporous zeolite, nitrogen fertilizers, zinc nanofertilizers, potash fertilizers, nanopesticides and nanoherbicides [251, 252].

2.7.6 NANOPARTICLES IN FOOD INDUSTRY

Green nanoparticles are found to be very impactful in the production, packaging, storage and transportation of a large number of food products due to their use as nanoadditives, nanocomposites, anticaking agents, nanocarriers and more [253].

Nanochelates are used to increase the nutrient delivery and absorption of food nutrients without altering the taste, color and flavor of the food. Food supplements based on nanoparticles are found to be more effective due to their ability to react with human cells. A gelatin-based nanocomposite film with silver increases its antimicrobial activity. This combination shows excellent antibacterial activity against foodborne bacteria, that is, both gram-positive and gram-negative bacteria [254]. Recent applications of nanoparticles in the food industry are important because they have the ability to form emulsion bilayers, reverse micelles, surfactant micelles and nano-emulsions, among others. Food ingredients with a nanotexture claim to offer better taste, texture and acceptability [255]. Green nanoparticles proved to be very helpful in the early stages of food manufacturing, nutrition, smart packing and quality control.

2.8 SUMMARY

Metal-based nanoparticles obtained by green synthesis have replaced the harmful and toxic nanoparticles synthesized by using hazardous solvents and chemicals. They are also found to show potential and extraordinary capabilities to use in broad areas, such as the fields of pharmaceuticals, optics, food, textiles and many more. Green nanoparticles are able to attract the interest of researchers due to their ease of preparation in less time by consuming less toxic and economical materials.

REFERENCES

1. Gurunathan S, Kalishwaralal K, Vaidyanathan R, Venkataraman D, Pandian SR, Muniyandi J, Hariharan N, Eom SH. Biosynthesis, purification and characterization of silver nanoparticles using escherichia coli. Colloids and Surfaces B: Biointerfaces. 2009 Nov 1;74(1):328–335.
2. Castro-Longoria E, Vilchis-Nestor AR, Avalos-Borja M. Biosynthesis of silver, gold and bimetallic nanoparticles using the filamentous fungus neurospora crassa. Colloids and Surfaces B: Biointerfaces. 2011 Mar 1;83(1):42–48.
3. Philip D. Green synthesis of gold and silver nanoparticles using hibiscus rosa sinensis. Physica E: Low-Dimensional Systems and Nanostructures. 2010 Mar 1;42(5):1417–1424.
4. Mukherjee P, Senapati S, Mandal D, Ahmad A, Khan MI, Kumar R, Sastry M. Extracellular synthesis of gold nanoparticles by the fungus fusarium oxysporum. ChemBioChem. 2002 May 3;3(5):461–463.
5. Indira K, Mudali UK, Rajendran N. Corrosion behavior of electrochemically assembled nanoporous titania for biomedical applications. Ceramics International. 2013 Mar 1;39(2):959–967.
6. Indira K, Ningshen S, Mudali UK, Rajendran N. Effect of anodization parameters on the structural morphology of titanium in fluoride containing electrolytes. Materials Characterization. 2012 Sep 1;71:58–65.
7. Indira K, Ningshen S, Kamachi Mudali U, Rajendran N. Structural features of self-organized. TiO_2 nanopore arrays formed by electrochemical anodization of titanium. Journal of the Electrochemical Society of India. 2011;60:145–147.
8. Indira K. *Development of titanium nanotube arrays for orthopaedic applications* (Doctoral dissertation, Anna University).
9. Tian C, Mao B, Wang E, Kang Z, Song Y, Wang C, Li S, Xu L. One-step, size-controllable synthesis of stable Ag nanoparticles. Nanotechnology. 2007 Jun 15;18(28):285607.

10. Srivastava SK, Hasegawa T, Yamada R, Ogino C, Mizuhata M, Kondo A. Green synthesis of Au, Pd and Au@ Pd core–shell nanoparticles via a tryptophan induced supramolecular interface. RSC Advances. 2013;3(40):18367–18372.

11. Kalabegishvili TL, Murusidze IG, Prangishvili DA, Kvachadze LI, Kirkesali EI, Rcheulishvili AN, Ginturi EN, Janjalia MB, Tsertsvadze GI, Gabunia VM, Frontasyeva M. Gold nanoparticles in sulfolobus islandicus biomass for technological applications. Advanced Science, Engineering and Medicine. 2014 Dec 1;6(12):1302–1308.

12. Kumar CG, Mamidyala SK, Sreedhar B, Reddy BV. Synthesis and characterization of gold glyconanoparticles functionalized with sugars of sweet sorghum syrup. Biotechnology Progress. 2011 Sep;27(5):1455–1463.

13. Kalpana D, Lee YS. Synthesis and characterization of bactericidal silver nanoparticles using cultural filtrate of simulated microgravity grown Klebsiella pneumoniae. Enzyme and Microbial Technology. 2013 Mar 5;52(3):151–156.

14. Logeswari P, Silambarasan S, Abraham J. Synthesis of silver nanoparticles using plants extract and analysis of their antimicrobial property. Journal of Saudi Chemical Society. 2015 May 1;19(3):311–317.

15. Ahmed S, Ahmad M, Swami BL, Ikram S. A review on plants extract mediated synthesis of silver nanoparticles for antimicrobial applications: a green expertise. Journal of Advanced Research. 2016 Jan 1;7(1):17–28.

16. Gopinathan P, Ashok AM, Selvakumar R. Bacterial flagella as biotemplate for the synthesis of silver nanoparticle impregnated bionanomaterial. Applied Surface Science. 2013 Jul 1;276:717–722.

17. Ahmad R, Mohsin M, Ahmad T, Sardar M. Alpha amylase assisted synthesis of TiO_2 nanoparticles: structural characterization and application as antibacterial agents. Journal of Hazardous Materials. 2015 Feb 11;283:171–177.

18. Raliya R, Biswas P, Tarafdar JC. TiO_2 nanoparticle biosynthesis and its physiological effect on mung bean (Vigna radiata L.). Biotechnology Reports. 2015 Mar 1;5:22–26.

19. Indira K, KamachiMudali U, Rajendran N. In vitro bioactivity and corrosion resistance of Zr incorporated TiO_2 nanotube arrays for orthopaedic applications. Applied Surface Science. 2014 Oct 15;316:264–275.

20. Indira K, Mudali UK, Rajendran N. In-vitro biocompatibility and corrosion resistance of strontium incorporated TiO_2 nanotube arrays for orthopaedic applications. Journal of Biomaterials Applications. 2014 Jul;29(1):113–129.

21. Ajitha B, Reddy YA, Reddy PS. Green synthesis and characterization of silver nanoparticles using lantana camara leaf extract. Materials Science and Engineering: C. 2015 Apr 1;49:373–381.

22. Feng A, Wu S, Chen S, Zhang H, Shao W, Xiao Z. Synthesis of silver nanoparticles with tunable morphologies via a reverse nano-emulsion route. Materials Transactions. 2013 Jul 1;54(7):1145–1148.

23. Moreira dos Santos M, João Queiroz M, Baptista PV. Enhancement of antibiotic effect via gold: silver-alloy nanoparticles. Journal of Nanoparticle Research. 2012;14(5):859–866.

24. Aritonang HF, Onggo D, Ciptati C, Radiman CL. Synthesis of platinum nanoparticles from K2PtCl4 solution using bacterial cellulose matrix. Journal of Nanoparticles. 2014;1–6.

25. Aritonang HF, Onggo D, Radiman CL. Insertion of platinum particles in bacterial cellulose membranes from PtCl4 and H2PtCl6 precursors. Macromolecular Symposia. 2015 Jul;353(1):55–61.

26. Aritonang HF, Kamu VS, Ciptati C, Onggo D, Radiman CL. Performance of platinum nanoparticles/multiwalled carbon nanotubes/bacterial cellulose composite as anode catalyst for proton exchange membrane fuel cells. Bulletin of Chemical Reaction Engineering & Catalysis. 2017;12(2):287.

27. Raut RW, Nikam T, Kashid SB, Malghe YS. Rapid biosynthesis of platinum and palladium metal nanoparticles using root extract of asparagus racemosus linn. Advanced Materials Letters. 2013 Aug 1;4(8):650–654.

28. Vazquez-Muñoz R, Borrego B, Juárez-Moreno K, García-García M, Morales JD, Bogdanchikova N, Huerta-Saquero A. Toxicity of silver nanoparticles in biological systems: does the complexity of biological systems matter?. Toxicology Letters. 2017 Jul 5;276:11–20.

29. Kim J, Kwon S, Ostler E. Antimicrobial effect of silver-impregnated cellulose: potential for antimicrobial therapy. Journal of Biological Engineering. 2009 Dec;3(1):1–9.

30. Rai M, Yadav A, Gade A. Silver nanoparticles as a new generation of antimicrobials. Biotechnology Advances. 2009 Jan 1;27(1):76–83.

31. Zielińska A, Skwarek E, Zaleska A, Gazda M, Hupka J. Preparation of silver nanoparticles with controlled particle size. Procedia Chemistry. 2009 Nov 1;1(2):1560–1566.

32. Sondi I, Salopek-Sondi B. Silver nanoparticles as antimicrobial agent: a case study on E. coli as a model for gram-negative bacteria. Journal of Colloid and Interface Science. 2004 Jul 1;275(1):177–182.

33. Singh A, Jain D, Upadhyay MK, Khandelwal N, Verma HN. Green synthesis of silver nanoparticles using argemone mexicana leaf extract and evaluation of their antimicrobial activities. Digest Journal of Nanomaterials and Biostructures. 2010 Jul;5(2):483–489.

34. Abbaszadegan A, Ghahramani Y, Gholami A, Hemmateenejad B, Dorostkar S, Nabavizadeh M, Sharghi H. The effect of charge at the surface of silver nanoparticles on antimicrobial activity against gram-positive and gram-negative bacteria: a preliminary study. Journal of Nanomaterials. 2015 Oct;16:15.

35. Jirovetz L, Buchbauer G, Shafi MP, Leela NK. Analysis of the essential oils of the leaves, stems, rhizomes and roots of the medicinal plant Alpinia galanga from southern India. Acta Pharmaceutica-Zagreb. 2003 Jun 1;53(2):73–82.

36. Chandran SP, Chaudhary M, Pasricha R, Ahmad A, Sastry M. Synthesis of gold nanotriangles and silver nanoparticles using aloevera plant extract. Biotechnology Progress. 2006;22(2):577–583.

37. Shekhawat MS, Manokari M, Kannan N, Revathi J, Latha R. Synthesis of silver nanoparticles using cardiospermum halicacabum L. leaf extract and their characterization. Journal of Phytopharmacology. 2013;2:15–20.

38. Becker L, Poreda RJ, Bada JL. Extraterrestrial helium trapped in fullerenes in the Sudbury impact structure. Science. 1996 Apr 12;272(5259):249–252.

39. Heymann D, Jenneskens LW, Jehlička J, Koper C, Vlietstra E. Terrestrial and extraterrestrial fullerenes. Fullerenes, Nanotubes and Carbon Nanostructures. 2003 Dec 1;11(4):333–370.

40. Banfield JF, Zhang H. Nanoparticles in the environment. Reviews in Mineralogy and Geochemistry. 2001 Jan 1;44(1):1–58.

41. Waychunas GA, Kim CS, Banfield JF. Nanoparticulate iron oxide minerals in soils and sediments: unique properties and contaminant scavenging mechanisms. Journal of Nanoparticle Research. 2005 Oct;7(4):409–433.

42. Anastasio C, Martin ST. Atmospheric nanoparticles. Reviews in Mineralogy and Geochemistry. 2001 Jan 1;44(1):293–349.

43. Hochella Jr MF, Madden AS. Earth's nano-compartment for toxic metals. Elements. 2005 Sep 1;1(4):199–203.

44. Zereini F, Wiseman C, Alt F, Messerschmidt J, Müller J, Urban H. Platinum and rhodium concentrations in airborne particulate matter in Germany from 1988 to 1998. Environmental Science & Technology. 2001 May 15;35(10):1996–2000.

45. Thilgen C, Diederich F. The higher fullerenes: covalent chemistry and chirality. Fullerenes and Related Structures. 1999;135–171. ISBN: 978-3-540-64939-7

46. Mokhtari A, Karimi-Maleh H, Ensafi AA, Beitollahi H. Application of modified multiwall carbon nanotubes paste electrode for simultaneous voltammetric determination of morphine and diclofenac in biological and pharmaceutical samples. Sensors and Actuators B: Chemical. 2012 Jul 5;169:96–105.

47. Bevilacqua AC, Köhler MH, Piquini PC. Corrole–Fullerene Dyads: stability, photophysical, and redox properties. The Journal of Physical Chemistry C. 2019 Aug 8;123(34):20869–20876.

48. Giles J. Top five in physics. Nature. 2006 May 1;441(7091):265.

49. Dai H. Carbon nanotubes: synthesis, integration, and properties. Accounts of Chemical Research. 2002 Dec 17;35(12):1035–1044.

50. Bianco A, Prato M. Can carbon nanotubes be considered useful tools for biological applications?. Advanced Materials. 2003 Oct 16;15(20):1765–1768.

51. Goldberg ED. Black carbon in the environment: Properties and distribution. 1986 Nov; 28(9).

52. Blackford DB, Simons GR. Particle size analysis of carbon black. Particle & Particle Systems Characterization. 1987;4(1–4):112–117.

53. Sirisinha C, Prayoonchatphan N. Study of carbon black distribution in BR/NBR blends based on damping properties: influences of carbon black particle size, filler, and rubber polarity. Journal of Applied Polymer Science. 2001 Sep 23;81(13):3198–203.

54. Buffle J. The key role of environmental colloids/nanoparticles for the sustainability of life. Environmental Chemistry. 2006 Jul 10;3(3):155–158.

55. Shields SP, Richards VN, Buhro WE. Nucleation control of size and dispersity in aggregative nanoparticle growth: a study of the coarsening kinetics of thiolate-capped gold nanocrystals. Chemistry of Materials. 2010 May 25;22(10):3212–3225.

56. Filho PI, Carmalt CJ, Angeli P, Fraga ES. Mathematical modeling for the design and scale-up of a large industrial aerosol-assisted chemical vapor deposition process under uncertainty. Industrial & Engineering Chemistry Research. 2020 Jan 6;59(3):1249–1260.

57. Koziara JM, Lockman PR, Allen DD, Mumper RJ. In situ blood–brain barrier transport of nanoparticles. Pharmaceutical Research. 2003 Nov;20(11):1772–1778.

58. Kim JY, Cohen C, Shuler ML, Lion LW. Use of amphiphilic polymer particles for in situ extraction of sorbed phenanthrene from a contaminated aquifer material. Environmental Science & Technology. 2000 Oct 1;34(19):4133–4139.

59. Baer DR, Tratnyek PG, Qiang Y, Amonette JE, Linehan J, Sarathy V, Nurmi JT, Wang C, Antony J. Synthesis, characterization, and properties of zero-valent iron nanoparticles. Environmental Applications of Nanomaterials: Synthesis, Sorbents and Sensors. 2012 Jul 24:49–86.

60. Xu Y, Zhao D. Removal of copper from contaminated soil by use of poly (amidoamine) dendrimers. Environmental Science & Technology. 2005 Apr 1;39(7):2369–2375.

61. Morones JR, Elechiguerra JL, Camacho A, Holt K, Kouri JB, Ramírez JT, Yacaman MJ. The bactericidal effect of silver nanoparticles. Nanotechnology. 2005 Aug 26;16(10):2346.

62. Brust M, Kiely CJ. Some recent advances in nanostructure preparation from gold and silver particles: a short topical review. Colloids and Surfaces A: Physicochemical and Engineering Aspects. 2002 9;202:175–186.

63. Nowack B. Pollution prevention and treatment using nanotechnology. Nanotechnology. 2008;2:1–5.

64. Aitken RJ, Chaudhry MQ, Boxall AB, Hull M. Manufacture and use of nanomaterials: current status in the UK and global trends. Occupational Medicine. 2006 Aug 1;56(5):300–306.

65. Rittner MN. Market analysis of nanostructured materials. American Ceramic Society Bulletin. 2002;81:33e36.

66. Dong H, Li L, Wang Y, Ning Q, Wang B, Zeng G. Aging of zero-valent iron-based nanoparticles in aqueous environment and the consequent effects on their reactivity and toxicity. Water Environment Research. 2020 May;92(5):646–661.
67. Tortella GR, Rubilar O, Durán N, Diez MC, Martínez M, Parada J, Seabra AB. Silver nanoparticles: toxicity in model organisms as an overview of its hazard for human health and the environment. Journal of Hazardous Materials. 2020 May 15;390:121974.
68. Exbrayat JM, Moudilou EN, Abrouk L, Brun C. Apoptosis in amphibian development. Advances in Bioscience and Biotechnology. 2012 Oct 29;3(6):669.
69. Hammond SA, Carew A, Helbing C. Evaluation of the effects of titanium dioxide nanoparticles on cultured Rana catesbeiana tailfin tissue. Frontiers in Genetics. 2013 Nov 21;4:251.
70. Bacchetta R, Santo N, Fascio U, Moschini E, Freddi S, Chirico G, Camatini M, Mantecca P. Nano-sized CuO, TiO$_2$ and ZnO affect Xenopus laevis development. Nanotoxicology. 2012 Jun 1;6(4):381–398.
71. Wahid F, Ul-Islam M, Khan R, Khan T, Khattak WA, Hwang KH, Park JS, Chang SC, Kim YY. Stimulatory effects of zinc oxide nanoparticles on visual sensitivity and electroretinography b-waves in the bullfrog eye. Journal of Biomedical Nanotechnology. 2013 Aug 1;9(8):1408–1415.
72. Trickler WJ, Lantz SM, Murdock RC, Schrand AM, Robinson BL, Newport GD, Schlager JJ, Oldenburg SJ, Paule MG, Slikker Jr W, Hussain SM. Silver nanoparticle induced blood-brain barrier inflammation and increased permeability in primary rat brain microvessel endothelial cells. Toxicological Sciences. 2010 Nov 1;118(1):160–170.
73. Zhang Y, Ferguson SA, Watanabe F, Jones Y, Xu Y, Biris AS, Hussain S, Ali SF. Silver nanoparticles decrease body weight and locomotor activity in adult male rats. Small. 2013 May 27;9(9–10):1715–1720.
74. Oberdorster G, Oberdorster E, Oberdorster J. Nanotoxicology: an emerging discipline evolving from studies of ultrafine particles. Environmental Health Perspectives. 2005;113(7):823–839.
75. Exbrayat JM, Moudilou EN, Abrouk L, Brun C. Apoptosis in amphibian development. Advances in Bioscience and Biotechnology. 2012 Oct 29;3(6):669.
76. Hammond SA, Carew A, Helbing C. Evaluation of the effects of titanium dioxide nanoparticles on cultured Rana catesbeiana tailfin tissue. Frontiers in Genetics. 2013 Nov 21;4:251.
77. Bacchetta R, Santo N, Fascio U, Moschini E, Freddi S, Chirico G, Camatini M, Mantecca P. Nano-sized CuO, TiO$_2$ and ZnO affect Xenopus laevis development. Nanotoxicology. 2012 Jun 1;6(4):381–398.
78. Wahid F, Ul-Islam M, Khan R, Khan T, Khattak WA, Hwang KH, Park JS, Chang SC, Kim YY. Stimulatory effects of zinc oxide nanoparticles on visual sensitivity and electroretinography b-waves in the bullfrog eye. Journal of Biomedical Nanotechnology. 2013 Aug 1;9(8):1408–1415.
79. Unrine JM, Shoults-Wilson WA, Zhurbich O, Bertsch PM, Tsyusko OV. Trophic transfer of Au nanoparticles from soil along a simulated terrestrial food chain. Environmental Science & Technology. 2012 Sep 4;46(17):9753–9760.
80. Jo HJ, Choi JW, Lee SH, Hong SW. Acute toxicity of Ag and CuO nanoparticle suspensions against Daphnia magna: the importance of their dissolved fraction varying with preparation methods. Journal of Hazardous Materials. 2012 Aug 15;227:301–308.
81. Adam N, Vakurov A, Knapen D, Blust R. The chronic toxicity of CuO nanoparticles and copper salt to Daphnia magna. Journal of Hazardous Materials. 2015 Feb 11;283:416–422.
82. Gomes T, Chora S, Pereira CG, Cardoso C, Bebianno MJ. Proteomic response of mussels *Mytilus galloprovincialis* exposed to CuO NPs and Cu(2)(+): an exploratory biomarker discovery. Aquatic Toxicology. 2014;155:327–336.

83. Bayat N, Lopes VR, Schoelermann J, Jensen LD, Cristobal S. Vascular toxicity of ultra-small TiO$_2$ nanoparticles and single walled carbon nanotubes in vitro and in vivo. Biomaterials. 2015 Sep 1;63:1–3.

84. He X, Aker WG, Hwang HM. An in vivo study on the photo-enhanced toxicities of S-doped TiO$_2$ nanoparticles to zebrafish embryos (Danio rerio) in terms of malformation, mortality, rheotaxis dysfunction, and DNA damage. Nanotoxicology. 2014 Aug 31;8(supl):185–195.

85. Lin S, Wang X, Ji Z, Chang CH, Dong Y, Meng H, Liao YP, Wang M, Song TB, Kohan S, Xia T. Aspect ratio plays a role in the hazard potential of CeO2 nanoparticles in mouse lung and zebrafish gastrointestinal tract. ACS Nano. 2014 May 27;8(5):4450–4464.

86. Ma H, Williams PL, Diamond SA. Ecotoxicity of manufactured ZnO nanoparticles–a review. Environmental Pollution. 2013 Jan 1;172:76–85.

87. Bayat N, Lopes VR, Schoelermann J, Jensen LD, Cristobal S. Vascular toxicity of ultra-small TiO$_2$ nanoparticles and single walled carbon nanotubes in vitro and in vivo. Biomaterials. 2015 Sep 1;63:1–3.

88. Wehmas LC, Anders C, Chess J, Punnoose A, Pereira CB, Greenwood JA, Tanguay RL. Comparative metal oxide nanoparticle toxicity using embryonic zebrafish. Toxicology Reports. 2015 Jan 1;2:702–715.

89. Velzeboer I, Hendriks AJ, Ragas AM, Van de Meent D. Nanomaterials in the environment aquatic ecotoxicity tests of some nanomaterials. Environmental Toxicology and Chemistry: An International Journal. 2008 Sep;27(9):1942–1947.

90. Bhuvaneshwari M, Iswarya V, Archanaa S, Madhu GM, Kumar GS, Nagarajan R, Chandrasekaran N, Mukherjee A. Cytotoxicity of ZnO NPs towards fresh water algae Scenedesmus obliquus at low exposure concentrations in UV-C, visible and dark conditions. Aquatic Toxicology. 2015 May 1;162:29–38.

91. Fu L, Hamzeh M, Dodard S, Zhao YH, Sunahara GI. Effects of TiO$_2$ nanoparticles on ROS production and growth inhibition using freshwater green algae pre-exposed to UV irradiation. Environmental Toxicology and Pharmacology. 2015 May 1;39(3):1074–1080.

92. Yang L, Watts DJ. Particle surface characteristics may play an important role in phytotoxicity of alumina nanoparticles. Toxicology Letters. 2005 Aug 14;158(2):122–132.

93. Lin D, Xing B. Root uptake and phytotoxicity of ZnO nanoparticles. Environmental Science & Technology. 2008 Aug 1;42(15):5580–5585.

94. Rana S, Kalaichelvan PT. Antibacterial activities of metal nanoparticles. Antibacterial Activities of Metal Nanoparticles. 2011 Aug;11(2):21–23.

95. Manjumeena R, Duraibabu D, Sudha J, Kalaichelvan PT. Biogenic nanosilver incorporated reverse osmosis membrane for antibacterial and antifungal activities against selected pathogenic strains: an enhanced eco-friendly water disinfection approach. Journal of Environmental Science and Health, Part A. 2014 Aug 24;49(10):1125–1133.

96. Elumalai K, Velmurugan S, Ravi S, Kathiravan V, Ashokkumar S. Retracted: facile, eco-friendly and template free photosynthesis of cauliflower like ZnO nanoparticles using leaf extract of Tamarindus indica (L.) and its biological evolution of antibacterial and antifungal activities. Spectrochimica Acta Part A: Molecular and Biomolecular Spectroscopy. 2015:1052–1057.

97. Kumar NA, Rejinold NS, Anjali P, Balakrishnan A, Biswas R, Jayakumar R. Preparation of chitin nanogels containing nickel nanoparticles. Carbohydrate Polymers. 2013 Sep 12;97(2):469–474.

98. Mortimer M, Kasemets K, Kahru A. Toxicity of ZnO and CuO nanoparticles to ciliated protozoa Tetrahymena thermophila. Toxicology. 2010 Mar 10;269(2–3):182–189.

99. Kasemets K, Ivask A, Dubourguier HC, Kahru A. Toxicity of nanoparticles of ZnO, CuO and TiO$_2$ to yeast Saccharomyces cerevisiae. Toxicology in Vitro. 2009 Sep 1;23(6):1116–1122.

100. Kim S, Ryu DY. Silver nanoparticle-induced oxidative stress, genotoxicity and apoptosis in cultured cells and animal tissues. Journal of Applied Toxicology. 2013 Feb;33(2):78–89.
101. Hadrup N, Loeschner K, Mortensen A, Sharma AK, Qvortrup K, Larsen EH, Lam HR. The similar neurotoxic effects of nanoparticulate and ionic silver in vivo and in vitro. Neurotoxicology. 2012 Jun 1;33(3):416–423.
102. McNeil PL, Boyle D, Henry TB, Handy RD, Sloman KA. Effects of metal nanoparticles on the lateral line system and behaviour in early life stages of zebrafish (Danio rerio). Aquatic Toxicology. 2014 Jul 1;152:318–323.
103. Navarro E, Baun A, Behra R, Hartmann NB, Filser J, Miao AJ, Quigg A, Santschi PH, Sigg L. Environmental behavior and ecotoxicity of engineered nanoparticles to algae, plants, and fungi. Ecotoxicology. 2008 Jul;17(5):372–386.
104. Suresh AK, Pelletier DA, Doktycz MJ. Relating nanomaterial properties and microbial toxicity. Nanoscale. 2013;5(2):463–474.
105. Seaton A, Godden D, MacNee W, Donaldson K. Particulate air pollution and acute health effects. The Lancet. 1995 Jan 21;345(8943):176–178.
106. Nel A, Xia T, Madler L, Li N. Toxic potential of materials at the nanolevel. Science. 2006 Feb 3;311(5761):622–627.
107. Donaldson K, Poland CA. Inhaled nanoparticles and lung cancer-what we can learn from conventional particle toxicology. Swiss Medical Weekly. 2012 Jun 18;142(2526).
108. Li N, Xia T, Nel AE. The role of oxidative stress in ambient particulate matter-induced lung diseases and its implications in the toxicity of engineered nanoparticles. Free Radical Biology and Medicine. 2008 May 1;44(9):1689–1699.
109. Donaldson K, Stone V, Tran CL, Kreyling W, Borm PJ. Nanotoxicology. Occupational and Environmental Medicine. 2004 Sep 1;61(9):727–728.
110. Capasso L, Camatini M, Gualtieri M. Nickel oxide nanoparticles induce inflammation and genotoxic effect in lung epithelial cells. Toxicology Letters. 2014 Apr 7;226(1):28–34.
111. Arora S, Jain J, Rajwade JM, Paknikar KM. Cellular responses induced by silver nanoparticles: in vitro studies. Toxicology Letters. 2008 Jun 30;179(2):93–100.
112. Yu M, Mo Y, Wan R, Chien S, Zhang X, Zhang Q. Regulation of plasminogen activator inhibitor-1 expression in endothelial cells with exposure to metal nanoparticles. Toxicology Letters. 2010 May 19;195(1):82–89.
113. Setyawati MI, Tay CY, Leong DT. Mechanistic investigation of the biological effects of SiO2, TiO$_2$, and ZnO nanoparticles on intestinal cells. Small. 2015 Jul;11(28):3458–3468.
114. Ristic BZ, Milenkovic MM, Dakic IR, Todorovic-Markovic BM, Milosavljevic MS, Budimir MD, Paunovic VG, Dramicanin MD, Markovic ZM, Trajkovic VS. Photodynamic antibacterial effect of graphene quantum dots. Biomaterials. 2014 May 1;35(15):4428–4435.
115. Isakovic A, Markovic Z, Todorovic-Markovic B, Nikolic N, Vranjes-Djuric S, Mirkovic M, Dramicanin M, Harhaji L, Raicevic N, Nikolic Z, Trajkovic V. Distinct cytotoxic mechanisms of pristine versus hydroxylated fullerene. Toxicological Sciences. 2006 May 1;91(1):173–183.
116. Arora S, Jain J, Rajwade JM, Paknikar KM. Interactions of silver nanoparticles with primary mouse fibroblasts and liver cells. Toxicology and Applied Pharmacology. 2009 May 1;236(3):310–318.
117. Lin W, Huang YW, Zhou XD, Ma Y. In vitro toxicity of silica nanoparticles in human lung cancer cells. Toxicology and Applied Pharmacology. 2006 Dec 15;217(3):252–259.
118. Lison D, Thomassen LC, Rabolli V, Gonzalez L, Napierska D, Seo JW, Kirsch-Volders M, Hoet P, Kirschhock CE, Martens JA. Nominal and effective dosimetry of silica nanoparticles in cytotoxicity assays. Toxicological Sciences. 2008;104:155–162.

119. Chen Y, Chen J, Dong J, Jin Y. Comparing study of the effect of nanosized silicon dioxide and microsized silicon dioxide on fibrogenesis in rats. Toxicology and Industrial Health. 2004 Feb;20(1–5):21–27.

120. Arora S, Jain J, Rajwade JM, Paknikar KM. Cellular responses induced by silver nanoparticles: in vitro studies. Toxicology Letters. 2008 Jun 30;179(2):93–100.

121. Trop M, Novak M, Rodl S, Hellbom B, Kroell W, Goessler W. Silver-coated dressing acticoat caused raised liver enzymes and argyria-like symptoms in burn patient. Journal of Trauma and Acute Care Surgery. 2006 Mar 1;60(3):648–652.

122. Nabeshi H, Yoshikawa T, Matsuyama K, Nakazato Y, Arimori A, Isobe M, Tochigi S, Kondoh S, Hirai T, Akase T, Yamashita T. Size-dependent cytotoxic effects of amorphous silica nanoparticles on Langerhans cells. Die Pharmazie-An International Journal of Pharmaceutical Sciences. 2010 Mar 1;65(3):199–201.

123. Hsin YH, Chen CF, Huang S, Shih TS, Lai PS, Chueh PJ. The apoptotic effect of nanosilver is mediated by a ROS-and JNK-dependent mechanism involving the mitochondrial pathway in NIH3T3 cells. Toxicology Letters. 2008 Jul 10;179(3):130–139.

124. Lademann J, Weigmann HJ, Rickmeyer C, Barthelmes H, Schaefer H, Mueller G, Sterry W. Penetration of titanium dioxide microparticles in a sunscreen formulation into the horny layer and the follicular orifice. Skin Pharmacology and Physiology. 1999;12(5):247–256.

125. Mironava T, Hadjiargyrou M, Simon M, Jurukovski V, Rafailovich MH. Gold nanoparticles cellular toxicity and recovery: effect of size, concentration and exposure time. Nanotoxicology. 2010 Mar 1;4(1):120–137.

126. Pernodet N, Fang X, Sun Y, Bakhtina A, Ramakrishnan A, Sokolov J, Ulman A, Rafailovich M. Adverse effects of citrate/gold nanoparticles on human dermal fibroblasts. Small. 2006 Jun;2(6):766–773.

127. Zhang XD, Wu HY, Di Wu YY, Chang JH, Zhai ZB, Meng AM, Liu PX, Zhang LA, Fan FY. Toxicologic effects of gold nanoparticles in vivo by different administration routes. International Journal of Nanomedicine. 2010;5:771.

128. Lankveld DP, Oomen AG, Krystek P, Neigh A, Troost–de Jong A, Noorlander CW, Van Eijkeren JC, Geertsma RE, De Jong WH. The kinetics of the tissue distribution of silver nanoparticles of different sizes. Biomaterials. 2010 Nov 1;31(32):8350–8361.

129. Xie G, Sun J, Zhong G, Shi L, Zhang D. Biodistribution and toxicity of intravenously administered silica nanoparticles in mice. Archives of Toxicology. 2010 Mar;84(3):183–190.

130. Shukla R, Bansal V, Chaudhary M, Basu A, Bhonde RR, Sastry M. Biocompatibility of gold nanoparticles and their endocytotic fate inside the cellular compartment: a microscopic overview. Langmuir. 2005 Nov 8;21(23):10644–10654.

131. Ferreira GK, Cardoso E, Vuolo FS, Michels M, Zanoni ET, Carvalho-Silva M, Gomes LM, Dal-Pizzol F, Rezin GT, Streck EL, Paula, MMS. Gold nanoparticle alters parameters of oxidative stress and energy metabolism in organs of adult rats. Biochemistry and Cell Biology. 2015;150710143811008. https://doi.org/10.1139/bcb-2015-0030.

132. Kim YS, Kim JS, Cho HS, Rha DS, Kim JM, Park JD, Choi BS, Lim R, Chang HK, Chung YH, Kwon IH. Twenty-eight-day oral toxicity, genotoxicity, and gender-related tissue distribution of silver nanoparticles in Sprague-Dawley rats. Inhalation Toxicology. 2008 Jan 1;20(6):575–583.

133. Kato S, Itoh K, Yaoi T, Tozawa T, Yoshikawa Y, Yasui H, Kanamura N, Hoshino A, Manabe N, Yamamoto K, Fushiki S. Organ distribution of quantum dots after intraperitoneal administration, with special reference to area-specific distribution in the brain. Nanotechnology. 2010 Jul 27;21(33):335103.

134. Fischer HC, Chan WC. Nanotoxicity: the growing need for in vivo study. Current Opinion in Biotechnology. 2007 Dec 1;18(6):565–571.

135. Gonzalez L, Lison D, Kirsch-Volders M. Genotoxicity of engineered nanomaterials: a critical review. Nanotoxicology. 2008 Jan 1;2(4):252–273.
136. IARC Working Group on the Evaluation of Carcinogenic Risks to Humans. Cobalt and cobalt compounds. In *Chlorinated drinking-water; chlorination by-products; some other halogenated compounds; cobalt and cobalt compounds.* International Agency for Research on Cancer, Lyon, 1999.
137. Arsenide G, Phosphide I, Pentoxide V. *IARC monographs on the evaluation of carcinogenic risks to humans.* IARC, Lyon.
138. Bolt HM. Highlight report: critical evaluation of key evidence on health hazards of the general European population by exposure to arsenic. Archives of Toxicology. 2015 Dec;89(12):2455–2457.
139. Mena S, Ortega A, Estrela JM. Oxidative stress in environmental-induced carcinogenesis. Mutation Research/Genetic Toxicology and Environmental Mutagenesis. 2009 Mar 31;674(1–2):36–44.
140. Singh N, Manshian B, Jenkins GJ, Griffiths SM, Williams PM, Maffeis TG, Wright CJ, Doak SH. NanoGenotoxicology: the DNA damaging potential of engineered nanomaterials. Biomaterials. 2009 Aug;30(23–24):3891–3914.
141. Beyersmann D, Hartwig A. Carcinogenic metal compounds: recent insight into molecular and cellular mechanisms. Archives of Toxicology. 2008 Aug;82(8):493–512.
142. Arita A, Costa M. Epigenetics in metal carcinogenesis: nickel, arsenic, chromium and cadmium. Metallomics. 2009 May;1(3):222–228.
143. Salnikow K, Zhitkovich A. Genetic and epigenetic mechanisms in metal carcinogenesis and cocarcinogenesis: nickel, arsenic, and chromium. Chemical Research in Toxicology. 2008 Jan 21;21(1):28–44.
144. Li Q, Ke Q, Costa M. Alterations of histone modifications by cobalt compounds. Carcinogenesis. 2009 Jul 1;30(7):1243–1251.
145. Oyoshi MK, He R, Li Y, Mondal S, Yoon J, Afshar R, et al. Leukotriene B4-driven neutrophil recruitment to the skin is essential for allergic skin inflammation. Immunity. 2012;37:747–758.
146. Sabella S, Carney RP, Brunetti V, Malvindi MA, Al-Juffali N, Vecchio G, Janes SM, Bakr OM, Cingolani R, Stellacci F, Pompa PP. A general mechanism for intracellular toxicity of metal-containing nanoparticles. Nanoscale. 2014;6(12):7052–7061.
147. Tschopp J, Schroder K. NLRP3 inflammasome activation: the convergence of multiple signalling pathways on ROS production?. Nature Reviews Immunology. 2010 Mar;10(3):210–215.
148. Franchi L, Eigenbrod T, Núñez G. Cutting edge: TNF-α mediates sensitization to ATP and silica via the NLRP3 inflammasome in the absence of microbial stimulation. The Journal of Immunology. 2009 Jul 15;183(2):792–796.
149. Manda-Handzlik A, Demkow U. Neutrophils: the role of oxidative and nitrosative stress in health and disease. Pulmonary Infection. 2015:51–60.
150. Konior A, Schramm A, Czesnikiewicz-Guzik M, Guzik TJ. NADPH oxidases in vascular pathology. Antioxidants & Redox Signaling. 2014 Jun 10;20(17):2794–2814.
151. Thannickal VJ, Toews GB, White ES, Lynch Iii JP, Martinez FJ. Mechanisms of pulmonary fibrosis. Annual Review of Medicine. 2004 Feb 18;55:395–417.
152. Xia T, Kovochich M, Brant J, Hotze M, Sempf J, Oberley T, Sioutas C, Yeh JI, Wiesner MR, Nel AE. Comparison of the abilities of ambient and manufactured nanoparticles to induce cellular toxicity according to an oxidative stress paradigm. Nano Letters. 2006 Aug 9;6(8):1794–1807.
153. Reisetter AC, Stebounova LV, Baltrusaitis J, Powers L, Gupta A, Grassian VH, Monick MM. Induction of inflammasome-dependent pyroptosis by carbon black nanoparticles. Journal of Biological Chemistry. 2011 Jun 17;286(24):21844–21852.

154. Shirasuna K, Usui F, Karasawa T, Kimura H, Kawashima A, Mizukami H, Ohkuchi A, Nishimura S, Sagara J, Noda T, Ozawa K. Nanosilica-induced placental inflammation and pregnancy complications: different roles of the inflammasome components NLRP3 and ASC. Nanotoxicology. 2015 Jul 4;9(5):554–567.
155. Yazdi AS, Guarda G, Riteau N, Drexler SK, Tardivel A, Couillin I, Tschopp J. Nanoparticles activate the NLR pyrin domain containing 3 (Nlrp3) inflammasome and cause pulmonary inflammation through release of IL-1α and IL-1β. Proceedings of the National Academy of Sciences. 2010 Nov 9;107(45):19449–19454.
156. Dostert C, Pétrilli V, Van Bruggen R, Steele C, Mossman BT, Tschopp J. Innate immune activation through Nalp3 inflammasome sensing of asbestos and silica. Science. 2008 May 2;320(5876):674–677.
157. Franchi L, Núñez G. The Nlrp3 inflammasome is critical for aluminium hydroxide-mediated IL-1β secretion but dispensable for adjuvant activity. European Journal of Immunology. 2008 Aug;38(8):2085–2089.
158. Riteau N, Gasse P, Fauconnier L, Gombault A, Couegnat M, Fick L, Kanellopoulos J, Quesniaux VF, Marchand-Adam S, Crestani B, Ryffel B. Extracellular ATP is a danger signal activating P2X7 receptor in lung inflammation and fibrosis. American Journal of Respiratory and Critical Care Medicine. 2010 Sep 15;182(6):774–783.
159. Anastas P, Eghbali N. Green chemistry: principles and practice. Chemical Society Reviews. 2010;39(1):301–312.
160. Bali R, Razak N, Lumb A, Harris AT. *The synthesis of metallic nanoparticles inside live plants.* 2006 International Conference on Nanoscience and Nanotechnology. IEEE, Brisbane, 2006 Jul 3.
161. Song JY, Jang HK, Kim BS. Biological synthesis of gold nanoparticles using Magnolia kobus and Diopyros kaki leaf extracts. Process Biochemistry. 2009 Oct 1;44(10):1133–1138.
162. Haefeli C, Franklin CH, Hardy K. Plasmid-determined silver resistance in Pseudomonas stutzeri isolated from a silver mine. Journal of bacteriology. 1984 Apr;158(1):389–392.
163. Bridges K, Kidson A, Lowbury EJ, Wilkins MD. Gentamicin-and silver-resistant pseudomonas in a burns unit. The British Medical Journal. 1979 Feb 17;1(6161):446–449.
164. Brock TD, Gustafson JO. Ferric iron reduction by sulfur-and iron-oxidizing bacteria. Applied and Environmental Microbiology. 1976 Oct;32(4):567–571.
165. Taylor DE. Bacterial tellurite resistance. Trends in Microbiology. 1999 Mar 1;7(3):111–115.
166. Lloyd JR, Ridley J, Khizniak T, Lyalikova NN, Macaskie LE. Reduction of technetium by Desulfovibrio desulfuricans: biocatalyst characterization and use in a flowthrough bioreactor. Applied and Environmental Microbiology. 1999 Jun 1;65(6):2691–2696.
167. Lovley DR, Stolz JF, Nord GL, Phillips EJ. Anaerobic production of magnetite by a dissimilatory iron-reducing microorganism. Nature. 1987 Nov;330(6145):252–254.
168. Sunkar S, Nachiyar CV. Biogenesis of antibacterial silver nanoparticles using the endophytic bacterium Bacillus cereus isolated from Garcinia xanthochymus. Asian Pacific Journal of Tropical Biomedicine. 2012 Dec 1;2(12):953–959.
169. Rai A, Singh A, Ahmad A, Sastry M. Role of halide ions and temperature on the morphology of biologically synthesized gold nanotriangles. Langmuir. 2006 Jan 17;22(2):736–741.
170. Wen L, Lin Z, Gu P, Zhou J, Yao B, Chen G, Fu J. Extracellular biosynthesis of monodispersed gold nanoparticles by a SAM capping route. Journal of Nanoparticle Research. 2009 Feb;11(2):279–288.
171. Saifuddin N, Wong CW, Yasumira AA. Rapid biosynthesis of silver nanoparticles using culture supernatant of bacteria with microwave irradiation. E-Journal of Chemistry. 2009 Jan 1;6(1):61–70.
172. Beveridge TJ, Murray RG. Sites of metal deposition in the cell wall of Bacillus subtilis. Journal of Bacteriology. 1980 Feb;141(2):876–887.

173. Southam G, Beveridge TJ. The in vitro formation of placer gold by bacteria. Geochimica et Cosmochimica Acta. 1994 Oct 1;58(20):4527–4530.
174. Cunningham DP, Lundie Jr LL. Precipitation of cadmium by Clostridium thermoaceticum. Applied and Environmental Microbiology. 1993 Jan;59(1):7–14.
175. Labrenz M, Druschel GK, Thomsen-Ebert T, Gilbert B, Welch SA, Kemner KM, Logan GA, Summons RE, Stasio GD, Bond PL, Lai B. Formation of sphalerite (ZnS) deposits in natural biofilms of sulfate-reducing bacteria. Science. 2000 Dec 1;290(5497):1744–1747.
176. Saifuddin N, Wong CW, Yasumira AA. Rapid biosynthesis of silver nanoparticles using culture supernatant of bacteria with microwave irradiation. E-Journal of Chemistry. 2009 Jan 1;6(1):61–70.
177. Zhang H, Li Q, Lu Y, Sun D, Lin X, Deng X, He N, Zheng S. Biosorption and bioreduction of diamine silver complex by Corynebacterium. Journal of Chemical Technology & Biotechnology: International Research in Process, Environmental & Clean Technology. 2005 Mar;80(3):285–290.
178. Sweeney RY, Mao C, Gao X, Burt JL, Belcher AM, Georgiou G, Iverson BL. Bacterial biosynthesis of cadmium sulfide nanocrystals. Chemistry & Biology. 2004 Nov 1;11(11):1553–1559.
179. Deplanche K, Macaskie LE. Biorecovery of gold by escherichia coli and desulfovibrio desulfuricans. Biotechnology and Bioengineering. 2008 Apr 1;99(5):1055–1064.
180. Mahanty A, Bosu RA, Panda P, Netam SP, Sarkar B. Microwave assisted rapid combinatorial synthesis of silver nanoparticles using E. coli culture supernatant. International Journal of Pharma and Bio Sciences. 2013;4(2):1030–1035.
181. Correa-Llanten DN, Mu noz-Ibacache SA, Castro ME, Munoz PA and Blamey JM, Gold nanoparticles synthesized by Geobacillus sp. strain ID17 a thermophilic bacterium isolated from deception island, Antarctica. Microbial Cell Factories. 2013;12(1), article 75.
182. Nair B, Pradeep T. Coalescence of nanoclusters and formation of submicron crystallites assisted by lactobacillus strains. Crystal Growth & Design. 2002 Jul 3;2(4):293–298.
183. Korbekandi H, Iravani S, Abbasi S. Optimization of biological synthesis of silver nanoparticles using Lactobacillus casei subsp. casei. Journal of Chemical Technology & Biotechnology. 2012 Jul;87(7):932–937.
184. Bai HJ, Zhang ZM, Gong J. Biological synthesis of semiconductor zinc sulfide nanoparticles by immobilized rhodobacter sphaeroides. Biotechnology Letters. 2006 Jul;28(14):1135–1139.
185. He S, Guo Z, Zhang Y, Zhang S, Wang J, Gu N. Biosynthesis of gold nanoparticles using the bacteria rhodopseudomonas capsulata. Materials Letters. 2007 Jul 1;61(18):3984–3987.
186. He S, Zhang Y, Guo Z, Gu N. Biological synthesis of gold nanowires using extract of Rhodopseudomonas capsulata. Biotechnology Progress. 2008 Mar;24(2):476–480.
187. Fawcett D, Verduin JJ, Shah M, Sharma SB, Poinern GE. A review of current research into the biogenic synthesis of metal and metal oxide nanoparticles via marine algae and seagrasses. Journal of Nanoscience. 2017;15.
188. Michalak I, Chojnacka K. Algae as production systems of bioactive compounds. Engineering in Life Sciences. 2015 Mar;15(2):160–176.
189. Sharma A, Sharma S, Sharma K, Chetri SP, Vashishtha A, Singh P, Kumar R, Rathi B, Agrawal V. Algae as crucial organisms in advancing nanotechnology: a systematic review. Journal of Applied Phycology. 2016 Jun;28(3):1759–1774.
190. Prasad R, Pandey R, Barman I. Engineering tailored nanoparticles with microbes: quo vadis?. Wiley Interdisciplinary Reviews: Nanomedicine and Nanobiotechnology. 2016 Mar;8(2):316–330.

191. Lengke MF, Fleet ME, Southam G. Biosynthesis of silver nanoparticles by filamentous cyanobacteria from a silver (I) nitrate complex. Langmuir. 2007 Feb 27;23(5): 2694–2699.
192. Dahoumane SA, Djediat C, Yéprémian C, Couté A, Fiévet F, Coradin T, Brayner R. Species selection for the design of gold nanobioreactor by photosynthetic organisms. Journal of Nanoparticle Research. 2012 Jun;14(6):1–7.
193. Blackwell M. The Fungi: 1, 2, 3 . . . 5.1 million species?. American Journal of Botany. 2011 Mar;98(3):426–438.
194. Sastry M, Ahmad A, Khan MI, Kumar R. Biosynthesis of metal nanoparticles using fungi and actinomycete. Current Science. 2003 Jul 25:162–170.
195. Castro-Longoria E, Moreno-Velasquez SD, Vilchis-Nestor AR, Arenas-Berumen E, Avalos-Borja M. Production of platinum nanoparticles and nanoaggregates using neurospora crassa. Journal of Microbiology and Biotechnology. 2012;22(7):1000–1004.
196. Siddiqi KS, Husen A. Fabrication of metal nanoparticles from fungi and metal salts: scope and application. Nanoscale Research Letters. 2016 Dec;11(1):1–5.
197. Volesky B, Holan ZR. Biosorption of heavy metals. Biotechnology Progress. 1995 May;11(3):235–250.
198. Mukherjee P, Ahmad A, Mandal D, Senapati S, Sainkar SR, Khan MI, Parishcha R, Ajaykumar PV, Alam M, Kumar R, Sastry M. Fungus-mediated synthesis of silver nanoparticles and their immobilization in the mycelial matrix: a novel biological approach to nanoparticle synthesis. Nano Letters. 2001 Oct 10;1(10):515–519.
199. Ahmad A, Senapati S, Khan MI, Kumar R, Ramani R, Srinivas V, Sastry M. Intracellular synthesis of gold nanoparticles by a novel alkalotolerant actinomycete, Rhodococcus species. Nanotechnology. 2003 Jun 6;14(7):824.
200. Marcato PD, De Souza GI, Alves OL, Esposito E, Durán N. Antibacterial activity of silver nanoparticles synthesized by Fusarium oxysporum strain. 4th MERCOSUR Congress on Process Systems Engineering 2nd MERCOSUR Congress on Chemical Engineering: Proceedings of ENPROMER. 2005 Aug:1–5.
201. Riddin TL, Gericke M, Whiteley CG. Analysis of the inter-and extracellular formation of platinum nanoparticles by Fusarium oxysporum f. sp. lycopersici using response surface methodology. Nanotechnology. 2006 Jun 20;17(14):3482.
202. Boroumand Moghaddam A, Namvar F, Moniri M, Md. Tahir P, Azizi S, Mohamad R. Nanoparticles biosynthesized by fungi and yeast: a review of their preparation, properties, and medical applications. Molecules. 2015 Sep 11;20(9):16540–16565.
203. Kumar D, Karthik L, Kumar G, Roa KB. Biosynthesis of silver nanoparticles from marine yeast and their antimicrobial activity against multidrug resistant pathogens. Pharmacologyonline. 2011;3:1100–1111.
204. Kowshik M, Ashtaputre S, Kharrazi S, Vogel W, Urban J, Kulkarni SK, Paknikar KM. Extracellular synthesis of silver nanoparticles by a silver-tolerant yeast strain MKY3. Nanotechnology. 2002 Dec 20;14(1):95.
205. Pimprikar PS, Joshi SS, Kumar AR, Zinjarde SS, Kulkarni SK. Influence of biomass and gold salt concentration on nanoparticle synthesis by the tropical marine yeast Yarrowia lipolytica NCIM 3589. Colloids and Surfaces B: Biointerfaces. 2009 Nov 1;74(1):309–316.
206. López-Lorente ÁI, Mizaikoff B. Recent advances on the characterization of nanoparticles using infrared spectroscopy. TrAC Trends in Analytical Chemistry. 2016 Nov 1;84:97–106.
207. Rostron P, Gaber S, Gaber D. Raman spectroscopy, review. Laser. 2016;21:24.
208. Pugsley AJ, Bull CL, Sella A, Sankar G, McMillan PF. XAS/EXAFS studies of Ge nanoparticles produced by reaction between Mg2Ge and GeCl4. Journal of Solid State Chemistry. 2011 Sep 1;184(9):2345–2352.
209. Jaber B, Laânab L. One step synthesis of ZnO nanoparticles in free organic medium: Structural and optical characterizations. Materials Science in Semiconductor Processing. 2014 Nov 1;27:446–451.

210. Utsunomiya S, Ewing RC. Application of high-angle annular dark field scanning transmission electron microscopy, scanning transmission electron microscopy-energy dispersive X-ray spectrometry, and energy-filtered transmission electron microscopy to the characterization of nanoparticles in the environment. Environmental Science & Technology. 2003;37(4):786–791.

211. Singh S, Singh A. Current status of nanomedicine and nanosurgery. Anesthesia, Essays and Researches. 2013 May;7(2):237.

212. Balan K, Qing W, Wang Y, Liu X, Palvannan T, Wang Y, Ma F, Zhang Y. Antidiabetic activity of silver nanoparticles from green synthesis using Lonicera japonica leaf extract. RSC Advances. 2016;6(46):40162–40168.

213. Prabhu S, Vinodhini S, Elanchezhiyan C, Rajeswari D. Retracted: Evaluation of antidiabetic activity of biologically synthesized silver nanoparticles using Pouteria sapota in streptozotocin-induced diabetic rats. Journal of Diabetes. 2018 Jan;10(1):28–42.

214. Guo Y, Jiang N, Zhang L, Yin M. Green synthesis of gold nanoparticles from Fritillaria cirrhosa and its anti-diabetic activity on Streptozotocin induced rats. Arabian Journal of Chemistry. 2020 Apr 1;13(4):5096–5106.

215. Umaralikhan L, Jamal Mohamed Jaffar M. Green synthesis of MgO nanoparticles and it antibacterial activity. Iranian Journal of Science and Technology, Transaction A: Science. 2018;42(19).

216. Vickers NJ. Animal communication: when I'm calling you, will you answer too?. Current Biology. 2017 Jul 24;27(14):R713–R715.

217. Prasanth R, Kumar SD, Jayalakshmi A, Singaravelu G, Govindaraju K, Kumar VG. Green synthesis of magnesium oxide nanoparticles and their antibacterial activity. Indian Journal of Geo Marine Sciences. 2019;48(8):1210–1215.

218. Gurunathan S, Han JW, Eppakayala V, Jeyaraj M, Kim JH. Cytotoxicity of biologically synthesized silver nanoparticles in MDA-MB-231 human breast cancer cells. BioMed Research International. 2013 Oct. DOI: 10.1155/2013/535796.

219. Firdhouse MJ, Lalitha P. Biosynthesis of silver nanoparticles using the extract of Alternanthera sessilis—antiproliferative effect against prostate cancer cells. Cancer Nanotechnology. 2013 Dec;4(6):137–143.

220. Liang CH, Chan LP, Ding HY, So EC, Lin RJ, Wang HM, Chen YG, Chou TH. Free radical scavenging activity of 4-(3, 4-dihydroxybenzoyloxymethyl) phenyl-O-β-D-glucopyranoside from Origanum vulgare and its protection against oxidative damage. Journal of Agricultural and Food Chemistry. 2012 Aug 8;60(31):7690–7696.

221. Shankar S, Jaiswal L, Aparna R, Prasad V, Kumar P, Murthy Manohara C. Wound healing potential of green synthesized silver nanoparticles prepared from Lansium domesticum fruit peel extract. Mater Express. 5: 159–164.

222. Arunachalam KD, Annamalai SK, Arunachalam AM, Kennedy S. Green synthesis of crystalline silver nanoparticles using indigofera aspalathoides-medicinal plant extract for wound healing applications. Asian Journal of Chemistry. 2013 Aug 2;25.

223. Shanmugasundaram T, Radhakrishnan M, Gopikrishnan V, Kadirvelu K, Balagurunathan R. In vitro antimicrobial and in vivo wound healing effect of actinobacterially synthesised nanoparticles of silver, gold and their alloy. RSC Advances. 2017;7(81):51729–51743.

224. Casson RJ, Chidlow G, Ebneter A, Wood JP, Crowston J, Goldberg I. Translational neuroprotection research in glaucoma: a review of definitions and principles. Clinical & Experimental Ophthalmology. 2012 May;40(4):350–357.

225. Dayem AA, Kim B, Gurunathan S, Choi HY, Yang G, Saha SK, Han D, Han J, Kim K, Kim JH, Cho SG. Biologically synthesized silver nanoparticles induce neuronal differentiation of SH-SY5Y cells via modulation of reactive oxygen species, phosphatases, and kinase signaling pathways. Biotechnology Journal. 2014 Jul;9(7):934–943.

226. Patra S, Mukherjee S, Barui AK, Ganguly A, Sreedhar B, Patra CR. Green synthesis, characterization of gold and silver nanoparticles and their potential application for cancer therapeutics. Materials Science and Engineering: C. 2015 Aug 1;53:298–309.

227. Patra CR, Mukherjee S, Kotcherlakota R. Biosynthesized silver nanoparticles: a step forward for cancer theranostics?. Nanomedicine. 2014 Jul;9(10):1445–1448.
228. Patra S, Mukherjee S, Barui AK, Ganguly A, Sreedhar B, Patra CR. Green synthesis, characterization of gold and silver nanoparticles and their potential application for cancer therapeutics. Materials Science and Engineering: C. 2015 Aug 1;53:298–309.
229. Anjum M, Miandad R, Waqas M, Gehany F, Barakat MA. Remediation of wastewater using various nano-materials. Arabian Journal of Chemistry. 2019 Dec 1;12(8):4897–4919.
230. Sharma S, Bhattacharya A. Drinking water contamination and treatment techniques. Applied Water Science. 2017 Jun;7(3):1043–1067.
231. Pedahzur R, Lev O, Fattal B, Shuval HI. The interaction of silver ions and hydrogen peroxide in the inactivation of E. coli: a preliminary evaluation of a new long acting residual drinking water disinfectant. Water Science and Technology. 1995 Jan;31(5–6):123–129.
232. Luo Y, Shen S, Luo J, Wang X, Sun R. Green synthesis of silver nanoparticles in xylan solution via Tollens reaction and their detection for Hg 2+. Nanoscale. 2015;7(2):690–700.
233. Khatoon N, Mazumder JA, Sardar M. Biotechnological applications of green synthesized silver nanoparticles. Journal of Nanoscience: Current Research. 2017;2(107):2572.
234. Sadegh H, Ali GA, Gupta VK, Makhlouf AS, Shahryari-Ghoshekandi R, Nadagouda MN, Sillanpää M, Megiel E. The role of nanomaterials as effective adsorbents and their applications in wastewater treatment. Journal of Nanostructure in Chemistry. 2017 Mar;7(1):1–4.
235. Santhosh C, Velmurugan V, Jacob G, Jeong SK, Grace AN, Bhatnagar A. Role of nanomaterials in water treatment applications: a review. Chemical Engineering Journal. 2016 Dec 15;306:1116–1137.
236. Dil EA, Ghaedi M, Asfaram A. The performance of nanorods material as adsorbent for removal of azo dyes and heavy metal ions: application of ultrasound wave, optimization and modeling. Ultrasonics Sonochemistry. 2017 Jan 1;34:792–802.
237. Bottino A, Capannelli G, D'asti V, Piaggio P. Preparation and properties of novel organic–inorganic porous membranes. Separation and Purification Technology. 2001 Mar 1;22:269–275.
238. Maximous N, Nakhla G, Wong K, Wan W. Optimization of Al2O3/PES membranes for wastewater filtration. Separation and Purification Technology. 2010 Jun 18;73(2):294–301.
239. Yurekli Y. Removal of heavy metals in wastewater by using zeolite nano-particles impregnated polysulfone membranes. Journal of Hazardous Materials. 2016 May 15;309:53–64.
240. Zodrow K, Brunet L, Mahendra S, Li D, Zhang A, Li Q, Alvarez PJ. Polysulfone ultrafiltration membranes impregnated with silver nanoparticles show improved biofouling resistance and virus removal. Water Research. 2009 Feb 1;43(3):715–723.
241. Patanjali P, Singh R, Kumar A, Chaudhary P. Nanotechnology for water treatment: a green approach. In *Green synthesis, characterization and applications of nanoparticle* (pp. 485–512). Elsevier, Amsterdam, 2019.
242. Anjum M, Al-Makishah NH, Barakat MA. Wastewater sludge stabilization using pre-treatment methods. Process Safety and Environmental Protection. 2016 Jul 1;102:615–632.
243. Ditta A. How helpful is nanotechnology in agriculture?. Advances in Natural Sciences: Nanoscience and Nanotechnology. 2012 May 29;3(3):033002.
244. Polash SA, Nadaf NY, Rahman MA, Shohael AM. Green synthesis of silver nanoparticles (AgNPs): agricultural applications and future vision. Journal of Biological &. Environmental Sciences. 2018;13(2):35–57.

245. Ashoka P, Meena RS, Gogoi N, Kumar S, Yadav GS, Layek J. Green nanotechnology is a key for eco-friendly agriculture. Journal of Cleaner Production. 2017;4(142).
246. Khodakovskaya M, Dervishi E, Mahmood M, Xu Y, Li Z, Watanabe F, Biris AS. Carbon nanotubes are able to penetrate plant seed coat and dramatically affect seed germination and plant growth. ACS Nano. 2009 Oct 27;3(10):3221–3227.
247. Zheng L, Hong F, Lu S, Liu C. Effect of nano-TiO$_2$ on strength of naturally aged seeds and growth of spinach. Biological Trace Element Research. 2005 Apr;104(1):83–91.
248. Yang F, Hong F, You W, Liu C, Gao F, Wu C, Yang P. Influence of nano-anatase TiO$_2$ on the nitrogen metabolism of growing spinach. Biological Trace Element Research. 2006 May;110(2):179–190.
249. Li J, Chang PR, Huang J, Wang Y, Yuan H, Ren H. Physiological effects of magnetic iron oxide nanoparticles towards watermelon. Journal of Nanoscience and Nanotechnology. 2013 Aug 1;13(8):5561–5567.
250. Irshad MA, Nawaz R, ur Rehman MZ, Adrees M, Rizwan M, Ali S, Ahmad S, Tasleem S. Synthesis, characterization and advanced sustainable applications of titanium dioxide nanoparticles: a review. Ecotoxicology and Environmental Safety. 2021 Apr 1;212:111978.
251. Rameshaiah GN, Pallavi J, Shabnam S. Nano fertilizers and nano sensors–an attempt for developing smart agriculture. International Journal of Engineering Research and General Science. 2015;3(1):314–320.
252. Maksimović M, Omanović-Mikličanin E. *Green internet of things and green nano-technology role in realizing smart and sustainable agriculture* (pp. 2290–2295). VIII International Scientific Agriculture Symposium AGROSYM, Jahorina, Bosnia and Herzgovina, 2017 Oct.
253. Ezhilarasi PN, Karthik P, Chhanwal N, Anandharamakrishnan C. Nanoencapsulation techniques for food bioactive components: a review. Food and Bioprocess Technology. 2013 Mar;6(3):628–647.
254. Kanmani P, Rhim JW. Physical, mechanical and antimicrobial properties of gelatin based active nanocomposite films containing AgNPs and nanoclay. Food Hydrocolloids. 2014 Mar 1;35:644–652.
255. Singh T, Shukla S, Kumar P, Wahla V, Bajpai VK, Rather IA. Application of nanotech-nology in food science: perception and overview. Frontiers in Microbiology. 2017 Aug 7;8:1501.

3 Economic Analysis and Environmental Aspects of Metal Oxide Materials

Saleem Ahmad Yatoo, G. N. Najar, Sajad Ahmad Rather, Rayees Ahmad Zargar, Ab Qayoom Mir, and S. A. Shameem

CONTENTS

3.1 INTRODUCTION

Nanotechnology is a novel technology that emphasizes the development of structures, devices, and systems using nanosized particles (1–100 nm) in several fields

DOI: 10.1201/9781003323464-3

of science like chemistry, physics, biotechnology and material science. As the nanotechnology industry has a strong connection to and plays a very important role in economic development. Presently, nanotechnology has a wide range of applications in many sectors; scientists, researchers, investors and policymakers at the global level are aware of its potential and effective power. It is a fact that industry is considered a driving force behind economic development, which contributes to the creation of more job opportunities, the encouragement of new technological innovations, output rate increases and the attainment of a higher standard of living [1].

The use of nanometer-scale material properties has significant implications in all aspects of life, particularly in manufacturing. By using nanotechnology in the manufacturing sector, it will be possible to direct the placement of atoms in the reaction in a specific way, resulting in the creation of more specific and precise, that redefined than traditional production techniques, which improves product quality while reducing energy consumption and lowering production costs [2].

Advanced nanoengineered materials and nano-based products have become increasingly popular in various industries, like healthcare, energy conversion, energy storage, electronics and consumer goods, over the last two decades [3]. In recent years, nanotechnology has made progress in every field including nanoparticles (NPs) and remediation techniques. Throughout the 21st century, many researchers have made significant developments in nanotechnology, and nanomaterials have received more attention for remediation [4, 5]. An increase in engineered nanomaterial (ENM) emissions into the environment has raised concerns about the harmful effects on human health and safety and the harmful effects on biological species (flora and fauna) and the natural environment [6]. In order to explore the full potential in terms of physical, chemical and biological transformations of ENMs in interaction with important environmental conditions, it is very necessary to characterize not only nanosized emissions but also the physical and chemical characteristics of the discharged particles [7]. NPs derived from metal oxides are one of the most commonly used NPs [8]. Fora couple of years, a large quantity of FeO_2, TiO_2, AlO_2 and SiO_2 have been manufactured. However, due to their large use, they have been synthesized at the nanoscale and are being found in consumer market applications, like ZnO in sunscreens [9]. Among such nanomaterials, metal oxide NPs(MONPs) such as ZnO, TiO_2, CuO and Fe_2O_3 are several that are commonly used in various fields [10–14].

3.2 NANOTECHNOLOGY: USES AND APPLICATIONS

Nanotechnology has evolved as a miracle science for the development of various materials at the nanoscale in the 21st century. An American physicist Richard Feynman commenced the idea or concept of nanoscience and nanotechnology with a famous talk titled "There's Plenty of Room at the Bottom". He revolutionize the new era, in which it is promising to change the properties of devices at the atomic, molecular or macromolecular level. Gradually, this technology advanced due to its significant improvements through numerous scientific researchers [15]. However, the word *nano* is derived from the Greek language, denoting dwarf or extremely small particles having a size less than 100 nm. Therefore, at present, the nanotechnology industry is a developing and growing science; thus, its market is flourishing. All branches of nanotechnology-based products have a profound effect on the market of innovative substances and change in consumer choices and planning in the near future.

Better nanotechnology products guarantee the long sustainability of industry. The use of waste products as raw material and the use of methodologies that consume a small amount of water and energy are necessary to protect human health and the environment [16]. Industrial technologies and advanced scientific applications have a firm position towards growing strategic sectors of countries economy and, specifically, economic growth. The applications of nanotechnology to traditional materials offer a chance for the industry to increase its innovative capacity [17].

In general, it is clear that materials at the nanoscale provide typical opportunities for developing our society; for example, they increase our information and provide better performing technological devices that make life simpler and subsequently result in a decrease in humans' physically completing hard tasks [18]. The use of nanotechnology for market products needs tough standards for reproducibility and safety. So, it has become significant to frame a common systematic standard for security and long-term sustainability in nanotechnology and continuous research and advances in nanotechnology [19] as nanotechnology is progressing and producing new products in all small and large industrial sectors. The following sections mentioned some examples of nanotechnology use in industrial applications.

3.3 NANOTECHNOLOGY APPLICATIONS IN BIOMEDICAL FIELDS

For some years, NPs have been applied to disease diagnosis, the detection of infection, drug delivery, cancer therapy treatment, the detection of pulmonary diseases and the prevention of some infections.

3.4 NANOTECHNOLOGY APPLICATIONS IN AGRICULTURAL AND FOOD PRODUCTION

A number of nanomaterials, mostly metal-based and carbon-based nanomaterials like graphene and fullerene have been exploited for their use in absorption, translocation and accumulation and are used to find the impact on the growth and development of crop plants [20].

3.5 NANOSIZED FERTILIZERS

Common fertilizers are gradually being replaced by nanofertilizers, an advanced method of release of nutrients in the soil gradually. Nanofertilizers have more benefits than commonly used fertilizers because they have exceptional features like ultra-high absorption, increased growth in production, an increased photosynthetic process and a substantial expansion in the leaf surface area. Nanotechnology applications in nanofertilizers have also enhanced the efficiency of the elements and decreased the toxicity of the soil [20].

3.6 NANO-BASED AGROCHEMICALS

In agriculture, pesticides are applied to improve crop yield and better efficiency. Industrial growth in nanotechnology has made a good technological improvement in the delivery of nano-based pesticide strategies to address the problems and issues in agriculture.

3.7 NANOTECHNOLOGY IN THE TEXTILE INDUSTRY

Nanotechnology has made great improvements in the textile industry compared to the traditional way of production due to the increased durability and efficiency of fabrics, healthy properties and reduced production cost. NPs provide extraordinary durability to fabrics as they have huge surface areas and high surface energies; for example, silver (Ag) NPs are being used in socks and sports clothing because of their capability of killing bacteria and preventing undesirable smells [20].

3.8 NANOTECHNOLOGY APPLICATIONS IN THE OIL INDUSTRY

For the oil industry, nanotechnology has created nanotubes that are being used to create very lightweight, durable, strong and more corrosion-resistant materials in the oil industry. The application of nanotechnology has assisted in improving oil and gas production, making it easy to separate oil and gas in the reservoir [20]. Nanotechnology has been used in various environmental applications, including wastewater treatment and soil treatment, pollution remediation, sensors and energy

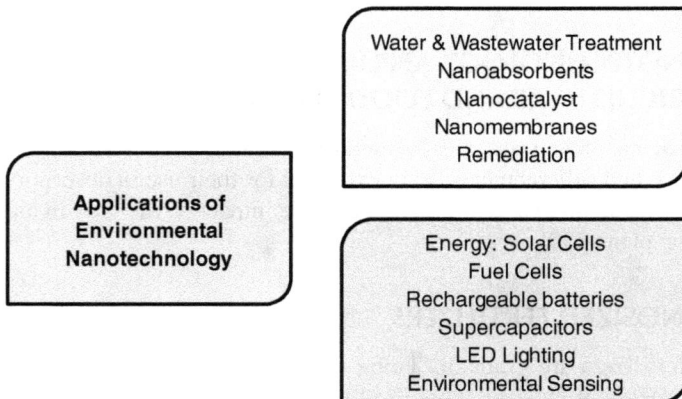

FIGURE 3.1 Applications of environmental nanotechnology in industry.

storage. The basic focus in environmental nanotechnology includes the safe design of nanomaterials having great environmental benefits, and this promotes the sustainable development of those materials [21] and introduces more resistant and more efficient materials [22].

3.9 SUSTAINABLE ENERGY

There has been a great demand for using renewable sources of energy that are sustainable and environmentally friendly, which are being explored globally. Nanotechnology has become a great way, because of its capacity, to improve energy efficiency, clean and eco-friendly energy production, energy conversion, energy storage and energy usage and saving. Therefore, potential nanomaterials can be used for producing more efficient of solar and fuel cells that are clean and environmentally friendly [21].

3.10 ENERGY CONVERSION

Pollution-free energy that is green can be formed by means of nanotechnology to create solar and some fuel cells which can be potential and efficient sources of commercially consumable alternate forms of unpolluted energy sources with significantly less cost. Therefore, this nanotechnology is economically efficient for large-scale power generation.

The economic impacts of nanotechnology may be of short, medium and long term on society, and in time, economic aspects can be noticed, like the availability of research facilities with instrumentation, the supply-chain structure having skilled graduates possessing knowledge of research and development and new technology platforms having infra-technologies producing the commercialization of new products. In the long run, wide-ranging industry and national economic benefits in terms of return on investment and gross domestic product influences at the national level (Table 3.1)

TABLE 3.1
Some Short-, Medium- and Long-Term Socioeconomic Impacts of Nanotechnologies

Short Term	Medium Term	Long Term
Partnership structures and strategic alliances organized New research facilities and instrumentation in place Initial research objectives met/ increased stock of technical knowledge	Supply-chain structure established New-skilled graduates produced Compression of R&D cycle New technology platforms and infra-technologies produced Commercialization of new products, new processes, licensing	Broad industry and national economic benefits: 1. Return on investment 2. Gross domestic product impacts

3.11 SOME METAL OXIDE MATERIALS AND THEIR APPLICATIONS

A selected number of applications of the typical metal oxide materials, characterized by different nanostructures is shown in Table 3.2. The list of oxides in Table 3.2 is not comprehensive but gives a broader concept of metal oxide applications. One thing is clear from the table: that structures of both core and functional materials are compulsory in modulating the bandgap as well as surface adsorption properties of

TABLE 3.2

Applications of Some Representative Metal Oxide–Based Materials with Various Structures

Metal Oxide-Based Material	Structural Features	Synthesis Method	Application	Reference
ZnO nanosheets	3D hierarchical flower-like architectures	Solvothermal	Adsorption of triphenylmethane dyes	[31]
Fe_3O_4UiO-66 composite	Cubical nanoparticles (NPs) arranged	Sonication	Adsorption	[32]
Fe_3O_4MIL-100 (Fe) Core–Shell Bionanocomposites	Core-shell structure with Fe_3O_4 core, immobilized on *P. putida*	Sonication followed by attaching the NPs on bacteria	Adsorption	[33]
ZnO–TiO_2/clay	TiO_2 and ZnO NPs mounted on clay surface	Sol-gel method	Degradation of MG	[34]
Cu/ZnO/Al_2O_3	Cu- and ZnO-impregnated Al_2O_3	Impregnation method	CO removal from reformed fuel	[35]
CO_2+, Ni_2+ doped Fe_3O_4	NPs Cubic lattice	Co-precipitation method	Photodegradation of carbol fuchsin	[36]
Ce/Fe bimetallic oxides (CFBO)	Flowerlike 3D hierarchical architecture	No-template hydrothermal method	As(V) and Cr(VI) remediation	[37]
Perovskite titanates ($ATiO_3$, A = Sr, Ca and Pb)	Leaf-architectured 3D-hierarchical structure	Combination of biosynthesis from cherry blossom, heating, grinding and photo de position	Artificial photosynthetic system for photoreduction of CO_2	[38]
TiO_2 polypyrrole	Core–shell nanowires (NWs)	Seed-assisted hydrothermal method	Flexible supercapacitors (SCs) on carbon cloth	[39]
Fe_3O_4/WO_3	Hierarchical core–shell structure	Solvothermal growth +oxidation route	Photodegradation of organic-dye materials	[40]

these heterogeneous materials photocatalysts. It is a fact that bandgap modulating, microstructure and optoelectronics properties are within the basic range of metal oxide applications [23–29]. There isa range of applications of metal oxide nanomaterials, including those of photocatalytic applications used for environmental remediation, decarbonization and energy sustainability [30].

Catalysts used in metal oxide–based reactions can alter any reaction type, and among the two ways that catalysts can change the reaction are, first, decreasing the activation energy and, second, providing another pathway; thus, metal oxide–based photocatalysts are applicable for both dye and organic pollution degradation. This degradation process occurs in three important steps:

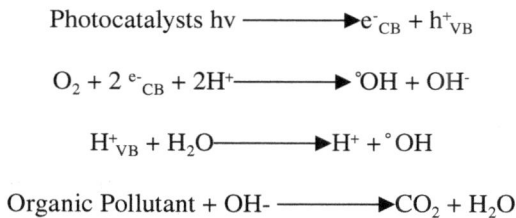

$$\text{Photocatalysts hv} \longrightarrow e^-_{CB} + h^+_{VB}$$

$$O_2 + 2\ e^-_{CB} + 2H^+ \longrightarrow {}^\circ OH + OH^-$$

$$H^+_{VB} + H_2O \longrightarrow H^+ + {}^\circ OH$$

$$\text{Organic Pollutant} + OH^- \longrightarrow CO_2 + H_2O$$

3.12 RECENT APPLICATIONS OF METAL OXIDE MATERIALS IN DIFFERENT FIELDS

In recent years, metal oxides have played a chief role in the fields of physics, chemistry and material science principally because of the exclusive physical and chemical properties of metal oxide nanostructures that lead to a wide scope of applications, including poisonous gas sensing, biomedical applications, textile coating for wearable electronic devices, photocatalytic degradation of organic contaminants and others [41, 42]. Engineered MONPs are among the extensively used manufactured nanomaterials because of their exceptional properties. The properties of nanophase structures that make them unique include higher ductility at elevated temperatures than the coarse-grained ceramics; unique catalytic, super-paramagnetic behavior; sensitivity; and selective activity [43]. Some of the major applications of metal oxide materials in the different fields are summarized in the following sections.

3.12.1 REMOVAL/ADSORPTION OF HEAVY METALS

Heavy metal pollution is a major problem in our environment; thus, its removal is necessary to protect the environment from degradation. Metal oxides at present are playing a vital role in environmental remediation as well as pollutant sensing. Heavy metal removal depends on the type, size and morphology of metal oxide–based materials. Three types of processes are mainly responsible for the metal oxide–based materials to work—reduction, oxidation and adsorption—in the remediation of heavy metals. The reaction between heavy metals mostly takes place during the adsorption process, which leads to the formation of inner-sphere complexes [44]. As far as the adsorption of heavy metals is concerned, surface area is an important factor to be considered. Moreover, these oxides are environmentally friendly and suitable since they do not cause secondary pollution [45–47]. A number of heavy metals, such as Pb, Hg, Zn, Cu, Cd and others, are adversely harmful and carcinogenic to human health and they may bring a noticeable threat to human health in the long run because they perhaps are biologically accumulated in food chains. For removal of heavy metals, nano-SiO_2 is being used in microwave-assisted solid-phase extraction like Pb(II), Cu(II), Cu(II), Cd(II) and Hg(II) are uptake by adsorbents of very less time 5–20s by applying this technique. Thus, this is an efficient method for quickly removing toxic heavy metals [48]. Zinc oxide (ZnO) NPs are among the most stable and efficient catalysts for degrading a varied range of pollutants. Moreover, a cork synthesis of ZnO-based materials has been developed that is considered a multifunction biomorphic system for environmental remediation.

Carbon nanotubes, graphene and fullerene may adsorb numerous heavy metals from water [49]. Nanomaterials, such as SWCNTs-COOH, SWCNTs-OH, SWCNTs-NH_2, MWCNTs-FeO_4, MWCNTs-ZrO_2, MWCNTs-MnO_2, MWCNTs-Fe_3O_4 etc [50], have heavy metal adsorption capacities, and they all are capable for the remediation of heavy metal from polluted water. From natural plants or their products, TiO_2NPs are being synthesized from *Jatropha curcas L.* through a green and eco-friendly procedure that is mostly economical, resulting in multiple benefits, namely, the removal of chemical oxygen demand (COD) and heavy metal like chromium ions from tannery wastewater and removal rates are; 82% and 76.48%, respectively, for COD and chromium could be achieved at the anatase phase of these TiO_2NPs under solar irradiation [51]. Cost of raw materials, bandgap, nanostructure and adsorption are considered some basic factors that determine the suitability of metal oxides as photocatalysts for organic pollution degradation and pure metal oxides often are found lacking in one or more of these parameters.

3.12.2 DECARBONIZATION/CARBON CAPTURE

It is a fact that most of the fossil fuels used in industry produce high levels of carbon dioxide (CO_2) during combustion that is released into the atmosphere, consequently causing global climate change. Nanotechnologies have received great attention for CO_2 capture due to their unique properties [22]. The increase in carbon dioxide concentration is a major problem in our environment. Its reduction is an important process because of its impact on the environment. Fortunately, sunlight-initiated breakdown of CO_2 might be also a cost-effective, readily achievable and better method if the right kind of catalyst is designed. Optimizing crystal structures and more accurately engineering the band gaps of metal-based photocatalysts. Materials like niobates and ferrite materials are widely used metal oxides, with these intentions for quite some time now [52]:

a. Adsorption of reductant and CO_2 onto the metal oxide catalyst surface
b. Adsorption of the sunlight on through perovskite leading to electron–hole separation

c. Transfer of photogenerated charges to the surface and reaction with adsorbed CO_2 and water

d. Desorption products

Besides the previously mentioned uses, a good reducing agent can enhance the efficiency of the valance bond, considering that its maximum is less than the reduction potential of water that is necessary for this type of process. Apart from environmental remediation, the photocatalytic reduction of CO_2 is emerging as a green tool to produce chemicals and fuels by closing the carbon cycle and realizing zero carbon emissions. Apart from this, CaO-base solid sorbents for carbon capture and its storage [53, 54], perovskite-based sodium tantalate nanocubes (Vo-NaTaON) [55], perovskite titanates [56], nanostructured $NaNbO_3$ [57], TiO2 [58], WO_3, ZrO_2, MgO, Ga_2O_3 and Al_2O_3, have been explored for CO_2 photocatalytic reduction under UV/Vis light [59].

3.12.3 Removal/Adsorption

Surfactants are considered one of the dangerous communal environmental problems and mostly find their way and get mixed up with the aqueous environment [60]. Graphene oxide (GO), as well as reduced graphene oxide (rGO) are being applied to nonionic surfactants such as TX-100 [61]. Through carbon nanotubes, aromatic cationic surfactants are also effortlessly eliminated. Titanium oxide is the most popular nanomaterial utilized a significant adsorbent for the elimination of surfactants from polluted water [62].

3.12.4 Metal Oxide Use in Semiconductors

The semiconductors manufactured from metal oxides are economically and environmentally friendly and are considered materials with high stability. They have been widely used as photoelectrodes in photovoltaics (PVs) in dye solar cells and for the development of metal oxide p–n ju Ritchie notions [63]. Their widespread use in new-generation PVs is due to their unusually flexible properties, low cost and easily scalable methods. Due to very large bandgap energy range and high tuneability, these metal oxides are used as photo-harvesters. Thin film semiconductor structures have widespread application potential in energy conversion and storage, sensing, electronics, and photonics. Metal oxide semiconductors are promising alternatives for large-scale applications due to their abundance in nature, nontoxicity and low cost. Cuprous oxide (Cu_2O), a good photoactive metal oxide with these advantages, is a p-type semiconductor with a band gap of about 2 eV, a large absorption coefficient ($>10^5$ cm^{-1}) in the visible region and high carrier mobility; therefore, it is used as a thin film active layer in solar cell applications [64]

Among all the metal oxide compounds available, copper oxides (Cu–O) are the most predominant materials to date. The good p-type semiconductor, cuprous oxide (Cu_2O), is used as PV material due to its low cost, good mobility, high absorption coefficient, nontoxicity and plentiful availability. Although Cu_2O has been used since 1920 in electronic materials, since 1970, its use has shown a manifold increase [65].

3.12.5 Metal Oxides for Energy Conversion

Nanomaterials, especially metal oxides, play an important role in the manufacture and application of sustainable energy sources. The efficiency of energy source utilization is significantly dependent on the properties of the nanomaterials and methods of utilization. Various innovative nanomaterials are hence explored for different methods of energy production and application, such as thermocatalysis, photoelectrocatalysis and photocatalysis. For the fabrication of these devices, a nanocrystalline mesoporous titanium dioxide (TiO_2) film, with a monolayer of the charge transfer dye attached to its surface, is pasted on a transparent conductive substrate [66, 67]. The large nanomaterial surface area for dye chemisorption and the short charge migration length underlie their power conversion efficiency [68]. Also, semiconductor nanomaterials like TiO_2 and cadmium sulfide nanostructures have been investigated as efficient catalysts for water conversion into oxygen and hydrogen [69].

3.12.6 Metal Oxides for Energy Storage

Metal oxides have many applications in other fields like industrial, energy production, conversion and storage. Energy is a basic need for development at present in many sectors. The growing demand for energy is tremendously increasing and a challenging factor for economic growth and climate change at present times. Nonrenewable resources like coal, petroleum and natural gas are limited in quantity but have dominated energy supplies in many past decades at a global level. Supercapacitors, a novel variety of energy storage device, reveals short charging time, high power density, low maintenance cost and long cycle life [70]. Supercapacitors can be efficiently used to prepare power supply systems, energy storage devices and many electronic devices [71].

Metal/metal oxides are common supercapacitors electrode materials, providing high power density and long cycle life. In combination, graphene and metal oxides exhibited the combination of the excellent cycle stability of graphene and the high-capacity characteristics of metal oxides which extraordinarily improve the comprehensive properties of nanocomposites [72]. Cobalt oxide/iron oxide and copper oxide/cobalt oxide as binary metal oxide systems appear to be promising as thermo-chemical storage materials [73].

3.12.7 Metal Oxides for Water Cleanup Technologies

The application of metal oxide nanomaterials allows for the development of non-reagent green technologies for water purification, which are considered more efficient and less energy-consuming; both the said properties are important for sustainable water treatment technologies. Improvement in the quality of water, air and soil is a vast obstacle of present times. Therefore, recognition and remediation of environmental pollutants and their prevention is a very necessary step in the conservation of the environment. Carbon nanotubes (SWCNTs) and fullerene may be applied as water purification membranes to rectify water quality by converting wastewater into drinking water (such as nano-sponge filtration), and adsorbents of polar and nonpolar organic chemicals, and to remove other toxic contaminants from water. A photocatalyst consisting of single-walled carbon nanotubes-titanium dioxide ($SWCNTs-TiO_2$) is significantly applied to the purification of water from oil [74, 75].

Nanotechnology-enabled metal oxides are used in water and wastewater treatment; they promise not only to overcome major challenges met by existing treatment technologies but also to convey new treatment potential. This could allow the economic utilization of alternative water sources to expand the water supply [76]. The deep removal of sulfur from liquid fuels, like diesel and others, is an extremely important issue to solve the issue of environmental pollution by sulfur-containing compounds [77, 78]. Remarkable applications may include the incorporation of functional nanomaterials, such as metaloxide nanomaterials such as aluminum oxide, TiO_2 and zeolite [79], and photocatalytic nanomaterials such as bimetallic nanomaterials, TiO_2 across the membranes to improve their permeability, biofilm control, fouling resistance, thermal and mechanical stability and offer pollutant degradation and self-cleaning ability [80].

3.12.8 ADSORPTION OF NUMEROUS HARMFUL DYES

At the present time, dyes play major roles in the paint industry and textile- and color pigment–fabricating industries. The sewage of organic dyes plus nitroarenes are recognized as some new harmful substances due to the fact that being unacceptable dyes, nitro compounds both are dangerous to humans and both to flora and fauna. Embellished titania on aerogel of SWCNTs and fabricated $MWCNTs-TiO_2$ are used for the elimination of methylene blue dye from water [50].

3.12.9 METAL OXIDES FOR HEALTH AND SAFETY CONSIDERATIONS

The unique properties of nano–metal oxide materials have made them striking for a number of innovative, sustainable and green applications. Metal oxides, such as TiO_2, Bi_2O_3, ZnO [13, 14], FeO, MnO_2, CuO, Ag_2O and Al_2O_3, show antimicrobial activities and play important roles in various medical applications. TiO_2 is being used against the transmission of various infectious diseases [81, 82]. Due to their antibacterial activity, ZnO NPs are attaining worldwide attention. It is now understood to be a biosafe material and is being used to carry out photocatalysis and photo-oxidation reactions on biological and chemical species [83]. Similarly, Al_2O_3 nanoparticles have vast practical applications and exhibited antimicrobial behavior [84]. Because of the excellent chemical properties reported for the CuO complex of *Ficus religiosa* NPs, it has been used as an anticancer agent in biomedicine [85]. Zinc-doped titania NPs have shown good proangiogenic properties, which might be advantageous in diverse applications [86]. Antimicrobial properties have been revealed from doped copper/TiO_2 NPs with carbon-based allotropes. To degrade various microbial species CuO has been successfully applied as an antimicrobial agent [87].

Similarly, different metal oxides and doped metal/metal composites, such as silver oxide, calcium oxide, copper oxide, titanium dioxide magnesium oxide and zinc oxide, have numerous distinctive properties and spectral activity and have exhibited outstanding antimicrobial ability [88]. Furthermore, the physicochemical properties and morphology of nanomaterials have been proven for their antimicrobial activities. It was recognized that the nanosized metal particles carry a potent bactericidal effect to resist the bacteria [89].

3.13 SUMMARY AND CONCLUSION

Gradually, nanotechnology and the economic system have recognized this important need for technological advancements. The idea of exploring how technology creates economic growth has been comprehensively explored. Nanotechnology has evolved as a miracle science for the development of various materials at the nanoscale in the 21st century that have benefited and increased economic development for humankind in various fields for the last five decades. New techniques and technologies are being continuously developed for numerous applications. Systematic, scientific and strong technological base is very important for better long-term economic development. Metal oxides are one of the promising nanomaterials for environmental applications tackling environmental challenges that presently exist. Over last the 20 years, tremendous growth has been seen in technological exploitation and applications of MONPs as evidenced by the many patent applications in the field. It is one of the most important nanomaterials used in various applications and particularly in pollution remediation of the environment. As one of the most exclusive fields of research due to scope of structural and functional variation that can be produced, metal oxides continue to be explored for environmental remediation, solving the energy crisis and for decarbonization and solving agriculture problems, oil industry issues, nanofertilizers, carbon capture, and so on. Thus being, economical and often long lasting without being much negative impact on the environment, nanotechnology is of great importance in the industrial sector and the environmental field.

REFERENCES

1. Reck, B.K. and Graedel, T.E. (2012) Challenges in metal recycling. *Science* 337(6095), 690–695.
2. Morganti, P. (2011) Chitin nanofibrils and their derivatives as cosmeceuticals. In: S.K. Kim (ed). *Chitin, Chitosan, Oligosaccharides and Their Derivatives: Biological Activities and Applications*, pp. 531–542. New York: CRC Press.
3. Gottschalk, F., Sun, T. and Nowack, B. (2013) Environmental concentrations of engineered nanomaterials: Review of modeling and analytical studies. *Environmental Pollution* 181, 287–300.
4. Yu, Z., Hu, L. and Lo, I.M.C. (2019a) Transport of the arsenic (As)-loaded nano zerovalent iron in groundwater-saturated sand columns: Roles of surface modification and as loading. *Chemosphere* 216, 428–436.
5. Yu, Z., Huang, J., Hu, L., Zhang, W. and Lo, I.M.C. (2019b) Effects of geochemical conditions, surface modification, and arsenic (As) loadings on as release from As-loaded nano zero-valent iron in simulated groundwater. *Chemical Engineering Journal* 332, 42–48.
6. Bystrzejewska-Piotrowska, G., Golimowski, J. and Urban, P.L. (2009) Nanoparticles: their potential toxicity, waste and environmental management. *Waste Management* 29, 2587–2595.

7. Handy, R.D., Von der Kammer, F., Lead, J.R., Hasserllöv, M., Owen, R. and Crane, M. (2008) The ecotoxicology and chemistry of manufactured nanoparticles. *Ecotoxicology* 17, 287–314.
8. Aitken, R.J., Chaudhry, M.Q., Boxall, A.B. and Hull, M. (2006) Manufacture and use of nanomaterials: Current status in the UK and global trends. *Occupational Medicine* 56(5), 300–306, Aug 1.
9. Rittner, M.N. (2002) Market analysis of nanostructured materials. *American Ceramic Society Bulletin* 81, 33e36.
10. Zargar, R.A., Kumar, K., Arora, M. and Shkir, M. (2022) Structural, optical, photoluminescence, and EPR behaviour of novel ZnO 80Cd0·200 thick films: An effect of different sintering temperatures. *Journal of Luminescence* 245, 118769.
11. Zargar, R.A., Kumar, K. and Shkir, M. (2022) Optical characteristics ZnO film: A metlab based computer calculation, under different thickness. *Physica B* 63, 414634.
12. Zargar, R.A., Arora, M., Alshahrani, T. and Shkir, M. (2021) Screen printed novel ZnO/MWCNTs nanocomposite thick film. *Ceramics International* 47, 6084–6093.
13. Zargar, R.A. (2019) ZnCdO thick film: A material for energy conversion devices. *Materials Research Express* 6, 095909.
14. Zargar, R.A. and Arora, M. (2018) Study of nanosized copper doped ZnO dilute magnetic semiconductor thick films for spintronic device applications. *Journal of Applied Physics* 124, 36.
15. Bleeker, E.A., de Jong, W.H., Geertsma, R.E., Groenewold, M., Heugens, E.H., Koers-Jacquemijns, M., van de Meent, D., Popma, J.R., Rietveld, A.G., Wijnhoven, S.W., Cassee, F.R. and Oomen, A.G. (2013). Considerations on the EU definition of nanomaterials: Science to support policy making. *Regulatory Toxicology Pharmacology* 65(1), 119–125.
16. Morganti, P. (2013) Innovation, nanotechnology and industrial sustainability by the use of natural underutilized byproducts. *Journal of Molecular Biochemistry* 2(3).
17. Oke, A.E., Aigbavboa, C.O. and Semenya, K. (2017) Energy savings and sustainable construction: Examining the advantages of nanotechnology. *Energy Procedia* 142, 3839–3843.
18. Mishra, Y.K., Murugan, N.A., Kotakoski, J. and Adam, J. (2017) Progress in electronics and photonics with nanomaterials. *Vacuum* 146.
19. Yeh, N.C. (2013) Nanotechnology for electronics & photonics. *Technovation* 33(4–5), 108.
20. Singh, N.A. (2017) Nanotechnology innovations, industrial applications and patents. *Environmental Chemistry Letters* 15(2), 185–191.
21. Pathakoti, K., Manubolu, M. and Hwang, H.M. (2018) Nanotechnology applications for environmental industry. In: *Handbook of Nanomaterials for Industrial Applications*, pp. 894–907. Amsterdam: Elsevier.
22. Delgado-Ramos, G.C. (2014) Nanotechnology in Mexico: Global trends and national implications for policy and regulatory issues. *Technology in Society* 37, 4–15.
23. Grilli, M.L., Chevallier, L., Di Vona, M.L., Licoccia, S. and Di Bartolomeo, E. (2005) Planar electrochemical sensors based on YSZ with WO3 electrode prepared by di_ erent chemical routes. *Sensors and Actuators B: Chemical* 111, 91–95.
24. Masetti, E., Grilli, M.L., Dautzenberg, G., Macrelli, G., Adamik, M. (1999) Analysis of the influence of the gas pressure during the deposition of electrochromic WO3 films by reactive r.f. sputtering of W and WO3 target. *Solar Energy Materials and Solar Cells* 56, 259–269.
25. Grilli, M.L., Kaabbuathong, N., Dutta, A., Di Bartolomeo, E. and Traversa, E. (2002) Electrochemical NO2 sensors with WO3 electrodes for high temperature applications. *Journal of the Ceramic Society of Japan* 110, 159–162.
26. Fujishima, A. and Honda, K. (1972) Electrochemical photolysis of water at a semiconductor electrode. *Nature Cell Biology* 238, 37–38.

27. Aydogan, S., Grilli, M.L., Yilmaz, M., Çaldiran, Z. and Kaçus, H. (2017) A facile growth of spray based ZnO films and device performance investigation for Schottky diodes: Determination of interface state density distribution. *Journal of Alloys and Compounds* 708, 55–66.

28. Yilmaz, M., Grilli, M.L. and Turgut, G. (2020) A bibliometric analysis of the publications on in doped ZnO to be a guide for future studies. *Metals* 10, 598.

29. Yilmaz, M. and Grilli, M.L. (2016) The modification of the characteristics of nanocrystalline ZnO thin films by variation of Ta doping content. *Philosophical Magazine* 96, 2125–2142.

30. USGS. (2020) *National Minerals Information Center.* Available online: www.usgs. gov/centers/nmic/iron-oxide-pigments-statistics-and-information (accessed on 20 March 2022).

31. Pei, C., Han, G., Zhao, Y., Zhao, H., Liu, B., Cheng, L., Yang, H. and Liu, S. (2016) Superior adsorption performance for triphenylmethane dyes on 3D architectures assembled by ZnO nanosheets as thin as 1.5 nm. *Journal of Hazardous Materials* 318, 732–741.

32. Zhan, X.Q., Yu, X.Y., Tsai, F.C., Ma, N., Liu, H.L., Han, Y., Xie, L., Jiang, T., Shi, D. and Xiong, Y. (2018) Magnetic MOF for AO7 removal and targeted delivery. *Crystals* 8, 250.

33. Fan, J., Chen, D., Li, N., Xu, Q., Li, H., He, J. and Lu, J. (2018) Adsorption and biodegradation of dye in wastewater with Fe3O4@MIL-100 (Fe) core–shell bio-nanocomposites. *Chemosphere* 191, 315–323.

34. Hadjltaief, H.B., Ben Zina, M., Galvez, M.E. and Da Costa, P. (2016) Photocatalytic degradation of methyl green dye in aqueous solution over natural clay-supported ZnO–TiO2 catalysts. *Journal of Photochemistry and Photobiology* 315, 25–33.

35. Tanaka, Y., Utaka, T., Kikuchi, R., Sasaki, K. and Eguchi, K. (2003) CO removal from reformed fuel over Cu/ZnO/Al2O3 catalysts prepared by impregnation and coprecipitation methods. *Applied Catalysis A: General* 238, 11–18.

36. Koli, P.B., Kapadnis, K.H. and Deshpande, U.G. (2019) Transition metal decorated ferrosoferric oxide (Fe3O4): An expeditious catalyst for photodegradation of Carbol Fuchsin in environmental remediation. *Journal of Environmental Chemical Engineering* 7.

37. Wen, Z., Ke, J., Xu, J., Guo, S., Zhang, Y. and Chen, R. (2018) One-step facile hydrothermal synthesis of flowerlike Ce/Fe bimetallic oxides for e_cient As(V) and Cr(VI) remediation: Performance and mechanism. *Chemical Engineering Journal* 343, 416–426.

38. Zhou, H., Guo, J., Li, P., Fan, T., Zhang, D. and Ye, J. (2013) Leaf-architectured 3D hierarchical artificial photosynthetic system of perovskite titanates towards CO2 photoreduction into hydrocarbon fuels. *Scientific Reports* 3, 01667.

39. Yu, M., Zeng, Y., Zhang, C., Lu, X., Zeng, C., Yao, C., Yang, Y. and Tong, Y. (2013) Titanium dioxide polypyrrole core-shell nanowires for all solid-state flexible supercapacitors. *Nanoscale* 5, 10806–10810.

40. Xi, G., Yue, B., Cao, J. and Ye, J. (2011) Fe3O4/WO3 hierarchical core-shell structure: High-performance and recyclable visible-light photocatalysis. *Chemistry—A European Journal* 17, 5145–5154.

41. Gautam, S., Agrawal, H., Thakur, M., Akbari, A., Sharda, H., Kaur, R. and Amini, M. (2020) Metal oxides and metal organic frameworks for the photocatalytic degradation: A review. *Journal of Environmental Chemical Engineering* 8(3), 103726.

42. Khin, M.M., Nair, A. S., Babu, V. J., Murugan, R. and Ramakrishna, S. (2012) A review on nanomaterials for environmental remediation. *Energy & Environmental Science* 5(8), 8075–8109.

43. Chavali, M.S. and Nikolova, M.P. (2019) Metal oxide nanoparticles and their applications in nanotechnology. *SN Applied Sciences* 1(6), 1–30.

44. Dizaj, S.M., Lotfipour, F., Barzegar-Jalali, M., Zarrintan, M.H. and Adibkia, K. (2014) Antimicrobial activity of the metals and metal oxide nanoparticles. *Materials Science and Engineering: C* 44, 278–284.
45. Dargahi, A., Gholestanifar, H., Darvishi, P., Karami, A., Hasan, S.H., Poormohammadi, A. and Behzadnia, A. (2016) An investigation and comparison of removing heavy metals (lead and chromium) from aqueous solutions using magnesium oxide nanoparticles. *Polish Journal of Environmental Studies* 25, 557–562.
46. Wang, X., Guo, Y., Yang, L., Han, M., Zhao, J. and Cheng, X. (2012) Nanomaterials as sorbents to remove heavy metal ions in wastewater treatment. *Journal of Environmental and Analytical Toxicology* 2.
47. Hua, M., Zhang, S., Pan, B., Zhang, W., Lv, L. and Zhang, Q. (2012) Heavy metal removal from water/wastewater by nanosized metal oxides: A review. *Journal of Hazardous Materials* 211, 317–331.
48. Xu, L., Xia, J., Wang, K., Wang, L., Li, H., Xu, H., Huang, L. and He, M. (2013) Ionic liquid assisted synthesis and photocatalytic properties of α-Fe2O3 hollow microspheres. *Dalton Transactions* 42, 6468–6477.
49. Mehndiratta, P., Jain, A., Srivastava, S. and Gupta, N. (2013) Environmental pollution and nanotechnology. *Environment and Pollution* 2(2), 49.
50. Baby, R., Saifullah, B. and Hussein, M.Z. (2019) Carbon nanomaterials for the treatment of heavy metal-contaminated water and environmental remediation. *Nanoscale Research Letters* 14(1), 341.
51. Reza, K.M., Kurny, A.S.W. and Gulshan, F. (2017) Parameters affecting the photocatalytic degradation of dyes using TiO2: A review. *Applied Water Science* 7, 1569–1578.
52. Kumar, A., Kumar, S. and Krishnan, V. (2019) Perovskite-based materials for photocatalytic environmental remediation. *Environmental Chemistry for a Sustainable World* 139–165.
53. Xu, P., Xie, M., Cheng, Z. and Zhou, Z. (2013) CO2 capture performance of CaO-based sorbents prepared by a solgel method. *Industrial & Engineering Chemistry Research* 52, 12161–12169.
54. Luisetto, I., Mancini, M.R., Della Seta, L., Chierchia, R., Vanga, G., Grilli, M.L. and Stendardo, S. (2020) CaO–CaZrO3 mixed oxides prepared by auto–combustion for high temperature CO2 capture: The effect of CaO content on cycle stability. *Metals* 10, 750.
55. Hou, J., Cao, S., Wu, Y., Liang, F., Ye, L., Lin, Z. and Sun, L. (2016) Perovskite-based nanocubes with simultaneously improved visible-light absorption and charge separation enabling efficient photocatalytic CO2 reduction. *Nano Energy* 30, 59–68.
56. Zhou, H., Guo, J., Li, P., Fan, T., Zhang, D. and Ye, J. (2013) Leaf-architectured 3D hierarchical artificial photosynthetic system of perovskite titanates towards CO2 photoreduction into hydrocarbon fuels. *Scientific Reports* 3, srep01667.
57. Shi, H., Wang, T., Chen, J., Zhu, C., Ye, J. and Zou, Z. (2010) Photoreduction of carbon dioxide over NaNbO3 nanostructured photocatalysts. *Catalysis Letters* 141, 525–530.
58. Chen, X. and Jin, F. (2019) Photocatalytic reduction of carbon dioxide by titanium oxide based semiconductors to produce fuels. *Frontiers in Energy* 13, 207–220.
59. Navalón, S., Dhakshinamoorthy, A., Álvaro, M., Garcia, H. (2013) Photocatalytic CO2 reduction using non-titanium metal oxides and sulfides. *ChemSusChem* 6, 562–577.
60. Ivanković, T. and Hrenović, J. (2010) Surfactants in the environment. *Arhiv za Higijenu Rada i Toksikologiju* 61(1), 95–109.
61. Prediger, P., Cheminski, T., de Figueiredo Neves, T., Nunes, W.B., Sabino, L., Picone, C.S.F. and Correia, C.R.D. (2018) Graphene oxide nanomaterials for the removal of non-ionic surfactant from water. *Journal of Environmental Chemical Engineering* 6, 1536–1545.
62. Abd El-Lateef, H.M., Ali, M.M.K. and Saleh, M.M. (2018) Adsorption and removal of cationic and anionic surfactants using zero-valent iron nanoparticles. *Journal of Molecular Liquids* 268, 497–505.

63. Jose, R., Thavasi, V. and Ramakrishna, S. (2009) Metal oxides for dye-sensitized solar cells. *Journal of American Ceramic Society* 92(2), 289–301.
64. Sinha, S., Nandi, D.K., Kim, S.H. and Heo, J. (2018) Atomic-layer-deposited buffer layers for thin film solar cells using earth-abundant absorber materials: A review. *Solar Energy Materials and Solar Cells* 176, 49–68.
65. Sullivan, I., Zoellner, B. and Maggard, P.A. (2016) Copper(I)-based p-type oxides for photoelectrochemical and photovoltaic solar energy conversion. *Chemistry of Materials* 28(17), 5999–6016.
66. Hagfeldt, A., Boschloo, G., Sun, L., Kloo, L. and Pettersson, H. (2010) Dye-sensitized solar cells. *Chemical Reviews* 110(11), 6595–6663.
67. Hagfeldt, A. (2012) Brief overview of dye-sensitized solar cells. *AMBIO* 41(2), 151–155.
68. Chen, X., Li, C., Grätzel, M., Kostecki, R. and Mao, S.S. (2012) Nanomaterials for renewable energy production and storage. *Chemical Society Reviews* 41(23), 7909–7937.
69. Wu, H. and Zhang, Z. (2011) High photoelectrochemical water splitting performance on nitrogen doped double-wall TiO2 nanotube array electrodes. *International Journal of Hydrogen Energy* 36(21), 13481–13487.
70. Sengupta, R., Bhattacharya, M., Bandyopadhyay, S. and Bhowmick, A.K. (2011) A review on the mechanical and electrical properties of graphite and modified graphite reinforced polymer composites. *Progress in Polymer Science* 36(5), 638–670.
71. Li, L., Seng, K.H., Liu, H., Nevirkovets, I.P. and Guo, Z. (2013) Synthesis of Mn3O4-anchored graphene sheet nanocomposites via a facile, fast microwave hydrothermal method and their supercapacitive behavior. *Electrochimica Acta* 87, 801–808.
72. Nandi, D., Mohan, V.B., Bhowmick, A.K. and Bhattacharyya, D. (2020) Metal/metal oxide decorated graphene synthesis and application as supercapacitor: A review. *Journal of Materials Science* 55(15), 6375–6400.
73. Block, T. and Schmücker, M. (2016) Metal oxides for thermochemical energy storage: A comparison of several metal oxide systems. *Solar Energy* 126, 195–207.
74. Gao, S.J., Shi, Z., Zhang, W.B., Zhang, F. and Jin, J. (2014) Photoinduced superwetting single-walled carbon nanotube/TiO2 ultrathin network films for ultrafast separation of oil-in-water emulsions. *ACS Nano* 8(6), 6344–6352.
75. Park, H.-A., Liu, S., Salvador, P.A., Rohrer, G.S. and Islam, M.F. (2016) High visible-light photochemical activity of titania decorated on single-wall carbon nanotubes aerogels. *RSC Advances* 6, 22285–22294.
76. Qu, X., Brame, J., Li, Q. and Alvarez, P.J. (2013) Nanotechnology for a safe and sustainable water supply: Enabling integrated water treatment and reuse. *Accounts of Chemical Research* 46(3), 834–843.
77. Liu, J.-F., Zhao, Z.S. and Jiang, G.B. (2008) Coating Fe3O4 magnetic nanoparticles with humic acid for high efficient removal of heavy metals in water. *Environmental Science & Technology* 42, 6949–6954.
78. Esmat, M., Farghali, A.A., Khedr, M.H. and El-Sherbiny, I.M. (2017) Alginate-based nanocomposites for e_cient removal of heavy metal ions. *International Journal of Biological Macromolecules* 102, 272–283.
79. Pendergast, M.T.M., Nygaard, J.M., Ghosh, A.K. and Hoek, E.M. (2010) Using nanocomposite materials technology to understand and control reverse osmosis membrane compaction. *Desalination* 261(3), 255–263.
80. Wu, L. and Ritchie, S.M. (2008) Enhanced dechlorination of trichloroethylene by membrane-supported Pd-coated iron nanoparticles. *Environmental Progress* 27(2), 218–224.
81. Janson, O., Gururaj, S., Pujari-Palmer, S., Ott, M.K., Strømme, M., Engqvist, H. and Welch, K. (2019) Titanium surface modification to enhance antibacterial and bioactive properties while retaining biocompatibility. *Materials Science and Engineering: C* 96, 272–279.

82. Yaqoob, A.A., Ahmad, H., Parveen, T., Ahmad, A., Oves, M., Ismail, I.M., Qari, H.A., Umar, K. and Mohamad Ibrahim, M.N. (2020) Recent advances in metal decorated nanomaterials and their various biological applications: A review. *Frontiers in Chemistry* 8, 341.

83. Sirelkhatim, A., Mahmud, S., Seeni, A., Kaus, N.H.M., Ann, L.C., Bakhori, S.K.M., Hasan, H. and Mohamad, D. (2015) Review on zinc oxide nanoparticles: antibacterial activity and toxicity mechanism. *Nano-Micro Letters* 7(3), 219–242.

84. Swaminathan, M. and Sharma, N.K. (2019) Antimicrobial activity of the engineered nanoparticles used as coating agents. In: *Handbook of Ecomaterials*, pp. 549–563. Cham: Springer International Publishing.

85. Sankar, R., Maheswari, R., Karthik, S., Shivashangari, K.S. and Ravikumar, V. (2014) Anticancer activity of Ficus religiosa engineered copper oxide nanoparticles. *Materials Science and Engineering: C* 44, 234–239.

86. Ebrahim-Saraie, H.S., Heidari, H., Rezaei, V., Mortazavi, S.M.J. and Motamedifar, M. (2018) Promising antibacterial effect of copper oxide nanoparticles against several multidrug resistant uropathogens. *Pharmaceutical Sciences* 24(3).

87. Nethi, S.K., Rico-Oller, B., Rodríguez-Diéguez, A., Gómez-Ruiz, S. and Patra, C.R. (2017) Design, synthesis and characterization of doped-titanium oxide nanomaterials with environmental and angiogenic applications. *Science of the Total Environment* 599, 1263–1274.

88. Dizaj, S.M., Lotfipour, F., Barzegar-Jalali, M., Zarrintan, M.H. and Adibkia, K. (2014) Antimicrobial activity of the metals and metal oxide nanoparticles. *Materials Science and Engineering: C* 44, 278–284.

89. Besinis, A., De Peralta, T. and Handy, R.D. (2014) The antibacterial effects of silver, titanium dioxide and silica dioxide nanoparticles compared to the dental disinfectant chlorhexidine on streptococcus mutans using a suite of bioassays. *Nanotoxicology* 8(1), 1–16.

4 Nanomaterials for Renewable Energy

Present Status and Future Demand

*Ajay Singh, Vishal Singh, Manju Arora, Balwinder
Kaur, Archana Sharma, and Sunil Sambyal*

CONTENTS

DOI: 10.1201/9781003323464-4

4.1   INTRODUCTION

Over the next 50 years, the world will have to face two serious challenges, namely, environmental and energy problems. These two problems are not separate but connected. Overconsumption of fossil fuels results in environmental pollution due to CO_2 emissions. The population of the world is expected to increase to about 9 billion by 2050 [1], which will be responsible for a linear increase in energy demand. The demand for energy will increase 50% by 2025 for the nations' development, which will result in the continuous depletion of fossil fuel reserves as about 80% of worldwide energy consumption is rooted in the chemical energy stored in it. Our fossil fuel consumption is escalating and could double by 2025. The overconsumption of fossil fuels results in environmental catastrophes such as the depletion of the ozone layer, global warming, the destruction of biospheres and geospheres, and ecological imbalances as the energy sector is mainly responsible for about 80% of CO_2 emissions in the world. It is therefore obvious that these two major problems regarding the energy problem and the environmental problem must be dealt together, and the simplest elucidation of both issues is adopting environmentally friendly renewable energy as more than 75% of the world's energy demand is met by nonrenewable sources. Fossil fuel potential is measured in MTOE, and the available fossil fuel in the world is about 5,500 MTOE, which includes 60% coal, 30% natural gas, and 10% crude oil. Insufficient fossil fuel supplies and environmental pollution by excessive gas emissions are the main issues facing the world today. It was reported that the available reserves of fossil fuel in the world will be exhausted by 2050 [2] as the consumption of petroleum is 10^5 times faster than its creation by nature and the energy demand globally is estimated to be about 30 and 46 TW by 2050 and 2100, respectively [3]. The major challenge our society is facing today is the energy requirements for the world's escalating population. Due to limited energy resources, there is worldwide competition, making energy an issue of security concerns throughout the world. Troubling oil prices, excessive emission of GHGs, limited nonrenewable resources reservoirs, and increasing energy demands have highlighted the need for alternative sources of energy, that is, renewable energy.

Due to the lesser availability of fossil fuels and the increasing demand for energy, there is a shift toward clean sources of energy called renewable energy sources. These potential renewable energy resources are inexhaustible and are environmentally friendly [4]. By 2040, it is estimated that renewable energy sources provide 50% of primary energy throughout the world. Nowadays, the scientific community is greatly interested in renewable energy sources as they will play a pivotal role in reducing the emission of harmful gases by about 70% till 2050.

The challenge of the present century involves evolving technologies for renewable energy production and storage and reducing energy demand. Energy supply and demand can be amalgamated in developing nanotechnology. Energy supplies mainly depend on nonrenewable sources such as coal-red plants, nuclear power stations, natural gas, and oil. There is some thriving beneficence from renewable resources such as wind, water, biofuels, and solar cells. The power required to run household appliances, offices, shops, industries, mines, transport people and goods, waste treatment plants, and farms falls under energy demand technologies. To resolve energy-related issues, nanotechnologies offer an extensive range of developing components in the nano-regime with important characteristics and functions that regulate the shape and size of matter to capture, store, and exchange energy. Furthermore, nanotechnology can play a major role in making efficient lighting and heating, improving electrical storage capacity, reducing pollution content in the environment, modifying materials' properties at the nano-scale, increasing materials' life spans, and increasing the efficiency of power systems.

This chapter examines the important utilization of nanomaterials in renewable energy systems. This chapter incorporates abstract and laboratory work concerned with materials properties and the future of renewable energies.

4.2 NANOTECHNOLOGY: APPLICATION IN THE RENEWABLE ENERGY SECTOR

Nanotechnology contributes to comprehensive energy security and supply. With the help of nanoscience and nanotechnology, functional nanomaterials with control physical and chemical properties are synthesized that show different strengths, conductivities, reflectivities, chemical reactivities, and more, in comparison to their bulk counterpart and have utmost importance in making new energy sources pragmatic on a profitable scale. In the arena of renewable energy, nanotechnology proved to be a cost-competitive technology. It is inexpensive, efficient, and environment-friendly technology [5]. For instance, solar energy, wind energy, hydrogen production, hydropower, bioenergy, geothermal energy clean coal, fuel cells, batteries, energy storage, and more, all benefited from the nanotechnology boom. Figure 4.1 shows nanotechnology application domains in renewable energy.

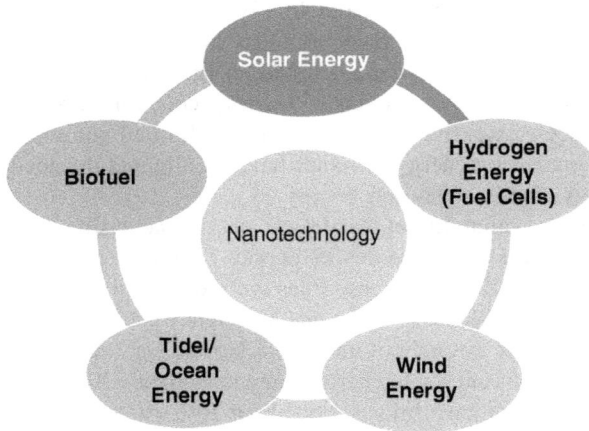

FIGURE 4.1 Nanotechnology application domains in renewable energy.

Examining and plucking properties of materials at a nano regime results in a reduction in the cost of PV and solar cells, power efficiencies and storage capacities of super-capacitors and batteries being improved, achieving a reduction of CO_2 to methanol by photocatalytic processes, commercializing technology of hydrogen generation from the photo-conversion of water and light, preventing current loss and thermal drop by designing quantum conductors and super-conductors based on nano-materials, economically extracting geothermal energy by developing nano-coating, and creating hydrogen-storage for reversible chemisorption by creating lightweight materials. Batteries based on nanotube ultra-capacitors have adequate power, have longer lives, perform excellently, and are rechargeable within a span of seconds. These batteries are used to run hybrid electric cars [6]. Nanotechnology plays a vital role in solar energy storage and generation by reducing assembly costs and enriching material efficiency [7]. There are four generations of PV cells, and nanostructured PV devices belong to the third generation. Third-generation PV shave a conducting layer made of oxide and a catalytic layer made of platinum that converts sunlight into electricity. PV technology provides only 0.04% of the total primary energy supply of the world. If solar energy accomplishes a 1% requirement of electricity in the world, 40 million tons of CO_2 emission may be eliminated per annum [8]. The nano-coatings used in solar cells modify the energy band gap, resulting in an amplification of the absorption path by multiple reflections and thus reducing recombination loss by decreasing absorption layer width. Another important aspect of nano-coatings is its self-cleaning feature [9]. Complex nanostructured lenses and mirrors are used for the large collection of solar energy.

The promising solution to sustainable energy and environment is the hydrogen fuel cell technologies. In a fuel cell, hydrogen and oxygen molecules combine to generate water, releasing heat and electricity as byproducts. Presently, nanomaterials are incorporated into fuel cells and electrode membranes. Nanoblades, nanofibers, and CNTs are used to hold huge volumes of hydrogen fuel [10]. For the conversion of

CO_2 to hydrocarbon fuels, such as methanol, technology based on nanocomposites is used [11]. Nanotechnology employing photocatalysis is used as an economic tool for hydrogen gas production from solar energy in an environmentally friendly manner [12].

Recently, nanotechnology is used for the production of biofuel from agricultural waste. Metal NPs are used as catalysts in the production of biofuel. Metal NPs are environmentally unfriendly; therefore, technology based on an amalgamation of nanocatalysts with biomass is used for biofuel production. A nanocatalyst is an eco-friendly technology as it provides high surface area that results in enhancing the reaction rate, increasing the efficiency of bioenergy generation. An iron nanocatalyst is utilized to produce biodiesel using Pongamia pinnata oil with methanol [13].

Efficient wind energy conversion can be achieved using nanotechnology. Nanoscience is used to develop coating, lubrication, and durable lightweight materials for wind energy conversion. To monitor stability or possible damage, sensing materials based on nanomaterials can be used. To check stiffness and bear weight-to-load proportions, efficient, longer, and stronger blades made up of nanocomposite materials are useful. To enhance wind turbine efficiency, nanolubricants and less frictional surface coatings are used. This will overcome the energy loss due to scuffing, micro-pitting, gearbox spalling, and wear. Nano-colloidal lubricant based on boron nitride additives is created for wear protection. CNTs are used to make light-weight wind turbine rotor blades to increase their conductivity and strength [14].

Another sustainable and clean renewable source of energy is ocean tidal energy. The power output that can be retrieved from ocean tidal energy is highest in India, 7000 MW, compared to the developed nations of the world [15]. Nanotechnology has made an exceptional advancement in ocean tidal energy. The TENG transforms rotational motion into electricity and is an exclusive way for large-scale power generation. The electromagnetic hybrid nanogenerator, due to its unique structure design, generates electricity from ocean tides and waves either by a rotation mode or a fluctuation mode. In this design, S-TENG is isolated from the environment by means of packaging and driven indirectly by the pairs of magnets, and then a wraparound electromagnetic generator can be easily alloyed. Finally, a nanotechnology-based single hybrid nanogenerator unit is used to run LEDs under distinct imitated conditions of water waves [16].

The characteristic properties of nanomaterials desired for various renewable harvesting is discussed in Section 4.3.

4.3 CHARACTERISTIC PROPERTIES OF NANOMATERIALS DESIRED FOR RENEWABLE ENERGY HARVESTING

The nanostructures that can be synthesized in order to utilize renewable energy sources are shown in Figure 4.2. Owing to their large surface area, faster diffusion rates, and short diffusion lengths, nanomaterials have enhanced device performance, particularly in solar cells, fuel cells, rechargeable LIBs, and more. The motion of the charge species is confined at the nano level, which enhances their mechanical, electrical, and optical properties [17, 18]. To improve devices' electrochemical performance, quantum dot, nanotubes, and nanosheets have been utilized.

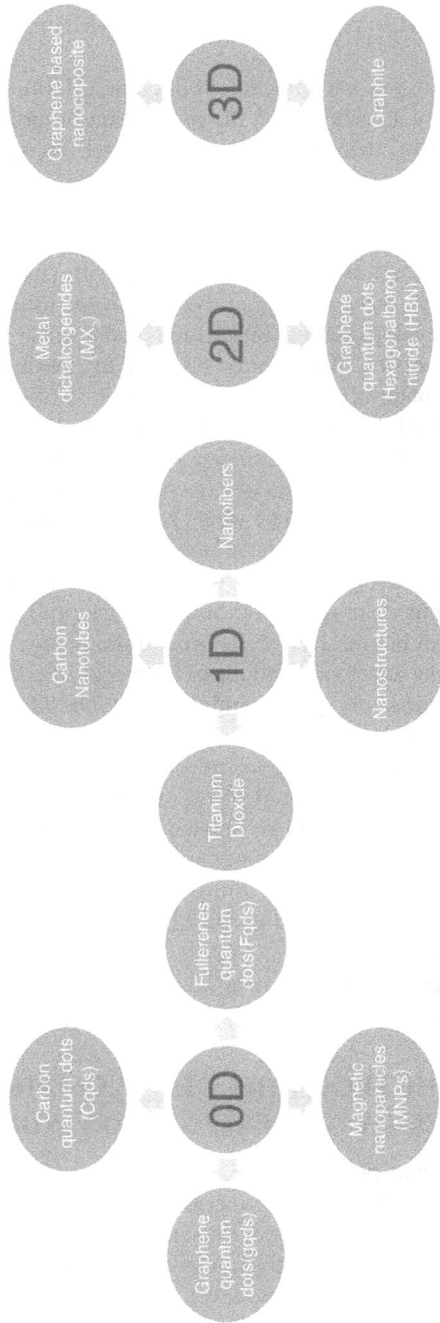

FIGURE 4.2   Nanostructures for energy storage devices.

PV solar cells using perovskite nanomaterials are promising substitutes for traditional sources of energy based on fossil fuel. Perovskite solar cells (PSCs) using nanomaterials in comparison show promising features like enhanced light conversion efficiency, improved absorption coefficient, high stability, and better transportation of charge carriers [19].

NPs of ZnO find applications in sensors [20], transistors [21], PV cells [22], hydrogen production [23], and dye degradation by photocatalysis [24]. These physicochemical, optical, piezoelectric, and catalytic properties [25, 26] are due to the crystallite size and various shapes of the NPs [27].

In hole-conduction-free PSC, carbon nanomaterials are preferred owing to their high conductivity, better stability, and lower cost. Recently, Wei et al. [28] reported an efficiency up to 12.03% of hole-conductor-free perovskite solar cells fabricated by hot pressing techniques.

In SOFCs, nanomaterials' properties, such as cost, catalytic activity, durability, chemical and thermal stability, and others, are employed for cathode, anode, and electrolyte fabrication [29]. To increase catalytic activity at the cathode, nanotubes, nanorods, nanowires, and others showed enhanced electrochemical properties are used [30, 31]. The porous structure and the fine particle size of nanomaterials are utilized for high conductivity and thermal stability, and, hence, increase the performance of the anode. From an application point of view, nanostructured materials are prepared and designed.

In LIBs, inorganic nanostructured materials have the following characteristics and advantages: (1) Nanostructured materials increase lithium insertion/removal rate and decrease intercalation time, (2) the large surface area of nanostructure increases lithium-ion flux cross-section coefficient, (3) tuning a material's size at the nanoscale modified the electrode potentials for ions and electrons, and (4) to prevent pulverization of electrode, nanostructured materials offer facile stain relaxation, large volume change tolerance during lithiation and delithiation.

The nanostructure materials show superiority in comparison to bulk materials because of improved kinetics and electrochemical activity.

4.4 NANOMATERIALS BASED ON METAL OXIDES USED FOR ENVIRONMENT POLLUTION MONITORING

Presently, environmental pollution is one of the top priorities in human and ecological systems owing to the high-speed development of scientific technologies and rapid industrialization. It is estimated that the consequence of death (i.e., 100 million people, 1 million seabirds, and 100,000 sea mammals) is highly probable to environmental pollution.

In this section, nanomaterials based on a variety of transient metal oxides such as Fe_2O_3, CuO, TiO_2, ZnO, and CeO_2, among others, are discussed with a focus on environment pollution monitoring.

4.4.1 METAL OXIDE NANOMATERIALS USED IN WATER PURIFICATION

Water pollution implies the occurrence of newly identified chemicals (e.g., insecticides, herbicides, and pesticides) and various well-identified hazardous pollutants (volatile organic compounds (VOCs), heavy metal ions, and chemical wastes). Per a 2003 report by the World Water Assessment Programme (WWAP), 2 million tons of sewage generated from industrial and agricultural wastewater is discharged every day into the world's water system. Nanomaterials based on metal oxides such as Fe_2O_3, CuO, TiO_2, ZnO, CeO_2, MgO, and others are utilized as economical, efficient adsorbents for wastewater purification. Photocatalysis using abundant natural sunlight is the potential technique for environmental remediation. The photocatalysis process in metal oxide semiconductors involves the production of photogenerated electrons and holes. An electron is shifted to an acceptor molecule, while holes are shifted to a donor molecule. Under ultraviolet (UV) illumination, both ZnO and TiO_2 show excellent photocatalytic activity since their band gap is about 3.2 eV. However, by defect engineering or doping, their photocatalytic activity can be achieved under visible illumination. Among various metal oxides, the magnetic NP shows excellent results owing to their interesting magnetic properties [32]. Magnetite (Fe_3O_4) synthesized/modified by using zirconium (IV)-metalloporphyrin at pH 5.5 is used to extract 92.0% fluoride ions with a contact time of 20 min [33]. Magnetite (Fe_3O_4) prepared using amino-functionalized (1, 6-hexadiamine) shows a maximum absorption capacity of 25.770 mg/g for Cu(II) at pH 6 and 298 K [34]. Polymer-modified Fe_3O_4 has been used to remove 95% of heavy ions like Cd(II), Zn(II), Pb(II), and Cu(II) at about 30 min at pH 5.5 [35]. Aerosol-assisted Fe_3O_4 synthesized by chemical vapor deposition shows 100% efficiency for arsenic (As) removal before 1 min of contact [36]. Amine functionalized Fe_3O_4 has been used to remove over 98% of Pb(II), Cd(II), and Cu(II) within 120 min at pH 7.0 [32]. NPs of mixed maghemite-magnetite (γ-Fe_2O_3-Fe_3O_4) shows about 40% Cd(II) removal within 5 min. [37]. A magnetic iron particle synthesized by tea water shows a high adsorption capacity of 188.690 mg g^{-1} for As(III) and As(V) [38]. Iron oxide-alumina (Fe_2O_3-Al_2O_3) shows maximum sorption capacity of 4.980 mg/g for Cu^{2+}, 32.360 mg/g for Ni^{2+}, 23.750 mg/g for Pb^{2+} and 63.690 mg/g for Hg^{2+} ions. The order of removal percentage is Cu^{2+} <Pb^{2+} <Ni^{2+} <Hg^{2+}[39]. The adsorption capacity of CeO_2, Fe_3O_4 and TiO_2NPs is 189.0 mg Pb/g, 83.0 mg Pb/g and 159.0 mg Pb/g [40]. MgO and ZnO show a maximum adsorption of Cu(II), and the adsorption capacity for MgO and ZnO for Cu(II) are, respectively, 593.0 mg/g and 226.0 mg/g at pH of 3–4. Thus, MgO has better adsorption for Cu(II) than ZnO [41]. Nano-ZnO synthesized by gel combustion shows almost complete Pb adsorption [42]. Amorphous zirconium oxide (am-ZrO_2) shows adsorption for phosphate (99.01 mg/g at pH 6.2) [43] and for As (III) and As (V) at pH 7 [44].

Thus, MO-NPs exhibit extraordinary multifunctional treatment options for wastewater that enhance pollution monitoring.

4.4.2 Metal Oxide Nanomaterials Used in Air Pollution Remediation

For increasing the performance of air pollution monitoring sensors, metal oxide nanomaterials are synthesized in different structures such as nanosheets, nanocomposites, nanotubes, and nanowires, among others.

Hydrogen sulfide (H_2S) can be removed from the air at increased temperatures by an adsorption mechanism using nanostructured ZnO [45]. Ultrasonic-assisted precipitation of zinc oxide NPs (U-ZnO) showed improved adsorption of H_2S (29.51 mg/g) in comparison to ZnO NPs prepared with no ultrasonic treatment (3.66 mg/g) [46]. NPs of ZnO synthesized by wire explosion technique have been used for the adsorption of carbon dioxide (CO_2) in air. ZnO synthesized by this technique showed enhanced CO_2 adsorption by decreasing temperature and increasing pressure [47].

Surface-fluorinated TiO_2 nanosheets synthesized by hydrothermal technique show 60% of ammonia gas separation from the air [48]. Nanorods of TiO_2 supported by CrO_x, FeO_x, MnO_x, and MnO_x-FeO_x-CrO_x catalysts show better efficiency for mercury (Hg^0) removal (80–83%) [49]. TiO_2 nanostructures, such as disks and rods, show a strong interaction with CO_2 [50]. TiO_2NPs are employed for manufacturing self-cleaning coatings that remove VOCs and nitrogen oxides into less poisonous species [51]. Different contaminants such as SO_2 and CO are degraded by ZnO and Al_2O_3 nanocatalysts. TiO_2/rGO film shows excellent selectivity to 1 ppm SO_2 gas at environmental temperatures [52]. For the detection of H_2S gas, sensors based on rGO/ferric oxide, rGO/zinc oxide, rGO/cupric oxide have been employed, and these sensors show good sensing ability.

4.4.3 CHALLENGES AND FUTURE PERSPECTIVES OF MO-NPs

The excessive use and production of MO-NPs pose a serious threat to the environment and human health as it increases the toxicity of soil. The nanofertilizers and pesticides available in the market do not mention the used concentration and types of NPs. Thus, before making them commercially available, these MO-NPs should be mentioned in safety evaluations and assessment standards.

Due to their specific physical and chemical properties and large surface–volume ratio, MO-NPs find applications in consumer products, cosmetics, electronics, environment remediation, fuel additives, paints, energy, and others. However, MO-NPs ending up in the soil disturb the soil's microbial functionality, plant growth, and, hence, human health through the food chain. Therefore, knowledge regarding the toxicological risks of MO-NPs in soil is emerging although with major gaps. Thus, it is important to develop a methodology for testing NPs toxicity in the natural environment. Future research must be conducted regarding MO-NPs, their use, their distribution, and their release in the environment with risk management.

4.5 NANOMATERIALS USED IN DIFFERENT RENEWABLE ENERGY PRODUCTION AND STORAGE DEVICES

In this section, the nanomaterials used for developing solar cells, LIBs, supercapacitors, fuel cells, and hydrogen storage are discussed.

4.5.1 NANOMATERIALS USED IN SOLAR CELLS

Solar cells based on sensitized dyes and quantum dots derived from nanomaterials have sufficient surface area for the adsorption of dye molecules and quantum dots as a monolayer to capture the maximum number of photons and minimize the interface charge recombination. They have a role in optical absorption antennas. The inorganic electrode materials must have exceptional charge mobility and a stable, long lifetime with light-scattering/photon-trapping properties. From a structural point of view, the inorganic nanomaterial has a perfect crystalline structure with minimum defects (surface and bulk) and low grain boundaries joining between individual nanostructures. Nanomaterials in solar cells should have the following characteristic properties; namely, quantum dot–based solar cells should have multiple exciton effects, DSCs should have a large internal surface area, one-dimensional nanostructure solar cells should have optical effects due to antireflection, metal NP–based solar cell should have surface plasmon resonance with NP agglomeration, and three-dimensional solar cell should have H-P-G core–shell spherical structures.

The semiconductors are available with Bohr radius of exciton in the 1–10-nm and 20–70-nm ranges. Silicon, CdS, CdSe, and ZnO have Bohr exciton radii of 4.2 nm, 3.1 nm, 6.1 nm, and 2.2 nm, respectively, while, PbS, PbSe, and InSb are examples of large Bohr radii semiconductors having values of 20.4 nm, 46, nm and 67.5 nm, respectively [53–55], and the characteristics of such semiconductor quantum dots are that their dimensions are proportionate to their excitonic Bohr radius, which develops a strong quantum confinement effect and is NP size–dependent. By optimizing the porous structure of the oxide film and the adsorption status of the quantum dots, the efficiency of the solar cell can be increased. The following synthesis methods have been used: (1) SILAR, (2) CBD, (3) ECD, (4) EP, and (5) linker-assisted binding or by combining two or more of these methods. Solar cells having an efficiency of about 4.92% are achieved in CdS and CdSe quantum dot–sensitized TiO_2 nanocrystalline film passivated by ZnS [56]. An increased efficiency of approximately 5.06% is achieved in an Sb_2S_3-sensitized TiO_2 nanocrystalline film [57]. In DSCs, nanowires/nanorods and nanotubes of ZnO have been used with reported values of efficiencies in the range of 1.0–2.5% [58–62], and its value is improved by modifying the ZnO nanowires' surface with a thin layer of TiO_2 [63, 64]. TiO_2 nanotubes are an example of a one-dimensional nanostructure in DSCs that have a maximum efficiency of 6.9%, which can be enhanced to about 7.4% with modified nanotubes. Although one-dimensional nanostructures have an advantage in electron conduction compared to their NPs, but the solar efficiencies of such DSCs are much less than that of TiO_2NPs, that is, about 11–12%.

In polymer solar cells (PSCs), particularly inverted structured polymer solar cells, one-dimensional nanostructures are useful. Normal PSCs consist of an indium-doped tin oxide (ITO) glass substrate, anode buffer layer (ABL) of poly(3,4-ethylenedioxythiophene) (PEDOT), a poly(styrenesulfonate) (PSS) as the hole transport layer, the active layer of conducting P3HT and PCBM, and a CBL of thin film made of oxide that acts as the layer of electron transport, and a conducting electrode consisting of a metal electrode. The excitons produced in the P3HT- and PCBM-conducting layer breaks into free electrons and holes. The holes move to PEDOT—the PSS and ITO sides—and electrons go toward the oxide and metal electrode sides. Conventional PSCs suffer from a severe stability problem due to the corrosion effect of PEDOT in contact with the ITO film and the top metal electrode (aluminum: Al), which drastically reduces cell performance, while in inverted structure PSCs, both these problems are solved. In inverted structure PSCs, the ITO glass substrate has been coated with a thin film made of oxide. On the oxide film, a blend of P3HT and

(6,6)--(phenyl-C61 butyric acid methyl ester) [PCBM], a layer of PEDOT, PSS, and a metal electrode are sequentially deposited. The PSCs' geometry restricts direct contact of the ITO with the PEDOT: PSS, and hence solves the tedious corrosion problem. Another demerit of PSCs, that is, the oxidation of the top metal electrode with air, is overcome in inverted structure PSCs by using a top electrode of metal that has a high work function, which transports the holes toward the metal electrode direction and the electrons to the oxide and ITO. These two modifications and the chemical stability of inverted PSCs have markedly improved their performance. The CBL's structure and material are important features that severely affect the performance of PSCs [65, 66]. It means the performance of solar cell is sensitive to the basic design and optimization of the CBL. In the inverted PSCs, ZnO and other oxides have been explored as CBLs. Nanocrystalline ZnO thin films exhibit a variety of morphologies, for example, nanowires/nanororods/spindles/needle/floral, among others, and have had a momentous impact on the performance of solar cells [67]. As a buffer layer in the inverted PSCs, it improves the electron transport and collection.

Three-dimensional structures, that is, an H-P-G schematic, have been used to investigate materials having high-electron mobility to improve the efficiency of DSCs. Instead of TiO_2, ZnO is used for DSCs due to its large electron mobility. However, ZnO treated with ruthenium-based dyes, such as N3, N719, and black dye, forms a Zn^{2+}–dye complex [68, 69], which is not active to injections of electrons and degrades the performance of DSCs compared to TiO_2. High-electron-mobility SnO_2 electrodes are also used in DSCs. But they suffer from problems of (1) lacking the availability of suitable surface bonds for adsorption of the ruthenium dye molecules and (2) a fast charge recombination rate, that is, 2–3 times faster, than TiO_2. The concept of a three-dimensional H-P-G structure has been proposed in which a high-electron-mobility material is used for DSCs [70]. SnO_2 or aluminum-doped ZnO (Al: ZnO) acts as a host (H) to form photoelectrode film for electron transport. To improve dye adsorption, a TiO_2 layer is deposited on the host (H); a TiO_2 coating shell forms the passivation (P) layer, which restricts the charge carrier recombination. The dye adsorption of the photoelectrode film is enhanced by filling TiO_2NPs within the gaps among the core–shell spheres. TiO_2NPs in the interstitial behave as guest (G) in the H–P–G structure. In the last, a $TiCl_4$ treatment further improves the photoelectrode film by improving its internal surface area and connectivity amid the core–shell spheres as well as with the NPs. The template method was employed to fabricate SnO_2 as a host using PSMSs of about 2.2 mm in diameter [70]. PSMSs were burned out to form hollow SnO_2 spheres having a thickness of about 90 nm. The atomic layer deposition technique was used to coat SnO_2 spheres by a TiO_2 passivation layer about 25 nm thick. The space between SnO_2–TiO_2 core–shell structured is filled with TiO_2NPs about 17nm in size. The studies were also made on Al: ZnO and TiO_2 as host materials to form H-P-G structured photoelectrodes. This setup does not require final $TiCl_4$ treatment because it dissolves the Al: ZnO. Using SnO_2 and Al: ZnO as hosts with photoelectrode films in DSCs have high efficiency compared to using a TiO_2 photoelectrode film as a host due to its higher electron mobility over the latter. SnO_2-host DSCs exhibit higher efficiency than Al: ZnO hosts. The open-circuit voltage of the photoelectrode with SnO_2 host was 842 mV significantly, and for TiO_2, the host was 791 mV. The higher open-circuit voltage value throws light on the effective decrease in charge recombination rate with the H-P-G structure. Such micro-core–shell spheres cause light scattering, which further improves the optical absorption of the photoelectrode.

4.5.2 LIBs

LIBs are made of an anode, a cathode, and an electrolyte and work on the principle of conversion of a chemical potential into electrical energy through a Faradaic reaction. While in its charging process, the electrical energy is reversed to chemical potential. The Faradaic reactions involve the transfer of mass and charge within the electrodes and dimension variations. In these batteries, the electrode's surface area and distance of migration of charges are prime factors that determine battery performance. The reaction rate and the transfer process are controlled by the composition, crystal structure, and morphology of the electrode material. By selecting appropriate materials for the electrode and the electrolyte, one can optimize the overall electrochemical performance of the battery [71]. The cathode and anode materials in commercially available LIBs are $LiCoO_2$ and graphite, respectively [72, 73]. The reported values for higher nominal voltage, specific energy/energy density, and cycle life values of commercial LIBs are 3.6 V, 125 Watt-hours kg^{-1} lit^{-1}, and more than 1000 cycles, respectively.

Economical rechargeable batteries based on lithium, which have the highest energy density, are widely employed in portable electronic devices, electric vehicles, and grid-scale energy storage. There has been fast growth in the field of nanotechnology over the last two decades; researchers have exhaustively explored many new and alternative materials for future-generation batteries. The advantages of nanotechnology application as compared to the earlier conventional methods in batteries are summarized as (1) a reduction in the dimensions of materials used in electrodes, (2) used in producing surface-coatings and functionalized layers on electrode materials by various chemical/physical routes to restrict side reactions and (3) affability to designing various component of batteries, for example, separator, current collector and others, with better functioning of batteries. The use of nanostructured electrode materials markedly improves the performance of LIBs, which enhances the intercalation capability [72]. The flow of Li ions between the electrode–electrolyte interface, in turn, improves the diffusion paths and modifies the reaction thermodynamics to permit phase transitions. It means that nanotechnology plays a key role in advanced LIBs setups.

Today, silicon nanowires/core–shell structured nanowires/nanospheres/mesoporous [74–79], lithium metal, carbon-based analogues (CNTs, graphene, mesoporous carbon) [80–84], intermetallics/nanocomposites [85–89], nano oxides [90–94] and thin films [95–99] have been tried as anode materials instead of earlier used lithium-insertion type anodes. A reversible capacity of 600 mAh/g was obtained using SWCNTs [100] while a higher reversible capacity, over 1700 mAh/g at room temperature with improved cycling performance was obtained in nanocomposites of Si [101]. Some nanosized transition-metal oxides used as anodes can deliver a lithium storage capacity of about 700 mAh/g, with 100% capacity retention for up to 100 cycles.

The cathode materials generally used in LIBs are layered $LiMO_2$ (e.g., $LiCoO_2$, $LiNiO_2$, and $LiMnO_2$) compounds, spinel-structured lithium manganese oxide ($LiMn_2O_4$), and other materials. In commercial LIBs, the $LiCoO_2$ compound is extensively used as cathode material due to its excellent electrochemical properties and long cycle life, with an acceptable capacity of about 140 mAh/g. But its high cost and the presence of toxic Co that pollutes the surrounding atmosphere after disposal are the two main disadvantages that restrict their use in LIBs. This problem is overcome by the substitution of Co with some other metal, for example, Ni, Cr, Al, and Mn, among others. $LiNi_xCo_{1-x}O_2$ compounds are found to be more suitable because their solid solution of any percentage does not disturb the inherent layered structure. Layered $LiMnO_2$ compounds are also used as cathode material in LIBs. $LiMnO_2$ compounds have a specific capacity of 285 mAh/g based on the transition of Mn^{3+} to Mn^{4+} state, that is, about two times more than the $LiMn_2O_4$ spinel. The polar carbon-based materials with a surface decorated with sulfur marginally enhance their electronic conductivity, cycle life, and system efficiency. The addition of nanomaterials in a solid polymer electrolyte matrix improves ionic conductivity due to the high aspect ratio and creates continuous ionic transport pathways.

The aim for improvement of future-generation lithium-based rechargeable batteries is really a very challenging problem at present. In-depth studies are required for the fabrication of high Li-to-host compound ratio or alloy electrode to improve their capacity. Commercially LIBs are successful and have reached the maximum limits of performance using the current electrode and electrolyte materials.

4.5.3 NANOMATERIALS FOR FUEL CELLS

W. Grove developed the first fuel cell to generate electricity [102, 103]. Experiments were conducted with a hydrogen fuel cell by Ludwig Mond (1839–1909) and Carl Langer that were able to produce 6 Amp/ft^2 at 0.73 V.F.W. Ostwald (1853–1932), presented different components of the fuel cells, such as electrodes, electrolytes and oxidizing and reducing agents (anions and cations) and the relationship between them. Then F.T. Bacon worked on fabricating high-pressure fuel cells with a maximum pressure of up to 3000 psi during the 1904–1992 period. In the early 1960s, the IFC established a power plant based on fuel cells for the Apollo spacecraft in Windsor, Connecticut, USA. In the 1970s, IFC developed a powerful alkaline-based fuel cell unit for NASA's space shuttle *Orbiter* [104, 105].

Fuel cells utilize hydrogen or other fuels to generate clean and efficient electricity. When hydrogen is the fuel, the final products are electricity, water, and heat. Fuel cells have many potential applications in transportation, industrial, commercial centers, residential buildings, and long-term energy storage for the grid in reversible setups for providing power to large power stations and small systems like laptop computers and others. The eco-friendly fuel cells have many advantages compared to coal/gas-based fuel combustion technologies due to their better efficiencies, that is, with a more than 60% transformation of fuel chemical energy into electrical energy with negligible or zero toxic emissions or air pollutants, and their being quite silent during operation. As is well known, air pollutants are involved in smog generation and raise many health issues in human beings.

Fuel cells work on the principle of batteries which do not drain or require recharging because their electricity and heat production rely on the supply of fuel. The fuel cell has an anode, a cathode, and a dense electrolyte between them. The following nanomaterials are used in anodes: $Sc_{0.1}Y_{0.1}Zr_{0.6}Ti_{0.2}O_{1.9}$, $La_{0.8}Sr_{0.2}Fe_{0.8}Cr_{0.2}O_3$, $(La_{0.7}Sr_{0.3})_{1-x}Ce_xCr_{1-x}Ni_xO_3$, $La_{0.8}Sr_{0.2}Cr_{0.95}Ru_{0.05}O_3$, $Sr_{0.88}Y_{0.08}TiO_3$, $CrTi_2O_5$, $Ti_{0.34}Nb_{0.66}O_2$, Ni-YSZ, Ni-SDC, Ni-GDC, $LaSrTiO_2$, Cu-GDCCrTi$_2$O$_5$ [106]. The cathode materials should have a high degree of porosity, be stable in an oxygen atmosphere, and have a good thermo-mechanical matching with electrolyte. Lanthanum manganite doped with rare earth elements [107, 108], Co, Ce, or Sr [109, 110] have been used as cathode materials. The following are some examples of cathode materials: $La_{1-x}Sr_xMnO_3$ (x =0.2, 0.3, 0.4), $Pr_{0.6}Sr_{0.4}MnO_3$, $La_{1-x}Sr_xCoO_3$ (x =0.2, 0.4), $La_{1-x}Sr_xFeO_3$ (x =0.2, 0.4, 0.5), $Pr_{1-x}Sr_xFeO_3$ (x =0.2, 0.5), $La_{0.7}Sr_{0.3}Fe_{0.8}Ni_{0.2}O_3$, $La_{1-x}Sr_xCo_{0.2}Fe_{0.8}O_3$ (x = 0.2, 0.4, 0.6), $La_{0.8}Sr_{0.2}Co_{0.2}Ni_{0.8}O_3$, $La_{0.8}Sr_{0.2}Co_{0.2}Mn_{0.2}O_3$, $La_{0.6}Sr_{0.4}Co_{0.8}Fe_{0.2}O_3$, and others. Zirconia, ceria, and lanthanum-based compounds, for example, YSZ/SSZ/CaSZ or GDC/SDC/YDC, are normally used as electrolytes in fuel cells. These pure and mixed nanomaterials involve in mixed ionic and electrical conduction. Hydrogen is used as a fuel and air is supplied to the anode and cathode, respectively. In such fuel cells, hydrogen molecules are broken into protons and electrons at the anode by a catalyst. They follow separate ways to reach the cathode. The flow of electron produces electricity in external circuit, while the protons transit through the electrolyte to pursue the cathode and react with oxygen and the electrons to generate heat and water. Several types of fuel cells are available in the market, and their categorization depends on the nature of the electrolyte. As per application demand, the fuel cell needs particular materials and fuels. The various types of fuel cells reported so far are proton exchange membrane fuel cell (PEMFC), DMFCs, PAFCs, AFCs, MCFCs, and SOFCs. Today, work is ongoing with SOFCs due to their better electrical efficiency, and the use of natural gas, biogas, or CH_4 as a fuel means fuel flexibility with high performance while being environmentally friendly [111, 112]. Such advantages encouraged their usage in many practical applications.

Nanoscale materials help in reducing the operating temperature of fuel cells. For instance, an Ni/YSZ/Pt [113] fuel cell at 600°C with dry H_2 and air as fuel for 10 hours showed a power density of 23.3 mW/cm^2, while a Pt/YSZ/Pt [114] fuel cell at 550°C with humidified H_2:N_2 (1:4) as fuel has reported 150 mW/cm^2 output power. Today, to achieve the highest performance, research is ongoing regarding the exploration of new nanomaterials for the latest smart technological applications.

In the following section, the key components of fuel cells are discussed.

4.5.3.1   Key Components of a Fuel Cell

4.5.3.1.1   PEM

The PEM is also known as a proton exchange membrane having a thickness of less than 20 µm. It allows the conduction of positively charged ions between the anode and cathode only and inhibits the electrons' movement. The PEM permits only selective ions to pass amid the anode and cathode, which is the key to fuel cell technology.

4.5.3.1.2 Catalyst Layers

The catalyst layer is placed on both sides of the membrane, that is, on one side on the anode and the other layer on the cathode. Nanosized Au, Ag, and Pt noble metals have been used as catalyst layers in fuel cells. Pt NPs have a dendrite morphology, and their different compositions show more electrocatalytic activities for small molecules oxidation and oxygen reduction reactions compared to commercially available catalysts [115–117]. The catalyst layers consist of nanosized platinum particles decorated on a carbon support. At the anode surface, the Pt catalyst splits hydrogen molecules into protons and electrons. The Pt catalyst reacts with the proton to produce water at the cathode by the process of reduction.

4.5.3.1.3 GDLs

AGDL sheet consists of carbon paper in which carbon fibers are partially coated with microporous PTFE. Gases pass through GDL pores. The PTFE helps keep the pores open and restricts excessive water buildup to maintain conductivity and the diffusion of hydrogen and oxygen into the electrodes.

4.5.3.1.4 Hardware

Hardware is required for the effective functioning of an MEA of a fuel cell where power is produced.

4.5.3.1.5 Bipolar Plates

Each MEA produces <1 V under normal operation, but higher voltages are required for desired applications. For that, a number of MEAs are stacked on each other to achieve required output voltage for the device/gadget. The stacked cell is placed amid two bipolar plates to isolate it from adjacent cells. These plates consist of metal, carbon, or composites, and their purpose is to add physical support to the stacked structure and permit electrical conduction between cells. The surfaces of the plates have a set of channels on their surface for gases flow on the MEA. The extra channels inside each plate have been used for liquid coolant circulation.

4.5.3.1.6 Gaskets

An MEA in a fuel cell stack is placed amid two bipolar plates. The polymer rubber gaskets are used all over the MEA boundaries for a gas-tight seal. Fuel cell technologies [118] are well commercialized for portable power source distribution and electrical energy generation in remote areas. To produce low-temperature fuel cells (T < 200°C), nanomaterials are used, the dispersion of nanosized precious and nonprecious metal/compound catalysts, reformation of fuel and storage of hydrogen, and the compilation of MEA. PEMFC has drawn much attention from researchers recently for applications as small portable power sources and in transportation. However, for stationary applications, PAFCS, SOFCs, and MCFCs are preferable. For low-temperature fuel cells using hydrogen, reformate, or methanol as fuel, platinum-based catalysts are preferred. To cut production costs, the precious platinum consumption should be reduced by either alloying it with economic transition metal or decreasing its particle size to maintain or enhance the MEA's performance by improving its electrocatalytic activity.

4.5.4 SUPERCAPACITORS

Supercapacitors are one of the recent energy storage solutions that bridge the gap between batteries and capacitors owing to their quick charging/discharging capacity, high specific power/energy, and good service life (cyclic stability) and have proved to be a very promising substitute for the present, as well as future, energy storage/redistribution systems and hybrid electric vehicles. Becker [119] first patented the concept of an ES in 1957.But this was not able to draw attention due to low energy storage and high cost in comparison to batteries. However, the evolution of new advanced nanomaterials and the progress in nanotechnology for ES electrode design have again emerged at the forefront of this field. Based on the charge storage mechanism and the active materials employed in fabrication, ESs are classified into three types: (1) EDLCs, (2) pseudo-capacitors or redox capacitors, and (3) hybrid electrochemical capacitors [120–122].

4.5.4.1 EDLCs

EDLCs are electrochemical capacitors, also known as Helmholtz layers, that store electric energy in a double layer. The double-layer concept is further extended in Gouy-Chapman model and Stem model [120]. The charge separation takes place on polarization at the interfaces amid solid electrode materials and the liquid electrolyte solution in the fine pores of the electrode.

EDLC capacitance is comparable to a normal electrolytic parallel plate capacitor and double-layer capacitance is given as

$$C_T = \mu_o \mu_r \frac{A}{d},$$

where ε_0 is the permittivity of free space, ε_r is the relative dielectric constant in the double layer, d is the effective thickness and A is the surface area. The thickness for an EDLC is in the 5–10 Å range, that is, much smaller than the distance between the electrolytic or dielectric capacitor plates.

Because the charge storage is mainly electrostatic in nature, the electric field in an EDLC is about 10^6 V cm^{-1}. Due to the non-transfer of charge across the interface (non-Faradaic), fast charging/discharging and a very long cycle life are achieved in EDLCs. For further enhancing the capacitance, the surface area of the electrode can be enhanced by employing porous materials, namely, carbon, with large internal effective surfaces, that is, more than 1000 m^2 g^{-1}. It means the combination of a highly accessible, large specific surface area and charge separation at the atomic scale results in a very high capacitance of the order of Farad (F) and increased electric energy in EDLCs.

In EDLCs, the complete cell setup consists of two capacitors connected in series. The cell capacitance (C_{Cell}) can be written as [121]

$$\frac{1}{C_{cell}} = \frac{1}{C_1} + \frac{1}{C_2},$$

where C_1, and C_2 are the capacitance of two similar electrodes.

4.5.4.2 Polymer Solar Cells (PSCs)

PSCs store Faradaic electrical energy by electron charge transfer between electrode and electrolyte through fast redox reaction and intercalation process. Pseudo-capacitance ($C\varphi$) is produced via the passage of charge 'q' originated from Faradaic charge transfer at solid electrode material surface and is a function of potential V. The derivative dq/dV is a Faradaic capacitance (pseudo-capacitance, PC) [123] given by the following equation:

$$C\varphi = dq/dV.$$

Three types of charge transfer processes take place during pseudo-capacitance [124, 125]: (1) ions adsorption from the electrolyte, (2) redox reactions, and (3) doping/de-doping of active conducting polymers. The first two processes are very sensitive to the electrode material's surface area while the last process pertains to the bulk process [126]. The second redox reaction exhibits reversible charge/discharge behavior in cyclic voltammetry with 10–100 times higher capacitance than those for carbon double-layer systems [127]. These two different storage mechanisms coexist in the supercapacitors system. Generally, one of the storage mechanisms dominates in comparison to the other one (e.g., 2% to 5% less). To distinguish between these two systems, a comparison of EDLCs and PCs is listed in Table 4.1 [128].

4.5.4.3 HSs

The use of EDL capacitance and pseudo capacitance materials for different electrodes in asymmetrical configuration form hybrid-type electrochemical capacitors. HSs have gained attention in recent years [129]. HSs have high specific energy

TABLE 4.1
Comparison of Double-Layer Capacitance vs. Pseudo-Capacitance

S. No.	Double-Layer Capacitance	Pseudo-Capacitance
1.	Involves non-Faradaic process	Involves Faradaic process
2.	20.0–50.0 μF cm^{-2}	2000 μF cm^{-2} for single-state process; 200–500 μF cm^{-2} for multistate, overlapping processes
3.	Capacitance 'C' fairly constant with potential, except through the point of zero charge	Capacitance 'C' fairly constant with potential for RuO_2; for 3 single-state process, exhibits marked maximum
4.	Highly reversible charging/ discharging	Can exhibit several maxima for overlapping, multistate processes, as for H at Pt
5.	Has restricted voltage range (contrast nonelectrochemical electrostatic capacitor)	Quite reversible but has intrinsic electrode-kinetic rate limitation determined by R_f
6.	Exhibits mirror-image voltammogram	Has restricted voltage range; exhibits mirror-image voltammogram

compared to electrochemical capacitors having two similar electrodes. The principle of the HS design is to use a double-layer capacitance electrode combined with a PC electrode. There are different modes by which two electrodes are charged: One is electrostatically charged-discharged, while the other is pseudo-capacitive; it goes through a Faradaic process, starting at defined electrode potentials results in PC. In HS, the increase in both the working voltage and the specific energy was achieved. The highest operating voltage and the lower ESR were additional beneficial factors, which allowed momentous enhancement in the overall performance of the supercapacitors for the HEV applications. Recently, a lot the work is going on the HSs, which are categorized on the basis of their electrode configurations: composite, asymmetric, and battery types [130].

4.5.4.4   Energy Storage Parameters

The nature of energy storage devices is defined by two important parameters: (1) energy density and (2) power density. Energy stored per unit mass is the energy density and is proportional to voltage squared. It is computed by the following expression:

$$E = \frac{1}{2} C \, V^2,$$

where C is the specific capacitance and V is the voltage.

The amount of energy delivered per unit mass is power density. It can be evaluated from the equation given as

$$P = iV,$$

where i is the current. The maximum power delivery is derived from the simple series circuit [131]:

$$P_{max} = \frac{1}{4R_s} V_i^2,$$

where R_s is the ESR and V_i is the initial voltage.

4.5.4.5   Characteristic Properties of Capacitive Storage Materials

The new multifunctional capacitive storage materials are now in demand because they can perform more than one function, for example, chemical as well as surface charge storage, and electrolytes can provide the fast removal of ions. It should also take care of the solvation of charging ions. The main challenge in any energy storage device is that the electrons insertion or removal should be fast. This can be achieved by obtaining a continuous electronic conductive pathway in the electrode. This is very important for high-power PCs. The nanosized porous materials have the inherent property of very large surface areas and potentially enhanced chemical redox behavior [132].

4.5.4.6 Electrode Materials for Supercapacitors

In supercapacitors, the electrode materials used are carbon-based materials AC [132, 133], CNTs [134–138], graphene [139–144], conducting polymers, and metal oxides/hydroxides. The material based on carbon are utilized in EDLCs as active electrode materials, while metal oxides such as NiO, MnO_2, IrO_2, and RuO_2, among others [145–152], and conducting polymers, for example, PANI, polypyrrole, and polythiophene [152–156], are used in PCs as active electrode materials. In EDLCs, the carbon-based material's surface area stores charge and shows a high power output, less energy density, and better cycling ability. In PSCs, charges are stored on the surface of the redox active material and within the subsurface layer. The electrode materials used in supercapacitors have various merits and demerits; for example, electrode materials based on carbon have high power density and longer life cycle but small specific capacitance in EDLCs. Metal oxide–based electrodes possess a wide range of charge/discharge potential but suffer from a small surface area and poor cycle life. Conducting polymer–based electrodes possess high capacitance, good conductivity, low cost, and ease of fabrication, but their low mechanical strength and cycle life are major disadvantages. By utilizing different nanoscale capacitive materials, electroactive nanocomposites are produced, which is a better approach to enhance supercapacitors' performance. Nanocomposite electrode properties depend on their respective materials, morphology, and interfacial characteristics. A lot of progress has been made on the development of electroactive nanocomposite materials for supercapacitors, but still the challenges are there to overcome.

4.5.4.7 Synthesis of Nanomaterials and Nanocomposite Active Electrode Materials for Supercapacitors Synthesis Methods

4.5.4.7.1 AC

AC is produced by a two-step reaction: (1) carbonization and (2) the chemical or physical activation of precursors [157]. In the first carbonization step, the precursor is treated with CO_2, steam, and air in the 700–1200°C temperature range to produce amorphous carbon in an inert atmosphere. The second chemical activation step is conducted in a low temperature range, that is, between 400 and 700°C, in the presence of NaOH, KOH, $ZnCl_2$, and H_3PO_4 as activating agents to have a porous network in the carbon particle bulk [10]. Activation improves the surface area, the porous structure, and the pore size distribution of charcoal. The maximum specific surface area of AC after activation obtained is about 3000 m²/g with a pore size distribution such as microspores (less than 2 nm), mesopores (2–50 nm), and macropores (more than 50 nm).

4.5.4.7.2 CNTs

CNTs are available in SWCNTs or MWCNTs form and exhibit good electrical conductivity. CNTs are synthesized through the catalytic decomposition of some hydrocarbons by the chemical vapor deposition technique. The synthesis parameters are optimized to obtain good-quality nanostructures in various conformations with controlled crystalline structures [132]. CNTs can be functionalized by refluxing them in

H_2SO_4 and HNO_3 acid solutions or activated in KOH solution. They have interconnected mesopores that permit a continuous charge distribution over the accessible surface area.

4.5.4.7.3 Graphene

Owing to its many fascinating properties like high electrical conductivity, a large surface area of about 2630 m^2/g, and chemical stability. The various methods reported for the preparation of graphene are chemical vapor deposition, micromechanical exfoliation, arch discharge method, unzipping of CNTs, epitaxial growth, electrochemical, chemical methods, intercalation methods in graphite, and so on [155, 158].

4.5.4.7.4 Metal Oxides/Hydroxides

Metal oxides/hydroxides, generally transition metal oxides/hydroxides, are used as electrode material as well as mild electrolytes in pseudo-capacitors. In nanoform, these oxides are easily produced by chemical coprecipitation, sol-gel, hydrothermal, green chemistry, and other methods under optimized conditions to produce the desired structure, size, and morphology of NPs. Their mixed oxide nanocomposites are synthesized by solid-state reactions, chemical coprecipitation, and anodic ECD methods [159–163]. Magnetite (Fe_3O_4) is used as an inexpensive electrode material. The capacitive behavior of Fe_3O_4 was investigated in sodium sulfite and sodium sulfate electrolytes.

4.5.4.7.5 Conducting Polymers

Conducting polymers like PANI, polypyrrole, polythiophene and poly [3,4-ethylenedioxythiophene] had been studied extensively due their low cost of production, environmental stability, and high electrical conductivity. They can be used as pseudo-material in supercapacitors to interact with electrolyte. These polymers are synthesized easily by chemical/electrochemical routes. Due to a lack of mechanical strength, nanocomposites are recommended.

4.5.4.7.6 Nanocomposite of a Conducting Polymer with Metal Oxides and CNTs with Metal Oxides

Nanocomposites of CNTs with metal oxides were synthesized either by mechanical mixing or by MOCVD or ECD or wet chemical precipitation. For instance, IrO_2-CNTs nanocomposites were prepared by the MOCVD process, MnO_2-CNTs nanocomposites were synthesized by ECD and RuO_2-CNTs were formed by wet chemical precipitation.

Nanocomposites of conducting polymer with metal oxide were synthesized by in situ polymerization. For instance, CNTs-PANI nanocomposites were synthesized using CNTs and an aniline monomer on adding $(NH_4)_2S_2O_8$ as an oxidizing agent. Graphene-PEDOT was prepared by oxidative polymerization of EDOT on adding ammonium peroxydisulfate [$(NH_4)_2S_2O_8$)] and $FeCl_3$ as oxidizing agents.

4.5.4.7.7 Carbon Aerogels

Carbon aerogels (CAGs) are prepared by sol-gel process with subsequent pyrolysis of the gel-based organic precursor (e.g. resorcinol-formaldehyde or phenol-furfural resinous gel, etc.) to replace the liquid component of the gel with a gas. With an ultra-low density, high porosity, good electrical properties, and controllable pore structure, CAGs are a potential alternative for supercapacitor electrode material. The dominance of mesopores (>2 nm) is responsible for their low capacitance, and an additional activation process is to be needed to enhance the accessible specific surface area (SSA) by the introduction of microporosity.

4.6 STRUCTURE, ELECTRICAL AND CHEMICAL COMPOSITION, AND SURFACE AREA CHARACTERIZATIONS

The crystalline structure, crystallite size, lattice parameters, morphology analysis of nanocomposites, stoichiometry, and surface area of pure and nanocomposite materials are inferred by powder XRD/XRF (X-ray fluorescence spectroscopy), SEM, energy dispersive analysis (EDX), HRTEM and Brunauer–Emmett–Teller. IR, Fourier transform infrared, and Raman spectra are powerful tools for identifying the vibrational peaks pertaining to different molecules present in the nanocomposites. Four probe technique is employed to obtain the electrical conducting of nanocomposites. TGA and DTA were used to study the weight loss and thermal decomposition of nanocomposite materials.

4.7 DRAWBACK OF PRESENT MATERIALS AND HOW TO OVERCOME THEM

The cost of some components, catalysts, electrodes, and electrolytes are very expensive materials and some pollute the environment. These are the main limitations for large-scale production and new alternative advanced materials are required. Researchers are now developing new coated electrode materials, electrolytes with membranes, and metal-electrolyte interfaces [164–167] to control undesired parasitic reactions. Recently, a fluoroethylene carbonate additive has been tried to improve the cycle life of Li-ion batteries with 60-nm Si NPs and restrict the formation of metastable c-$Li_{15}Si_4$ phase. The use of existing and new nanomaterials quantum confinement effects for the preparation of new electrolytes compatible with other materials of energy storage devices.

4.8 SUMMARY

The efficiency of perovskite-based solar cells, fuel cells, and Li-ion batteries is improved by employing nanostructured materials. Materials in nano-dimensional regime provide stability and enhanced performance of energy storage and conversion devices. The nanostructure materials show superiority in comparison to bulk materials because of improved kinetics and electrochemical activity. Thermodynamic stability, side reaction, and handling of "nano" materials are the few challenges that can be addressed by collective efforts to ensure clean, sustaining energy storage, and conversion devices for the future.

LIST OF ABBREVIATIONS IN THIS CHAPTER

AC	activated carbon
AFCs	alkaline fuel cells
CaSZ	calcium stabilized zirconia
CBD	chemical bath deposition
CBL	cathode buffer layer
CNTs	carbon nanotubes
DMFCs	direct methanol fuel cells
DSCs	dye-sensitized solar cells
DTA	differential thermal analysis
ECD	electrochemical deposition
EDLCs	electrochemical double layer supercapacitors
EDOT	ethylene dioxythiophene
EDX	energy dispersive analysis
ES	electrochemical supercapacitors
ESR	equivalent series resistance
GDC	gadolium-doped ceria
GDLs	gas diffusion layers
GHG	greenhouse gas
HEV	hybrid electrical vehicle
H-P-G	host–passivation–guest
HRTEM	high-resolution transmission electron microscopy
HS	hybrid supercapacitors
IFC	International Fuel Cells
IR	infrared spectra
LIBs	lithium-ionbatteries
MCFCs	molten carbonate fuel cells
MEA	membrane electrode assembly
MOCVD	metal–organic chemical vapor deposition
MO-NPs	metaloxide nanoparticles
MTOE	million tons of oil equivalent
MWCNTs	multiwalled carbon nanotubes
NPs	nanoparticles
P3HT	poly(3-hexylthiophene)
PAFCs	phosphoric acid fuel cells
PANI	polyaniline
PCBM	[6,6]--(phenyl-C61 butyric acid methyl ester)
PEDOT	poly 3,4-ethylenedioxythiophene
PEM	polymer electrolyte membrane
PEMFC	polymer electrolyte membrane fuel cell
PEMFC	proton exchange membrane fuel cell
PSC	pseudo-supercapacitor
PSCs	polymer solar cells
PSMS	polystyrene macrospheres
PTFE	polytetrafluoroethylene

PV	photovoltaic
SDC	strontium-doped ceria
SEM	scanning electron microscopy
SILAR	successive ionic layer adsorption and reaction
SOFCs	solid oxide fuel cells
SSZ	strontium-stabilized zirconia
S-TENG	spiral triboelectrification–based nanogenerator
SWCNTs	single-walled carbon nanotubes
TENG	triboelectrification-based nanogenerator
TGA	thermogravimetric analysis
TW	terawatt
XRD	X-ray diffraction
YDC	yttrium-doped ceria
YSZ	yttria-stabilized zirconia

REFERENCES

1. Zekić, E.; Vuković, Z.; Halkijević, I.; Zelić, E.; Hidrokon, M.; Ivan Halkijević, A. Application of Nanotechnology in Wastewater Treatment Authors: Subject Review. *Nano-Adsorbens*, **2018**, 4, 315–323.
2. Demirbas, A. Global Renewable Energy Projections. *Energy Sources Part B*, **2009**, 4 (2), 212–224.
3. Sahaym, U.; Norton, M. G. Advances in the Application of Nanotechnology in Enabling a 'Hydrogen Economy'.*J. Mater. Sci.*, **2008**, 43 (16), 5395–5429.
4. Deng, J.; Lu, X.; Liu, L.; Zhang, L.; Schmidt, O. G. Introducing Rolled-Up Nanotechnology for Advanced Energy Storage Devices. *Adv. Energy Mater.*, **2016**, 6 (23), 1600797.
5. Shafiei, S.; Salim, R. A. Non-Renewable and Renewable Energy Consumption and CO$_2$ Emissions in OECD Countries: A Comparative Analysis. *Energy Policy*, **2014**, 66, 547–556.
6. Pandey, G. Nanotechnology for Achieving Green-Economy Through Sustainable Energy. *Rasayan J. Chem.*, **2018**, 11 (3), 942–950.
7. Raghav, S. B.; Dinesh, V. Recent Developments on Nanotechnology in Solar Energy. *Int. J. Eng. Comput. Sci.*, **2016**, 5 (2), 15829–15834.
8. Hussein, A. K. Applications of Nanotechnology in Renewable Energies—A Comprehensive Overview and Understanding. *Renew. Sustain. Energy Rev.*, **2015**, 42, 460–476.
9. Kadirgan, F. Electrochemical Nano-Coating Processes in Solar Energy Systems. *Int. J. Photoenergy*, **2006**, 1–5.
10. Matteo, C.; Candido, P.; Vera, R.; Francesca, V. Current and Future Nanotech Applications in the Oil Industry. *Am. J. Appl. Sci.*, **2012**, 9 (6), 784–793.
11. Kamat*, P. V. Meeting the Clean Energy Demand: Nanostructure Architectures for Solar Energy Conversion. *J. Phys. Chem. C*, **2007**, 111 (7), 2834–2860.
12. Jang, J. S.; Kim, H. G.; Joshi, U. A.; Jang, J. W.; Lee, J. S. Fabrication of CdS Nanowires Decorated with TiO$_2$ Nanoparticles for Photocatalytic Hydrogen Production under Visible Light Irradiation. *Int. J. Hydrogen Energy*, **2008**, 21 (33), 5975–5980.
13. Ahmadi, M. H.; Ghazvini, M.; Nazari, M.A.; Ahmadi, M. A.; Pourfayaz, F.; Lorenzini, G.; Ming, T.Renewable Energy Harvesting with the Application of Nanotechnology: A Review. *Int. J. Energy Res.*, **2019**, 43 (4), 1387–1410.

14. Raina, N.; Sharma, P.; Slathia, P. S.; Bhagat, D.; Pathak, A. K. Efficiency Enhancement of Renewable Energy Systems Using Nanotechnology. *Nanotechnol. Life Sci.*, **2020**, 271–297.

15. Khan, N.; Kalair, A.; Abas, N.; Haider, A. Review of Ocean Tidal, Wave and Thermal Energy Technologies. *Renew. Sustain. Energy Rev.*, **2017**, 72, 590–604.

16. Abdalla, A. M.; Elnaghi, B. E.; Hossain, S.; Dawood, M.; Abdelrehim, O.; Azad, A. K. Nanotechnology Utilization in Energy Conversion, Storage and Efficiency: A Perspective Review. *Adv. Energy Conversion Mater.*, **2020**, 1(1), 30.

17. Hirscher, M. Nanoscale Materials for Energy Storage. *Mater. Sci. Eng. B*, **2004**, 1–2 (108), 1.

18. Nazar, L. F.; Goward, G.; Leroux, F.; Duncan, M.; Huang, H.; Kerr, T.; Gaubicher, J. Nanostructured materials for energy storage. *Int. J. Inorg. Mater.*, **2001**, 3, 191–200.

19. Park, N. G. Perovskite Solar Cells: An Emerging Photovoltaic Technology. *Mater. Today*, **2015**, 18 (2), 65–72.

20. Kumar, R.; Al-Dossary, O.; G. Kumar, G.; Umar, A. Zinc Oxide Nanostructures for NO_2 Gas-Sensor Applications: A Review. *Nano-Micro Lett.*, **2015**, 7, 97–120.

21. Hoffman, R. L.; Norris, B. J.; Wager, J. F. ZnO-Based Transparent Thin-Film Transistors. *Appl. Phys. Lett.*, **2003**, 82 (5), 733.

22. Huang, J.; Yin, Z.; Zheng, Q. Applications of ZnO in Organic and Hybrid Solar Cells. *Energy Environ. Sci.*, **2011**, 4 (10), 3861–3877.

23. Steinfeld, A. Solar Hydrogen Production via a Two-Step Water-Splitting Thermochemical Cycle Based on Zn/ZnO Redox Reactions. *Int. J. Hydrogen Energy*, **2002**, 27 (6), 611–619.

24. Jang, E. S.; Won, J.-H.; Hwang, S.-J.; Choy, J.-H. Fine Tuning of the Face Orientation of ZnO Crystals to Optimize Their Photocatalytic Activity. *Adv. Mater.*, **2006**, 18 (24), 3309–3312.

25. Wang, Z. L. ZnO Nanowire and Nanobelt Platform for Nanotechnology. *Mater. Sci. Eng. R Reports*, **2009**, 64 (3–4), 33–71.

26. Wang, H.; Li, G.; Jia, L.; Wang, G.; Tang, C. Controllable Preferential-Etching Synthesis and Photocatalytic Activity of Porous ZnO Nanotubes. *J. Phys. Chem. C*, **2008**, 112 (31), 11738–11743.

27. Wu, Q.; Chen, X.; Zhang, P.; Han, Y.; Chen, X.; Yan, Y.; Li, S. Amino Acid-Assisted Synthesis of ZnO Hierarchical Architectures and Their Novel Photocatalytic Activities. *Cryst. Growth Des.*, **2008**, 8 (8), 3010–3018.

28. Wei, H.; Xiao, J.; Yang, Y.; Lv, S.; Shi, J.; Xu, X.; Dong, J.; Luo, Y.; Li, D.; Meng, Q. Free-Standing Flexible Carbon Electrode for Highly Efficient Hole-Conductor-Free Perovskite Solar Cells. *Carbon N. Y.*, **2015**, 93, 861–868.

29. Abdalla, A. M.; Hossain, S.; Azad, A.T.; Petra, P. M. I; Begum, F.; Eriksson, S. G.; Azad, A. K. Nanomaterials for Solid Oxide Fuel Cells: A Review. *Renew. Sustain. Energy Rev.*, **2018**, 82, 353–368.

30. Martín, G.B.; Joaquín, G.S.; Diego, G.L.; Ana, G.L.; De Reca, N. E. W. High-Performance Solid-Oxide Fuel Cell Cathodes Based on Cobaltite Nanotubes. *J. Am. Chem. Soc.*, **2007**, 129 (11), 3066–3067.

31. Zhao, E.; Ma, C; Yang, W.; Xiong, Y.; Li, J.; Sun, C. Electrospinning $La_{0.8}Sr_{0.2}Co_{0.2}Fe_{0.8}O_{3-\delta}$ Tubes Impregnated with $Ce_{0.8}Gd_{0.2}O_{1.9}$ Nanoparticles for an Intermediate Temperature Solid Oxide Fuel Cell Cathode. *Int. J. Hydrogen Energy*, **2013**, 38 (16), 6821–6829.

32. Xin, X.; Wei, Q.; Yang, J.; Yan, L.; Feng, R.; Chen, G.; Du, B.; Li, He. Highly Efficient Removal of Heavy Metal Ions by Amine-Functionalized Mesoporous Fe_3O_4 Nanoparticles. *J. Chem. Eng.*, **2012**, 184, 132–140.

33. Poursaberi, T.; Hassanisadi, M.; Torkestani, K.; Zare, M. Development of Zirconium (IV)-Metalloporphyrin Grafted Fe_3O_4 Nanoparticles for Efficient Fluoride Removal. *J. Chem. Eng.*, 2012, 117, 189–190.

34. Hao, Y.-M.; Man, C.; Hu, Z.-B. Effective Removal of Cu (II) Ions from Aqueous Solution by Amino-Functionalized Magnetic Nanoparticles. *J. Hazard Mater.*, 2010, 184, 392–399.
35. Zargar, R. A. ZnCdO Thick Film: A Material for Energy Conversion Devices. *Mater. Res. Express*, **2019**, 6, 095909.
36. Monárrez-Cordero, B.; Amézaga-Madrid, P.; Antúnez-Flores W.; Leyva-Porras, C.; Pizá-Ruiz, P.; Miki-Yoshida, M. Highly Efficient Removal of Arsenic Metal Ions with High Superficial Area Hollow Magnetite Nanoparticles Synthetized by AACVD Method. *J. Alloys Compd.*, **2014**, 586(Supplement 1), S520–S525.
37. Chowdhury, S. R.; Yanful, E. K. Kinetics of Cadmium(II) Uptake by Mixed Maghemite-Magnetite Nanoparticles. *J. Environ. Manag.*, **2013**, 129, 642–651.
38. Lunge, S.; Singh, S.; Sinha, A. Magnetic Iron Oxide (Fe3O4) Nano-Particles from Tea Waste for Arsenic Removal. *J. Magn. Magn. Mater.*, **2014**, 356, 21–31.
39. Mahapatra, A.; Mishra, B. G.; Hota, G. Electrospun Fe_2O_3–Al_2O_3 Nanocomposite Fibers as Efficient Adsorbent for Removal of Heavy Metal Ions from Aqueous Solution. *J. Hazard Mater.*, **2013**, 258–259, 116–123.
40. Recillas, S.; García, A.; González, E.; Casals, E.; Puntes, V.; Sánchez, A.; Font, X. Use of CeO_2, TiO_2 and Fe_3O_4 Nanoparticles for the Removal of Lead from Water: Toxicity of Nanoparticles and Derived Compounds. *Desalination.*, **2011**, 277, 213–220.
41. Rafiq, Z.; Nazir, R.; Durre, S.; Shah, M. R.; Ali, S. Utilization of Magnesium and Zinc Oxide Nano-Adsorbents as Potential Materials for Treatment of Copper Electroplating Industry Wastewater. *J. Environ. Chem. Eng.*, **2014**, 2, 642–651.
42. Venkatesham, V.; Madhu, G. M.; Satyanarayana, S. V.; Preetham, H. S. Adsorption of Lead on Gel Combustion Derived Nano ZnO. *Proc. Eng.*, **2013**, 51, 308–313.
43. Su, Y.; Cui, H.; Li, Q.; Gao, S.; Shang, J. K. Strong Adsorption of Phosphate by Amorphous Zirconium Oxide Nanoparticles. *Water Res.*, **2013**, 47, 5018–5026.
44. Cui, Y.; Kim, S. N.; Naik, R.; McAlpine, M. C. Biomimetic Peptide Nanosensors. *Acc. Chem. Res.*, **2012**, 45, 696–704.
45. Zargar, R. A., Kumar, K., Arora, M., Shkir, M. Structural, Optical, Photoluminescence, and EPR Behaviour of Novel Zn0·80Cd0·20O Thick Films: An Effect of Different Sintering Temperatures. *J. Luminescence*, **2022**, 245, 118769.
46. Nguyen, N. H.; Vo Thi, T. T.; Nguyen, H. T.; Nguyen, T. T.; Le, T. Q.; Tran, T. K.; Dang, V. T. Facile One-Step Synthesis of Zinc Oxide Nanoparticles by Ultrasonic-Assisted Precipitation Method and Its Application for H_2S Adsorption in Air. *J. Phys. Chem. Solids*, **2019**, 132, 99–103.
47. Zargar, R. A.; Arora, M. Study of Nanosized Copper Doped ZnO Dilute Magnetic Semiconductor Thick Films for Spintronic Device Application. *J. Appl. Phys–A*, **2018**, 124, 36.
48. Wu, H.; Ma, J.; Li, Y.; Zhang, C.; He, H. Photocatalytic Oxidation of Gaseous Ammonia Over Fluorinated TiO_2 with Exposed (0 0 1) Facets. *Appl. Catal. B Environ.*, **2014**, 152, 82–87.
49. Jampaiah, D.; Chalkidis, A.; Sabri, Y.; Mayes, E. L. H., Reddy, B. M.; Bhargava, S. Low-Temperature Elemental Mercury Removal Over TiO_2Nanorods-Supported MnOx-FeOx-CrOx. *Catal. Today*, **2019**, 324, 174–182.
50. Tumuluri, U.; Howe, J. D.; Mounfield III, W. P.; Li, M.; Chi, M.; Hood, Z. D.; Walton, K. S.; Sholl, D. S.; Dai, S.; Wu, Z. Effect of Surface Structure of TiO_2 Nanoparticles on CO_2 Adsorption and SO_2 Resistance. *ACS Sustain. Chem. Eng.*, **2017**, 5 (10), 9295–9306.
51. Shen W.; Zhang C.; Li Q.; Zhang W.; Cao L.; Ye, J. Preparation of Titanium Dioxide Nanoparticle Modified Photocatalytic Self-Cleaning Concrete. *J. Clean. Prod.*, **2016**, 87, 762–765.

52. Zhang, H.; Li, Q.; Huang, J.; Du, Y.; Ruan, S. C. Reduced Graphene Oxide/Au Nanocomposite for NO_2 Sensing at Low Operating Temperature. *Sensors*, **2016**, 16 (7), 1152–1160.

53. Fahlman, B. D. *Materials Chemistry*, 2nd ed. Springer Nature, Cham, Switzerland, **2011**.

54. Wise, F. W.; Kang, I. Electronic Structure and Optical Properties of PbS and PbSe Quantum Dots. *J. Opt. Soc. Am. B.*, **1997**, 14 (7), 1632–1646.

55. Zargar, R. A., Kumar, K., Shkir, M. Optical Characteristics ZnO Film: A Metlab Based Computer Calculation, Under Different Thickness. *Physica B.*, **2022**, 63, 414634.

56. Hu, L.; Huang, Y.; Zhang, F.; Chen, Q. CuO/Cu_2O Composite Hollow Polyhedrons Fabricated from Metal–Organic Framework Templates for Lithium-Ion Battery Anodes with a Long Cycling Life. *Nanoscale*, **2013**, 5 (10), 4186–4190.

57. Chang, J.A.; Rhee, J.H.; Im, S.H.; Lee, Y.H.; Kim, H.J.; Seok, S.I.; Nazeeruddin, M.K.; Gratzel, M. High-Performance Nanostructured Inorganic-Organic Heterojunction Solar Cells. *Nano Lett.*, **2010**, 10 (7), 2609–2612.

58. Law, M.; Greene, L. E.; Johnson, J. C.; Saykally, R.; Yang, P. Nanowire Dye-Sensitized Solar Cells. *Nat. Mater.*, **2005**, 4 (6), 455–459.

59. Yodyingyong, S.; Zhang, Q.; Park, K.; Dandeneau, C. S.; Zhou, X.; Triampo, D.; Cao, G. ZnO Nanoparticles and Nanowire Array Hybrid Photoanodes for Dye-Sensitized Solar Cells. *Appl. Phys. Lett.*, **2010**, 96 (7), 073115.

60. Baxter, J. B.; Walker, A. M.; Ommering, K. V.; Aydil, E. S. Synthesis and Characterization of ZnO Nanowires and Their Integration into Dye-Sensitized Cells. *Nanotechnol.*, **2006**, 17 (11), S304.

61. Qiu, J.; Li, X.; Zhuge, F.; Gan, X.; Gao, X.; He, W.; Park, S.-J.; Kim, H.-K.; Hwang, Y.-H. Solution-Derived 40 Mm Vertically Aligned ZnO Nanowire Arrays as Photoelectrodes in Dye-Sensitized Solar. *Nanotechnol.*, **2010**, 21 (19), 195602.

62. Xu, C.; Shin, P.; Cao, L.; Gao, D. Preferential Growth of Long ZnO Nanowire Array and Its Application in Dye-Sensitized Solar Cells. *J. Phys. Chem. C*, **2009**, 114 (1), 125–129.

63. Matt, L.; Lori, E. G.; Aleksandra, R.; Tevye, K.; Jan, L.; Peidong, Y. $ZnO-Al_2O_3$ and $ZnO-TiO_2$ Core–Shell Nanowire Dye-Sensitized Solar Cells. *J. Phys. Chem. B*, **2006**, 110 (45), 22652–22663.

64. Irannejad, A.; Janghorban, K.; Tan, O. K.; Huang, H.; Lim, C. K.; Tan, P. Y.; Fang, X.; Chua, C. S.; Maleksaeedi, S.; Hejazi, S. M. H.; Shahjamali, M.M. Effect of the TiO_2 Shell Thickness on the Dye-Sensitized Solar Cells with $ZnO-TiO_2$ Core–Shell Nanorod Electrodes. *Electrochim. Acta*, **2011**, 58 (1), 19–24.

65. Wiranwetchayan, O.; Zhang, Q.; Zhou, X.; Liang, Z.; Singjai, P.; Cao, G. Impact of the Morphology of Tio 2 Films as Cathode Buffer Layer on the Efficiency of Inverted-Structure Polymer Solar Cells. *Chalcogenide Lett.*, **2012**, 9 (4), 157–163.

66. Liang, Z.; Zhang, Q.; Wiranwetchayan, O.; Xi, J.; Yang, Z.; Park, K.; Li, C.; Cao, G. Effects of the Morphology of a ZnO Buffer Layer on the Photovoltaic Performance of Inverted Polymer Solar Cells. *Adv. Funct. Mater.*, **2012**, 22 (10), 2194–2201.

67. Zargar, R.A.; Arora, M.; Alshahrani, T.; Shkir, M. Screen Printed Novel ZnO/ MWCNTs Nanocomposite Thick Film. *Ceramic Int.*, **2021**, 47, 6084–6093.

68. Chou, T. P.; Zhang, Q.; Cao, G. Effects of Dye Loading Conditions on the Energy Conversion Efficiency of ZnO and TiO_2 Dye-Sensitized Solar Cells. *J. Phys. Chem. C*, **2007**, 111, 18804–18811.

69. Zhang, Q.; Dandeneau, C. S.; Zhou, X.; Cao, G. ZnO Nanostructures for Dye-Sensitized Solar Cells. *Adv. Mater.*, **2009**, 21 (41), 4087–4108.

70. Tétreault, N.; Arsenault, E.; Heiniger, L.-P.; Soheilnia, N.; Brillet, J.; Moehl, T.; Zakeeruddin, S.; Ozin, G. A.; Grätzel, M. High-Efficiency Dye-Sensitized Solar Cell with Three-Dimensional Photoanode. *Nano Lett.*, **2011**, 11 (11), 4579–4584.

71. Cao, G.; Wang, Y. *Nanostructures and Nanomaterials: Synthesis, Properties and Applications*, 2nd ed. World Scientific Publishing, Singapore, **2004**.
72. Imanishi, N.; Kashiwagi, H.; Ichikawa, T.; Takeda, Y.; Yamamoto, O.; Inagaki, M. Charge-Discharge Characteristics of Mesophase-Pitch-Based Carbon Fibers for Lithium Cells. *J. Electrochem. Soc.*, **1993**, 140 (2), 315.
73. Besenhard, J. O.; Wagner, M. W.; Winter, M.; Jannakoudakis, A. D.; Jannakoudakis, P. D.; Theodoridou, E. Inorganic Film-Forming Electrolyte Additives Improving the Cycling Behaviour of Metallic Lithium Electrodes and the Self-Discharge of Carbon—Lithium Electrodes. *J. Power Sources*, **1993**, 44 (1–3), 413–420.
74. Bruce, P. G.; Freunberger, S. A.; Hardwick, L. J.; Tarascon, J.-M. $Li–O_2$ and Li–S Batteries with High Energy Storage. *Nat. Mater.* **2011**, 11 (1), 19–29.
75. Su, X.; Wu, Q.; Li, J.; Xiao, X.; Lott, A.; Lu, W.; Sheldon, B. W.; J. Wu, J. Silicon-Based Nanomaterials for Lithium-Ion Batteries: A Review. *Adv. Energy Mater.*, **2014**, 4 (1), 1300882.
76. Chan, C. K.; Peng, H.; Liu, G.; McIlwrath, K.; Zhang, X. F.; Huggins, R. A.; Cui, Y. High-Performance Lithium Battery Anodes Using Silicon Nanowires. *Nat. Nanotechnol.*, **2007**, 3 (1), 31–35.
77. Cui, L.-F.; Ruffo, R.; Chan, C. K.; Peng, H.; Cui, Y. Crystalline-Amorphous Core–Shell Silicon Nanowires for High Capacity and High Current Battery Electrodes. *Nano Lett.*, **2008**, 9 (1), 491–495.
78. Yao, Y.; McDowell, M. T.; Ryu, I.; Wu, H.; Liu, N.; Hu, L.; Nix, W. D.; Cui, Y. Interconnected Silicon Hollow Nanospheres for Lithium-Ion Battery Anodes with Long Cycle Life. *Nano Lett.*, **2011**, 11 (7), 2949–2954.
79. Ikonen, T.; Kalidas, N.; Lahtinen, K.; Isoniemi, T.; Toppari, J. J.; Vázquez, E.; Herrero-Chamorro, M. A.; Fierro, J. L. G.; Kallio, T.; Lehto, V.-P. Conjugation with Carbon Nanotubes Improves the Performance of Mesoporous Silicon as Li-Ion Battery Anode. *Sci. Rep.*, **2020**, 10 (1), 1–8.
80. Geng, H.; Peng, Y.; Qu, L.; Zhang, H.; Wu, M. Structure Design and Composition Engineering of Carbon-Based Nanomaterials for Lithium Energy Storage. *Adv. Energy Mater.*, **2020**, 10 (10), 1903030.
81. Chen, T.; Qiu, L.; Cai, Z.; Gong, F.; Yang, Z.; Wang, Z.; Peng, H. Intertwined Aligned Carbon Nanotube Fiber Based Dye-Sensitized Solar Cells. *Nano Lett.*, **2012**, 12 (5), 2568–2572.
82. Chong, W. G.; Huang, J.-Q.; Xu, Z.-L.; Qin, X.; Wang, X.; Kim, J.-K. Lithium–Sulfur Battery Cable Made from Ultralight, Flexible Graphene/Carbon Nanotube/Sulfur Composite Fibers. *Adv. Funct. Mater.*, **2017**, 27 (4), 1604815.
83. Lee, J. S.; Jun, J.; Jang, J.; Manthiram, A. Sulfur-Immobilized, Activated Porous Carbon Nanotube Composite Based Cathodes for Lithium–Sulfur Batteries. *Small*, **2017**, 13 (12), 1602984.
84. Liu, H.; Zhang, S.; Zhu, Q.; Cao, B.; Zhang, P.; Sun, N.; Xu, B.; Wu, F.; Chen, R. Fluffy Carbon-Coated Red Phosphorus as a Highly Stable and High-Rate Anode for Lithium-Ion Batteries. *J. Mater. Chem. A*, **2019**, 7 (18), 11205–11213.
85. Park, M.-G.; Song, J. H.; Sohn, J.-S.; Lee, C. K.; Park, C.-M. Co–Sb Intermetallic Compounds and Their Disproportionated Nanocomposites as High-Performance Anodes for Rechargeable Li-Ion Batteries. *J. Mater. Chem. A*, **2014**, 2 (29), 11391–11399.
86. Lin, T. C.; Dawson, A.; King, S. C.; Yan, Y.; Ashby, D. S.; Mazzetti, J. A.; Dunn, B. S.; Weker, J. N.; Tolbert, S. H. Understanding Stabilization in Nanoporous Intermetallic Alloy Anodes for Li-Ion Batteries Using Operando Transmission X-Ray Microscopy. *ACS Nano*, **2020**, 14 (11), 14820–14830.
87. Yi, Z.; Wang, Z.; Cheng, Y.; Wang, L. Sn-Based Intermetallic Compounds for Li-Ion Batteries: Structures, Lithiation Mechanism, and Electrochemical Performances. *Energy Environ. Mater.*, **2018**, 1 (3), 132–147.

88. Kim, I.; Kumta, P. N.; Blomgren, G. E. Si/TiN Nanocomposites Novel Anode Materials forLi‑Ion Batteries. *Electrochem. Solid-State Lett.*, **2000**, 3 (11), 493.

89. Sun, W.; Wang, Y. Graphene-Based Nanocomposite Anodes for Lithium-Ion Batteries. *Nanoscale*, **2014**, 6 (20), 11528–11552.

90. Wu, H. Bin; Chen, J. S.; Hng, H. H.; Lou, X. W. Nanostructured Metal Oxide-Based Materials as Advanced Anodes for Lithium-Ion Batteries. *Nanoscale*, **2012**, 4 (8), 2526–2542.

91. Reddy, M. V.; Tse, L. Y.; Bruce, W. K. Z.; Chowdari, B. V. R. Low Temperature Molten Salt Preparation of Nano-SnO$_2$ as Anode for Lithium-Ion Batteries. *Mater. Lett.*, **2015**, 138, 231–234.

92. Zhang, J.; Yu, A. Nanostructured Transition Metal Oxides as Advanced Anodes for Lithium-Ion Batteries. *Sci. Bull.*, **2015**, 60 (9), 823–838.

93. Zhang, W.; Du, L.; Chen, Z.; Hong, J.; Yue, L. ZnO Nanocrystals as Anode Electrodes for Lithium-Ion Batteries. *J. Nanomater.*, **2016**, 1–7.

94. DeKrafft, K. E.; Wang, C.; Lin, W. Metal-Organic Framework Templated Synthesis of Fe2O3/TiO2 Nanocomposite for Hydrogen Production. *Adv. Mater.*, **2012**, 24 (15), 2014–2018.

95. Xiao, A.; Zhou, S.; Zuo, C.; Zhuan, Y.; Ding, X. Electrodeposited Porous Metal Oxide Films with Interconnected Nanoparticles Applied as Anode of Lithium Ion Battery. *Mater. Res. Bull.*, **2014**, 60, 864–867.

96. Poizot, P.; Laruelle, S.; Grugeon, S.; Dupont, L.; Tarascon, J.-M. Nano-Sized Transition-Metal Oxides as Negative-Electrode Materials for Lithium-Ion Batteries. *Nat.*, **2000**, 407 (6803), 496–499.

97. Yuan, Y. F.; Xia, X. H.; Wu, J. B.; Yang, J. L.; Chen, Y. B.; Guo, S. Y. Hierarchically Ordered Porous Nickel Oxide Array Film with Enhanced Electrochemical Properties for Lithium Ion Batteries. *Electrochem. commun.*, **2010**, 12 (7), 890–893.

98. Mosa, J.; García-García, F. J.; González-Elipe, A. R.; Aparicio, M. New Insights on the Conversion Reaction Mechanism in Metal Oxide Electrodes for Sodium-Ion Batteries. *Nanomater.*, **2021**, 11 (4), 966.

99. Zhao, B.; Dhara, A.; Dendooven, J.; Detavernier, C. Atomic Layer Deposition of SnO$_2$-Based Composite Anodes for Thin-Film Lithium-Ion Batteries. *Front. Energy Res.*, **2020**, 8, 609417.

100. Guo, Z. P.; Zhao, Z. W.; Liu, H. K.; Dou, S. X. Electrochemical Lithiation and De-Lithiation of MWNT–Sn/SnNi Nanocomposites. *Carbon*, **2005**, 43 (7), 1392–1399.

101. Li, H.; Huang, X.; Chen, L.; Wu, Z.; Liang, Y. A High CapacityNano Si Composite Anode Material for Lithium Rechargeable Batteries. *Electrochem. Solid-State Lett.*, **1999**, 2 (11), 547.

102. Singhal, S. C.; Kendall, K. *High-Temperature Solid Oxide Fuel Cells : Fundamentals, Design, and Applicatons*, 1st ed. Elsevier, Oxford, **2003**, 405.

103. Ghosh, A.; Azad, A.; Irvine, J. T. Study of Ga Doped LSCM as an Anode for SOFC. *ECS Trans.*, **2011**, 35 (1), 1337.

104. Cook, B. Introduction to Fuel Cells and Hydrogen Technology. *Eng. Sci. Educ. J.*, **2002**, 11 (6), 205–216.

105. Singhal, S. C. Advances in Solid Oxide Fuel Cell Technology. *Solid State Ion.*, **2000**, 135 (1–4), 305–313.

106. Shaikh, S. P. S.; Muchtar, A.; Somalu, M. R. A Review on the Selection of Anode Materials for Solid-Oxide Fuel Cells. *Renew. Sustain. Energy Rev.*, **2015**, 51, 1–8.

107. McCarthy, B. P.; Pederson, L. R.; Chou, Y. S.; Zhou, X. D.; Surdoval, W. A.; Wilson, L. C. Low-Temperature Sintering of Lanthanum Strontium Manganite-Based Contact Pastes for SOFCs. *J. Power Sources*, **2008**, 180 (1), 294–300.

108. Meixner, D. L.; Cutler, R. A. Sintering and Mechanical Characteristics of Lanthanum Strontium Manganite. *Solid State Ion.*, **2002**, 146 (3–4), 273–284.

109. Sun, C.; Hui, R.; Roller, J. Cathode Materials for Solid Oxide Fuel Cells: A Review. *J. Solid State Electrochem.*, **2009**, 14 (7), 1125–1144.

110. Khandale, A. P.; Lajurkar, R. P.; Bhoga, S. S. $Nd_{1.8}Sr_{0.2}NiO_{4-\delta}$:$Ce_{0.9}Gd_{0.1}O_{2-\delta}$ Composite Cathode for Intermediate Temperature Solid Oxide Fuel Cells. *Int. J. Hydrogen Energy*, **2014**, 39 (33), 19039–19050.

111. Menzler, N. H.; Tietz, F.; Uhlenbruck, S.; Buchkremer, H. P.; Stöver, D. Materials and Manufacturing Technologies for Solid Oxide Fuel Cells. *J. Mater. Sci.*, **2010**, 45 (12), 3109–3135.

112. Haile, S. M. Fuel Cell Materials and Components. *Acta Mater.*, **2003**, 51 (19), 5981–6000.

113. Cho, G. Y.; Lee, Y. H.; Cha, S. W. Multi-Component Nano-Composite Electrode for SOFCS via Thin Film Technique. *Renew. Energy*, **2014**, 65, 130–136.

114. Evans, A.; Bieberle-Hütter, A.; Galinski, H.; Rupp, J.L. M.; Ryll, T,; Scherrer, B.; Tölke, R.; Gauckler, L. J. Micro-Solid Oxide Fuel Cells: Status, Challenges, and Chances. *Monatshefte für Chemie—Chem. Mon.*, **2009**, 140 (9), 975–983.

115. Tian, N.; Zhou, Z.-Y.; Sun, S.-G.; Ding, Y.; Wang, Z. L. Synthesis of Tetrahexahedral Platinum Nanocrystals with High-Index Facets and High Electro-Oxidation Activity. *Sci.*, **2007**, 316 (5825), 732–735.

116. Wang, L.; Y. Yamauchi, Y. Block Copolymer Mediated Synthesis of Dendritic Platinum Nanoparticles. *J. Am. Chem. Soc.*, **2009**, 131 (26), 9152–9153.

117. Mazumder, V.; Lee, Y.; Sun, S. Recent Development of Active Nanoparticle Catalysts for Fuel Cell Reactions. *Adv. Funct. Mater.*, **2010**, 20 (8), 1224–1231.

118. Aricò, A. S.; Bruce, P.; Scrosati, B.; Tarascon, J. M.; Schalkwijk, W. V. Nanostructured Materials for Advanced Energy Conversion and Storage Devices. *Mater. Sustain. Energy A Collect. Peer-Reviewed Res. Rev. Artic. from Nat. Publ. Gr.*, **2010**, 148–159.

119. Becker, H. J. *Low Voltage Electrolytic Capacitor.* United States Patent Office, New York, **1954**.

120. Chapman, D. L. A Contribution to the Theory of Electrocapillarity. *Lond. Edinb. Dublin philos. Mag. J. Sci.*, **2010**, 25 (148), 475–481.

121. Grahame, D. C. The Electrical Double Layer and the Theory of Electrocapillarity. *Chem. Rev.*, **2002**, 41 (3), 441–501.

122. Gerischer, H.; Sparnaay, M. J. *The Electrical Double Layer; Vol. 4 of Properties of Interfaces; Topic 14 of The International Encyclopedia of Physical Chemistry and Chemical Physics.* Pergamon Press, Oxford, **1972**.

123. Zhang, Y.; Feng, H; Wu, X.; Wang, L.; Zhang, A.; Xia, T.; Dong, H.; Li, X.; Zhang, L. Progress of Electrochemical Capacitor Electrode Materials: A Review. *Int. J. Hydrogen Energy*, **2009**, 34 (11), 4889–4899.

124. Fedorov, M. V.; A. A. Kornyshev, A. A. Towards Understanding the Structure and Capacitance of Electrical Double Layer in Ionic Liquids. *Electrochim. Acta*, **2008**, 53 (23), 6835–6840.

125. Kötz, R.; Carlen, M. Principles and Applications of Electrochemical Capacitors. *Electrochim. Acta*, **2000**, 45 (15–16), 2483–2498.

126. Conway, B. E.; Birss, V.; Wojtowicz, J. The Role and Utilization of Pseudocapacitance for Energy Storage by Supercapacitors. *J. Power Sources*, **1997**, 66 (1–2), 1–14.

127. Peng, C.; Zhang, S.; Jewell, D.; Chen, G. Z. Carbon Nanotube and Conducting Polymer Composites for Supercapacitors. *Prog. Nat. Sci.*, **2008**, 18 (7), 777–788.

128. Burke, A. Ultracapacitors: Why, How, and Where Is the Technology. *J. Power Sources*, **2000**, 91 (1), 37–50.

129. Conway, B. E.; Pell, W. G. Double-Layer and Pseudocapacitance Types of Electrochemical Capacitors and Their Applications to the Development of Hybrid Devices. *J. Solid State Electrochem.*, **2003**, 7 (9), 637–644.

130. Halper, M. S.; Ellenbogen, J. C. *Supercapacitors: A Brief Overview.* The MITRE Corporation, McLean, VA, **2006**.

131. Karthikeyan, K.; Aravindan, V.; Lee, S. B.; Jang, I. C.; Lim, H. H.; Park, G. J.; Yoshio, M.; Lee, Y. S. Electrochemical Performance of Carbon-Coated Lithium Manganese Silicate for Asymmetric Hybrid Supercapacitors. *J. Power Sources*, **2010**, 195 (11), 3761–3764.

132. Pandolfo, A. G.; A. F. Hollenkamp, A. F. Carbon Properties and Their Role in Supercapacitors. *J. Power Sources*, **2006**, 157 (1), 11–27.

133. Zhang, L. L.; Zhao, X. S. Carbon-Based Materials as Supercapacitor Electrodes. *Chem. Soc. Rev.*, **2009**, 38 (9), 2520–2531.

134. Cheng, Q.; Tang, J.; Ma, J.; Zhang, H.; Shinya, N.; Qin, L.-C. Graphene and Carbon Nanotube Composite Electrodes for Supercapacitors with Ultra-High Energy Density. *Phys. Chem. Chem. Phys.*, **2011**, 13 (39), 17615–17624.

135. Tamilarasan, P.; Mishra, A. K.; Ramaprabhu, S. Graphene/Ionic Liquid Binary Electrode Material for High Performance Supercapacitor. *2011 Int. Conf. Nanosci. Technol. Soc. Implic. NSTSI11*, **2011**, 1–5.

136. Du, C.; Pan, N. Carbon Nanotube-Based Supercapacitors. *Nanotechnol. Law Busi.*, **2007**, 4, 569–576.

137. Li, J. Review of Electrochemical Capacitors Based on Carbon Nanotubes and Graphene. *Graphene*, **2012**, 1 (1), 1–13.

138. Marcano, D. C.; Kosynkin, D. V.; Berlin, J. M.; Sinitskii, A.; Sun, Z.; Slesarev, A.; Alemany, L. B.; Lu, W.; Tour, J. M. Improved Synthesis of Graphene Oxide. *ACS Nano*, **2010**, 4 (8), 4806–4814.

139. Liu, C.; Yu, Z.; Neff, D.; Zhamu, A.; Jang, B. Z. Graphene-Based Supercapacitor with an Ultrahigh Energy Density. *Nano Lett.*, **2010**, 10 (12), 4863–4868.

140. Cheng, H. M. Development of Graphene-Based Materials for Energy Storage. *Proc.— 2010 8th Int. Vac. Electron Sources Conf. Nanocarbon, IVESC 2010 NANOcarbon*, **2010**, 49.

141. Hualan, W.; Hao, Q.; Yang, X.; Lu, L.; Wang, X. A Nanostructured Graphene/Polyaniline Hybrid Material for Supercapacitors. *Nanoscale*, **2010**, 2 (10), 2164–2170.

142. Karthika, P.; Rajalakshmi, N.; Dhathathreyan, K. S. Functionalized Exfoliated Graphene Oxide as Supercapacitor Electrodes. *Soft Nanosci. Lett.*, **2012**, 2012 (04), 59–66.

143. Kim, T.; Jung, G.; Yoo, S.; Suh, K. S.; Ruoff, R. S. Activated Graphene-Based Carbons as Supercapacitor Electrodes with Macro- and Mesopores. *ACS Nano*, **2013**, 7 (8), 6899–6905.

144. Zhao, C.; Zheng, W. A Review for Aqueous Electrochemical Supercapacitors. *Front. Energy Res.*, **2015**, 3, 23.

145. Jung, K.-D.; Gujar, T. P.; Kim, W.-Y.; Puspitasari, I.; Joo, O.-S. Electrochemically Deposited Nanograin Ruthenium Oxide as a Pseudocapacitive Electrode. *J. Electrochem. Sci*, **2007**, 2, 666–673.

146. Wu, H.-Y.; Wang, H.-W. Electrochemical Synthesis of Nickel Oxide Nanoparticulate Films on Nickel Foils for High-Performance Electrode Materials of Supercapacitors. *Int. J. Electrochem. Sci*, **2012**, 7, 4405–4417.

147. Liu, F.; Song, S.; Xue, D.; Zhang, H. Selective Crystallization with Preferred Lithium-Ion Storage Capability of Inorganic Materials. *Nanoscale Res. Lett.*, **2012**, 7 (1), 1–17.

148. Acznik, I.; Lota, K.; Sierczynska, A.; Lota, G. Carbon-Supported Manganese Dioxide as Electrode Material for Asymmetric Electrochemical Capacitors. *Int. J. Electrochem. Sci*, **2014**, 9, 2518–2534.

149. Chen, S.; Zhu, J.; Wu, X.; Han, Q.; Wang, X. Graphene Oxide–MnO_2 Nanocomposites for Supercapacitors. *ACS Nano*, **2010**, 4 (5), 2822–2830.

150. Liu, C.; Gui, D.; Liu, J. Preparation of MnO_2/Graphene Nanocomposite for the Application of Supercapacitor. *Proc. Electron. Packag. Technol. Conf. EPTC*, **2014**, 177–182.

151. Wang, D.-W.; Li, F.; Zhao, J.; Ren, W.; Chen, Z.-G.; Tan, J.; Wu, Z.-S.; Gentle, I.; Lu, G. Q.; Cheng, H.-M. Fabrication of Graphene/Polyaniline Composite Paper via In Situ Anodic Electropolymerization for High-Performance Flexible Electrode. *ACS Nano*, **2009**, 3 (7), 1745–1752.

152. Sharma, P.; Bhatti, T. S. A Review on Electrochemical Double-Layer Capacitors. *Energy Convers. Manag.*, **2010**, 51 (12), 2901–2912.

153. Cheng, Q.; Tang, J.; Shinya, N.; Qin, L. C. Polyaniline Modified Graphene and Carbon Nanotube Composite Electrode for Asymmetric Supercapacitors of High Energy Density. *J. Power Sources*, **2013**, 241, 423–428.

154. Li, J.; Xie, H.; Li, Y.; Liu, J.; Li, Z. Electrochemical Properties of Graphene Nanosheets/ Polyaniline Nanofibers Composites as Electrode for Supercapacitors. *J. Power Sources*, **2011**, 196 (24), 10775–10781.

155. Feng, X.; Li, R.; Yan, Z.; Liu, X.; Chen, R.; Ma, Y.; Li, X.; Fan, Q.; Huang, W. Preparation of Graphene/Polypyrrole Composite Film via Electrodeposition for Supercapacitors. *IEEE Trans. Nanotechnol.*, **2012**, 11 (6), 1080–1086.

156. Kuilla, T.; Bhadra, S.; Yao, D.; Kim, N. H.; Bose, S.; Lee, J. H. Recent Advances in Graphene Based Polymer Composites. *Prog. Polym. Sci.*, **2010**, 35 (11), 1350–1375.

157. Fang, S.; Bresser, D.; Passerini, S. Transition Metal Oxide Anodes for Electrochemical Energy Storage in Lithium- and Sodium-Ion Batteries. *Adv. Energy Mater.*, **2020**, 10 (1), 1902485.

158. Conway, B. E. Transition from "Supercapacitor" to "Battery" Behavior in Electrochemical Energy Storage. *J. Electrochem. Soc.*, **1991**, 138 (6), 1539.

159. Zhang, Y.; Ma, Q.; Feng, K.; Guo, J.; Wei, X.; Shao, Y.; Zhuang, J.; Lin, T. Effects of Microstructure and Electrochemical Properties of $Ti/IrO_2–SnO_2–Ta_2O_5$ as Anodes on Binder-Free Asymmetric Supercapacitors with $Ti/RuO2–NiO$ as Cathodes. *Ceram. Int.*, **2020**, 46 (11), 17640–17650.

160. Guo, W.; Sun, W.; Wang, Y. Multilayer CuO@NiO Hollow Spheres: Microwave-Assisted Metal–Organic-Framework Derivation and Highly Reversible Structure-Matched Stepwise Lithium Storage. *ACS Nano*, **2015**, 9 (11), 11462–11471.

161. Mondal, I.; Pal, U. Synthesis of MOF Templated Cu/CuO@TiO2 Nanocomposites for Synergistic Hydrogen Production. *Phys. Chem. Chem. Phys.*, **2016**, 18 (6), 4780–4788.

162. Luo, J.-Y.; Cui, W.-J.; He, P.; Xia, Y.-Y. Raising the Cycling Stability of Aqueous Lithium-Ion Batteries by Eliminating Oxygen in the Electrolyte. *Nat. Chem.*, **2010**, 2 (9), 760–765.

163. Deyab, M. A. 1-Allyl-3-Methylimidazolium Bis(Trifluoromethylsulfonyl) Imide as an Effective Organic Additive in Aluminum-Air Battery. *Electrochim. Acta*, **2017**, 244, 178–183.

164. Wei, S.; Choudhury, S.; Xu, J.; Nath, P.; Tu, Z.; Archer, L. A. Highly Stable Sodium Batteries Enabled by Functional Ionic Polymer Membranes. *Adv. Mater.*, **2017**, 29 (12), 1605512.

165. Tikekar, M. D.; Choudhury, S.; Tu, Z.; Archer, L. A. Design Principles for Electrolytes and Interfaces for Stable Lithium-Metal Batteries. *Nat. Energy*, **2016**, 1 (9), 1–7.

166. Gao, H.; Xiao, L.; Plümel, I.; Xu, G.-L; Ren, Y.; Zuo, X.; Liu, Y.; Schulz, C.; Wiggers, H.; Amine, K.; Chen, Z. Parasitic Reactions in Nanosized Silicon Anodes for Lithium-Ion Batteries. *Nano Lett.*, **2017**, 17 (3), 1512–1519.

167. Wang, X.; Mathis, T. S.; Li, K.; Lin, Z.; Vlcek, L.; Torita, T.; Osti, N. C.; Hatter, C.; Urbankowski, P.; Sarycheva, A.; Tyagi, M. Influences from Solvents on Charge Storage in Titanium Carbide mXenes. *Nat. Energy*, **2019**, 4 (3), 241–248.

5 Evaluation of the Healthiness of Nanomaterials

Environmental and Human Health Approaches

*Km Madhuri Singh, Paulami Sahu,
and Saleem Ahmad Yatoo*

CONTENTS

DOI: 10.1201/9781003323464-5

5.1 INTRODUCTION

Over the last several years, nanotechnology has been effectively introduced into our day-to-day life. An American physicist, Richard Feynman, commenced the idea or concept of nanoscience and nanotechnology with a famous talk titled "There's Plenty of Room at the Bottom". He revolutionizes a new era in which it is possible to change the properties of devices at the atomic, molecular or macromolecular level. Since numerous scientific researchers have significantly improved this technology [1]. However, the word *nano* is derived from the Greek language denoting dwarf or extremely small particles having a size of less than 100 nm. They exhibit novel or enhanced size-dependent characteristics containing a large portion of the surface of any atoms or molecules considered to have high surface activity. Therefore, nanoscience and nanotechnology cover the study of extremely small things. Moreover, they can be used in a variety of science disciples including chemistry, physics, material science and biology. Therefore, nanotechnology is a multidisciplinary field encompassing science, engineering and technology with many applications and nanoparticles are the mandatory component of nanotechnology. All the abbreviations used in the present research work are tabulated in Table 5.1, and numerous terms related to NMs are given in Table 5.2.

TABLE 5.1
Abbreviations

Sr. No.	Abbreviated form	Full Form
1	NPs	Nanoparticles
2	NMs	Nanomaterials
3	CNTS	Carbon nanotubes
4	SWCNTs	Single-walled carbon nanotubes
5	MWCNTs	Multiwalled carbon nanotubes
6	GO	Graphene oxide
7	rGO	Reduced graphene oxide
8	nZVI	Nano-zero-valent iron
9	GRA	Graphene
10	DWNTs	Double-walled carbon nanotubes
11	CNT-CS	Carbon nanotube–chitosan
12	ZnO	Zinc oxide
13	Fe_2O_3	Ferric oxide
14	SiO_2	Silicon dioxide
15	TiO_2	Titanium dioxide
16	TX-100	Triton X-100
17	TX	Triton X
18	Cr^{6+}	Chromium (VI) ion
19	Pb^{2+}	Lead (II) ion
20	Cd^{2+}	Cadmium (II) ion
21	EB	Ethidium bromide
22	EOB	Eosin bluish
23	OG	Orange G
24	AR18	Acid red 18
25	AO	Acridine orange
26	ACF	Activated carbon fiber
27	MB	Methylene blue
28	CV	Crystal violet
29	RhB	Rhodamine B
30	ARS	Alizarin red S
31	SWCNT-COOH	Single-walled nanotube–carboxylic acid
32	SWCNT-OH	Single-walled carbon nanotube–hydroxide
33	$SWCNT-NH_2$	Single-walled carbon nanotube–azanide
34	$MWCNT-FeSO_4$	Multiwalled carbon nanotube–ferrous sulfate
35	$MWCNT-ZrO_2$	Multiwalled carbon nanotube–zirconium oxide
36	$MWCNT-MnO_2$	Multiwalled carbon nanotubes–manganese dioxide
37	$MWCNT-TiO_2$	Multiwalled carbon nanotubes–titanium dioxide
38	Fe_3O_4	Magnetic oxide
39	DR23	Direct red 23
40	HC	Hydrocarbon

(*Continued*)

TABLE 5.1 *(Continued)*
Abbreviations

Sr. No.	Abbreviated form	Full Form
41	PM	Particulate matter
42	TEM	Transmission electron microscopy
43	TNF	Tumor necrosis factor
44	GALT	Gut-associate dlymphoid tissue
45	LDH	Lactate dehydrogenase
46	ROS	Reactive oxygen species
47	RNS	Reactive nitrogen species

TABLE 5.2
Summary of Numerous Terms Related to NMs

Term	Description
Nanotechnology	Nanotechnology is the study of extremely small things in multidisciplinary fields at the nanoscale range.
Nanoscale	A scale consisting between 1–100 nm.
NPs	NPs are those ultrafine particles that have a size of less than 100 nm and consist of one or more external dimensions.
Nanotubes	Hollow nanofibers are called nanotubes.
Engineered NMs	Those nanoscale materials that are produced purposely are called engineered NMs.
NMs	An NM, as any material that has unique properties, has at least one external dimension at nanoscale assembling in the size range of 1–100 nm.

NPs are a subclassification of natural or anthropogenic ultrafine particles that can be produced naturally or accidentally. They are generally less toxic, biodegradable and more biocompatible. Natural NPs are those smaller particles originating in the natural environment, including abiotic components like soil, air, water or biotic component. In soils, NPs occur as NMs having a high density of surface functional groups and high reactivity, which directly control various soil processes such as sorption capacity, mobility and diffusion mass transfer. Examples of NPs are certain clay minerals (kaolinite, smectite, and vermiculite), Fe oxides (allophan, imogolite) and Mn oxyhydroxides or carbon-containing NPs. Abiotic NPs include sea salt aerosols, forest fires, fine dust storms (lunar dust), volcanic ash (allophane) and others, and biological NPs consist of living organisms like bacteria and fungi and semi-living organisms such as viruses. Natural NPs are of basic importance in the Earth's system, metal-binding biogeochemical cycle, ecotoxicity, weathering and bioavailability. In addition, NPs may also be produced unintentionally as products of mechanical or industrial processes through combustion and vaporization [2, 1].

Nano-based materials perhaps show distinct novel properties belonging to their non-nano configurations; in addition, these dissimilarities in their characteristics have raised queries regarding possible human robustness and environmental threat. According to the International Organization for Standardization (ISO) "nanoparticles" as nanosized entities accompanied by three exterior dimensions in the nanoscale range, where nanoscale is explained as the size dimensions being about 1–100 nm [3] and "nanomaterials" as having some exterior dimensions in the nanoscale or an inner or external conformation in the nanoscale [4]. Therefore, NMs can be defined as any material that has novel or unique properties due to the nanoscale assembling, having at least one exterior dimension in the size range of 1–100 nm. NMNs are formed by the compaction or structuring of NPs.

NMs demonstrate novel physical or chemical properties, and they grant advancement to engineered materials, through increased optical properties, greater magnetic features and upgraded electrical actions, as well as enhanced better integrity. NMs are present everywhere, and they can be generated naturally or anthropogenically. Scientists can engineer NMs from common materials to improve their utility. In the natural world, NMs are found in coral and marine plankton skeletons, the feathers and beaks of birds, spider webs, the hair and bone matrices of animals and humans and so on. In the engineered world, through micro-fabrication techniques, NMs are purposefully designed and synthesized for various applications such as specific optical, mechanical, electronic, medical, photocatalytic and enzymatic applications. Presently, nano-assembling materials are extensively utilized in a diversity of commodities including foodstuffs, medicine, personal care products or cosmetics, clothing, electronic devices and more.

5.2 CLASSIFICATION OF NMS

Nanoscientists are working around the clock to explore or develop new kinds of nanomaterials with distinct structural designs that can be utilized in new productive applications. Therefore, various schemes are considered to classify NMs. In this chapter, nanomaterials are classified on the basis of their origin (Figure 5.1), dimensions (Figure 5.2), chemical composition (Figure 5.3) and shape (Figure 5.4).

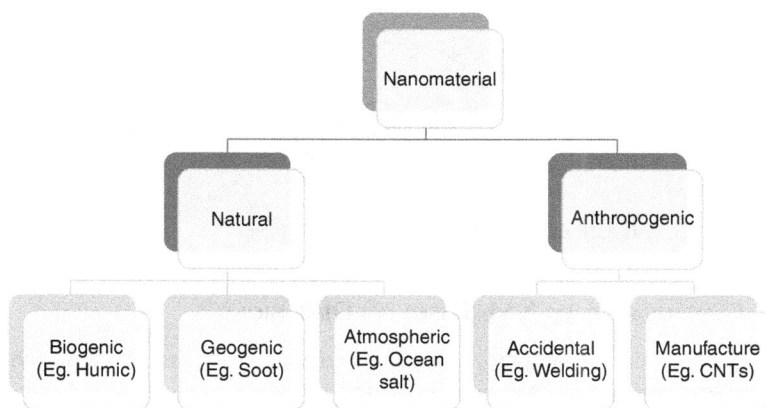

FIGURE 5.1   Classification of NMs based on origin.

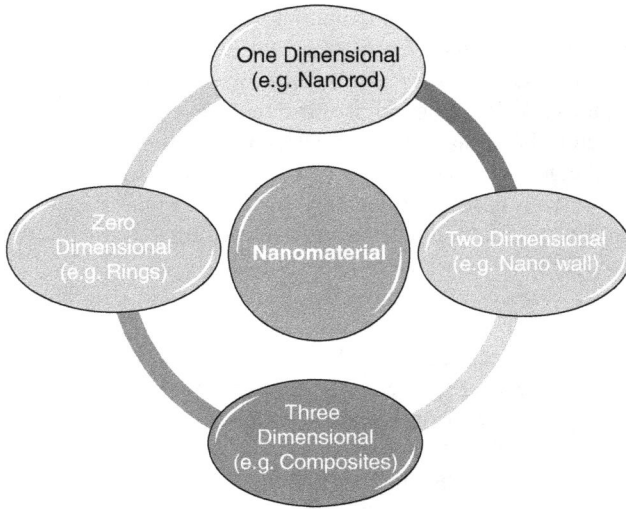

FIGURE 5.2 Classification of NMs based on dimension.

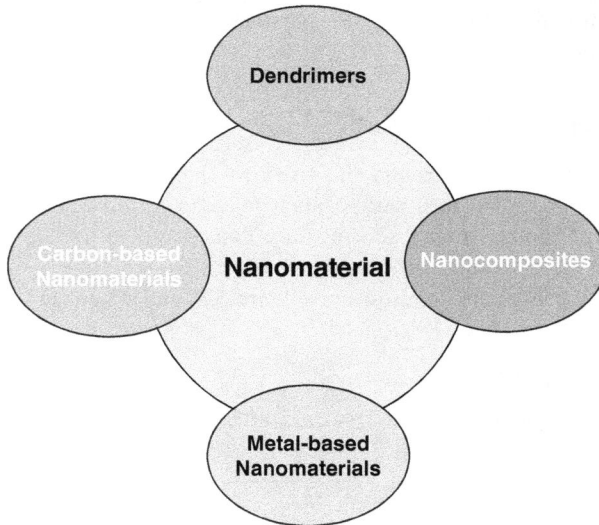

FIGURE 5.3 Classification based on chemical composition.

5.2.1 Classification of NMs Based on Dimensions

NMs' sizes can be affected by various parameters, such as pressure, synthesis method, time, temperature, pH and concentration.

They are categorized based on the foundation of dimensions, such as (1) zero-dimensional NMs having all the three dimensions less than 100 nm (e.g., rings,

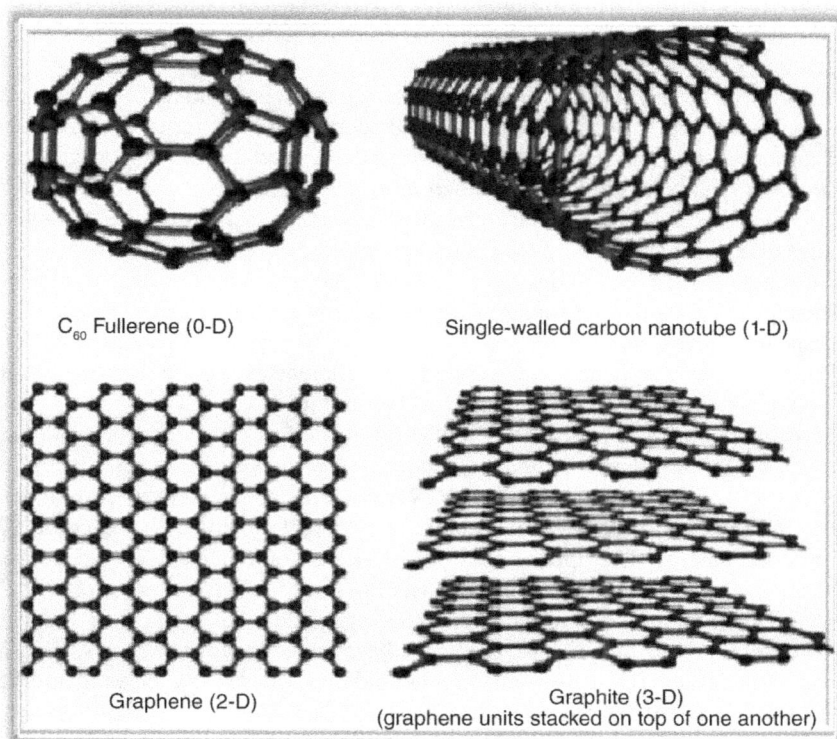

FIGURE 5.4 Different dimensions of carbon NMs adopted from Reference 13.

atomic clusters and quantum dots), (2) one-dimensional NMs having an individual dimension greater than 100 nm (e.g., CNTs, nanorods or wires), (3) two-dimensional having any two dimensions greater than 100 nm (e.g., nano walls and nanosheets). Those NMs with all three dimensions greater than 100 nm are known as three-dimensional materials (e.g., graphite and composites).

5.2.2 Classification of NMs Based on Chemical Composition

Classification of NMs based on chemical composition into further classes such as dendrimers, metal-based NMs and composites as well as carbon-based NMs [5].

The dendrimers are nanosized, radially symmetric, highly ordered and branched polymeric molecules. The synonymous terms used for dendrimers are *arborols* and *cascade*. They have effectively been described as analogous and are characterized by particles of uniform size in a dispersed phase. Dendrimers have a generally symmetric core, an external shell and an internal shell.

Metal-based NMs, such as gold and silver NPs, and other oxides, such as titanium oxides and platinum oxides, reveal a significant diversity of functioning characteristics, firmly determining their crystalline shape, morphological structure, constitution, intrinsic deficiency, doping and so on, which regulate their ocular, voltaic, catalytical and chemical properties.

A nanocomposite is a multiphase solid material, and it is found in nature (e.g., abalone shell and bone). The concept beyond nanocomposites is to use building blocks with dimensions in the nanoscale range to pattern, produce novel materials with extraordinary versatility and advance their morphological properties. Corresponding to their matrix substance, they are classified into three distinct classifications, for instance ceramic matrix, metal matrix and polymer matrix nanocomposites.

Carbon NMs are composed with carbon elements, and there is a minimum of one dimension in the nanoscale range of the latest nanosized materials, mostly GRA, fullerene (C_{60}) and CNTs.

The folding of graphene sheets around the fullerene gives rise to three distinct forms of CNTs, such as chiral, armchair and zigzag [6].

GRA is one member of the carbon NM family; it has come to light as a magic material, and it has also obtained tremendous prominence within only just a few years after its isolation from graphite. GRA contains sp^2 interbreed carbon atom planar sheets (two-dimensional), that is, firmly filled within honeycomb, for instance lattices [7]. The uppermost layer of GRA is about 2630 m^2 g^{-1}, is much higher than the fullerenes, graphite and CNTs. They are known as particular the finest substances, and they exhibit outstanding mechanical power. In the various literature, ample studies have mentioned graphene-based substances, like GO or graphenoids and rGO. Between graphenoids, GO is greatly investigated and described.

Fullerenes were discovered in 1985 and are an astonishing carbon allotrope. They are extremely uniform and confine sp^2 hybridization C atoms. Fullerenes are present in nature and interstellar space. Corresponding to the sum total of C atoms, fullerenes appear at different sizes, for example, C_{60}, C_{70}, C_{72}, C_{76}, C_{84} and C_{100}. Among all, the most superabundant and well-known example is C_{60} fullerene [6].

In the 21stcentury, it is identified as a pioneer of materials and the third carbon isomorphism, and its shape is analogous to a ball-shaped structure. As its tiny, globular and isotropous behavior, it is taken into consideration as an absolute zero-dimensional material.

CNTs circularly folded graphene layers covered with fullerene. The single-layer rolled sheets are sp^2 hybridization carbon atoms that are ordered in hexagons [6]. CNTs can be synthesized through various processes such as the chemical process of vapor deposition, laser ablation, the arc-discharge method and so on [8]. CNTs show captivating optical or electronic properties due to owing their novel one-dimensional nanostructures, which are, comparatively, differ from both other carbon NMs as well as NPs. In addition, CNTs have several characteristics such as small-scale configurations, suitable bio-affinities or surface operations, and they show higher reactivity. Depending on the graphitic ring's helicity and diameter, CNTs also display metallic or semiconducting behavior. The width and length depend on various synthesis routes, and many other factors so that they are not well defined in case of the CNTs. CNTs can persist stably at temperatures greater than 600° Celsius in air; these properties recommend that the CNTs have apparent, favorable features over graphite.

According to the folded graphene layers, CNTs can be further categorized into SWNTs, DWNTs and MWNTs.

5.3 APPLICATION OF NMS

5.3.1 NMs in Environmental Remediation

Upgrading the quality of soil, water and air is a vast obstacle in the present era. Recognition and remediation of environmental contaminants and their prevention are vital steps in the conservation of the environment.

SWCTNs and fullerene may be applied as a water purification membrane to rectify water quality by converting the wastewater into potable water (such as nano-sponge filtration) and adsorbents of polar and nonpolar organic chemicals and to remove other toxic contaminants from water.

Photocatalyst-based SWCNT-titanium dioxide (SWCNT-TiO_2) is significantly applied for purifying water from oil [9, 10]. CNTs can not only adsorb contaminants solitary, but they also integrate with further materials to fabricate complexes along with powerful adsorption capability [11, 12]. CNT-chitosan composite electrodes have a great electro-adsorption capability as well as a speedy electrosorption concentration, and their suitable mesoporosity enhances capacitative water desalination functions. Because of its high-quality electro adsorption performance, CNT-chitosan composite electrodes have been appropriately utilized in sanitization as well as treating sewage. CNTs refined by acid solution are productively used as

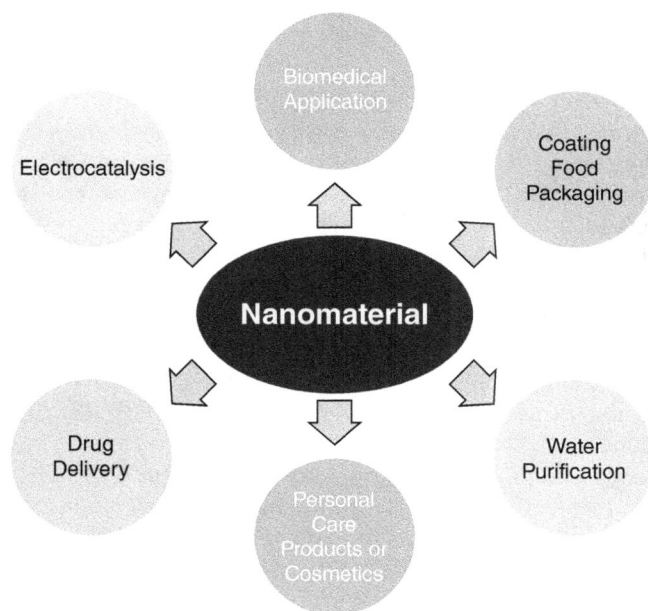

FIGURE 5.5   Various applications of NMs.

adsorption of carcinogenic tri-halomethanes from the water. CNTs have broad surface area highly used to manufacture composites to adsorb contaminants, which are eventually degenerated by electrochemical oxidation.

5.3.1.1 Adsorption of Numerous Dyes

Today, dyes play major roles in the paint industry and textile-and color pigment–fabricating industries. The effluents of organic dyes plus nitroarenes are acknowledged as a new hazardous substance due to the fact that the unacceptable dyes and nitro compounds both are harmful to humans, flora and fauna. *Embellished titania* on aerogel of SWCNTs and fabricated MWCNTs-TiO$_2$ are used for the elimination of MB from water [13].

MWCNTs are used for removing cationic dyes, namely, OG, EB, AO, EOB and others. MWCNTs are also used as adsorbents for eliminating AR18 from aqueous solutions [14]. It is effectively capable of eradicating both cationic and anionic dyes from aqueous solutions.

CNT-based composites such as CNT-chitosan, CNT-ACF, CNT-Fe$_3$O$_4$, CNT-dolomite, CNT-cellulose and CNT-GRA are widely used in dye adsorption [15–17]. Nanosized ZnO is used as nanocomposites together with chitosan (CS) for the dissociation of direct blue 78 plus acid black 26. Nanosized particles of ZnO have the capacity to lessen the unsafe dye MB by nearly 97% [18]. GRA-based nanosized adsorbent is one of the outstanding latest materials for the elimination of organic pollutants from the water. GO showed much better adsorption of MB and malachite green as ordinary organic dyes. It is applied for eradicating both cationic dyes, such as MB, RhB and CV, and anionic dyes, namely, DR23 and AO8, from aqueous solutions. The designed rGO-titanium dioxide, a photocatalyst, is effectively used for refractory dyes such as alizarin red S [13].

5.3.1.2 Adsorption of Various Heavy Metals

Several heavy metals such as lead, mercury, zinc, copper and others are immensely harmful and carcinogenic, and they may be a threat to human health as they perhaps biologically amassed in the food chain. CNTs, graphene and fullerene may adsorb numerous heavy metals from water [19]. NMs such as SWCNT-COOH, SWCNT-OH, SWCNT-NH$_2$, MWCNT-FeO$_4$, MWCNT-ZrO$_2$, MWCNT-MnO$_2$, MWCNT-Fe$_3$O$_4$ and others have heavy metal adsorption capacities, and they all are capable for the remediation of heavy metal from polluted water. SWCNTs and MWCNTs are also capable for adsorption of polycyclic aromatic HC, such as naphthalene and phenanthrene, plus pyrene. Oxidized MWCNTs have extreme adsorption capacity as well as removal capability for the Cr^{6+}, Pb^{2+} and Cd^{2+} ions from aqueous solutions [13].

GRA, widely used in photocatalysis, may decrease the heavy metals and may also be used in disinfecting wastewater as the absolute deterioration and mineralization of biological pollutants. The photocatalytic estimation of harmful heavy metal ions has a lower energy consumption, modest reactions and high-level proficiency conditions. It may remove chromium (VI) contamination from water. Moreover, GRA also exhibits excellent application possibilities in atmospheric remediation. It may ascertain the characteristics of an individual molecule or atom that can be applied to produce an extremely susceptible gas sensor. Superparamagnetic iron oxide NPs are effective sorbent materials for soft toxic arsenic, lead and cadmium materials.

5.3.1.3   Adsorption of the Surfactant

Following the surface activity characteristic, the surfactants considered as acute communal environmental problems due to surfactant could evenly disperse in the aqueous environment [20]. Various forms of graphene, namely, GO as well as rGO, have been utilized in the adsorption of nonionic surfactants such as TX-100 [21]. GO with phenyl tetraethyleneglycol is used to eliminate cationic and nonionic surfactants (DTAB & TX-100, respectively) from polluted water [22]. Through CNTs, aromatic cationic surfactants are also effortlessly eliminated.

Titanium oxide is the most popular NM that is utilized as a significant adsorbent for eliminating surfactants from polluted water. nZVI particles are powerful adsorbents for cationic surfactants together with anionic surfactants (such as hexadecyl pyridinium chloride and sodium dodecyl benzene sulfonate) from the dilute solutions [23].

5.3.1.4   Adsorption of Atmospheric Pollutants

GRA-based NMs can capture CO_2 and H_2 [13] from the atmosphere; it can also adsorb a few greenhouse gases along with other gases, namely, ammonia. After modification, GRA sheets have a profound adsorption capacity for gas, such as GRA altered by polyaniline, which has excessive adsorption of carbon dioxide. The C_{60} type of fullerene may react with additional materials during the adsorption of atmospheric contaminants, such as volatile organic compounds.

5.3.2   OTHER APPLICATIONS

Dendritic polymers have advantages in biomedical applications and anticancer drugs. Dendrimers are used in gene and drug delivery (e.g., in treating leishmaniasis).

Metal-based NMs are used for food packaging, electronics, textile engineering, biotechnology, sensors, imaging, cosmetics, environmental remediation, agricultural and biomedical applications.

Nanocomposites are used in finer film condensers for computer chips, food packaging and solidified synthetic polymer electrolyze for array and in auto-driven machine sections plus fuel tankers.

GRA is translucent and highly used in manufacturing light panels as well as touch screens. GRA-coated substances are considered excellent electrode commodities, used in power depository tools (supercapacitors).

GRA-coated electrodes enhance the functions of existing batteries, such as lithium-ion arrays, are supposedly convenient for evolving future-generation batteries such as sodium ion and lithium sulfur or lithium O_2 batteries. The salient characteristics of GRA have the guarantee to achieve innovation in the numerous sectors of catalysis, supercapacitors, field-effect transistors and solar cells, stretchable and flexible sensors, biosensors and membrane technology. Altogether, GRA is a tremendous material for advancing transparent and flexible devices.

Fullerenes are used in drug delivery and biomedicinal applications, electrochemical activities, the development of efficient solar cells and advanced lubricants, superconductors and so on [6].

Therefore, CNTs can be broadly utilized in the energy sector, biomedicine, electronics (rectifying diodes and field-effect transistors) and photoelectricity, analysis and catalysis.

5.4   EFFECT OF NMS ON THE ENVIRONMENT

Presently, concern regarding the lethal effects of NMs lies first and foremost in the fabrication and commercialization of such a wide range of utilizations of NMs. Consequently, the threat of increased production and usefulness of NM compounds will lead to an increase in different environmental exposure and becoming bioavailable at a very large scale. The hydrological ecosystem is the major pathway of exposure to NMs, as this kind of ecosystem is generally the last stop for nanocomposites. As routes of exposure, following the hydrological ecosystem, the atmosphere (troposphere), soil and sediment act in accordance with the sequence of priority. Certain organic and metallic NMs perhaps modified under anaerobic condition, likely in aquatic (benthic) sediments, may respond to natural organic substances and colloidal suspensions, suspended PMs, developing accumulation and probably sedimentation from solution. Sedimentation,

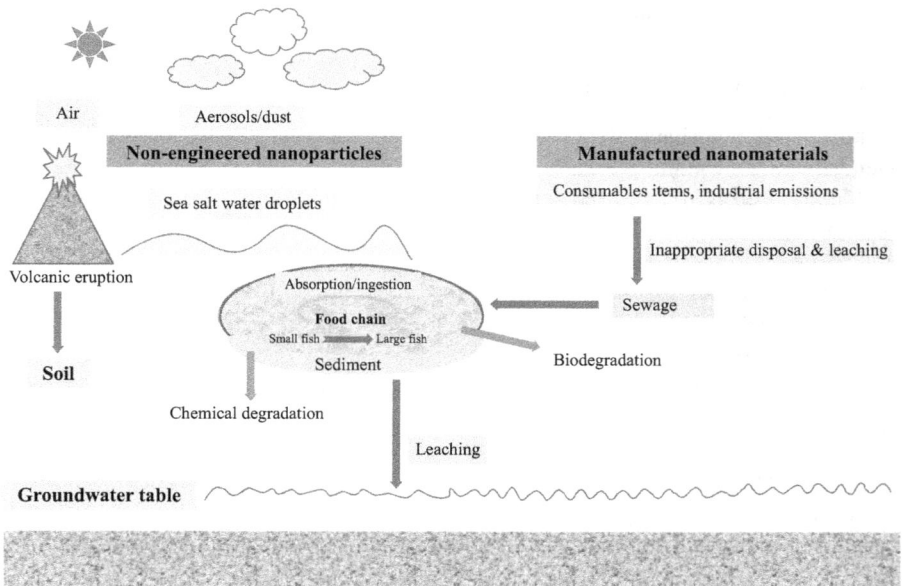

FIGURE 5.6   Major sources, way of exposure, and potential interactivity between NPs with the environment and organisms modified from Reference 26.

together with accumulation, may indicate a channel for the transmission of NPs from the water column to benthic sediments.

If nano-based materials are present in hydrological environments, they may be absorbed by cells, mainly through filtration by aquatic organisms, which may result in direct interference with their physiology, especially their ability to breathe and feed.

The NPs of ZnO, silver oxide (AgO) and copper oxide (CuO) are capable of generating lethal effects to algae bioindicators in a homogeneous route to particular solubilized metallic compounds. It was found that the metals assembled at different sites within the cells and that NPs mainly remained in the cell wall, while metals appeared predominantly in organelles, such as the endoplasmic reticulum.

NMs are released into the environment either intentionally or unintentionally. Intentional engineered NMs are disseminated either from a point source, for example, manufacturing with nanotechnology capabilities, landfills and wastewater treatment plants, or some from a nonpoint source, through wear from materials that contain NMs. During manufacture and transportation, NMs may become accidentally free through seepage from improper sealing or perhaps purposely released, as in the use of zero-valent iron NPs to remediate contaminated soil. Additionally, NMs may be getting into the environment either as individual fine particles or as embedded in a matrix, from where dissemination of nanosized materials will take place via the deterioration of matrix materials. Today, the major cause of unintentional NM release is the partial oxidation of manmade compounds through human actions, the burning of wood and fuel in cars with defective catalytic converters.

5.4.1 FATE AND BEHAVIOR IN AIR

Atmospheric nano-based material has three chief emission sources: (1) primary emission, or the first type of emission, which refers to exhaust gases emitted instantly from roadside traffic and industrial incineration; (2) secondary emission, or the second type, which is produced in the atmosphere during the condensation of small volatile vapors formed from the oxidation of atmospheric gases; and (3) the third type is created during diesel exhaust dilution. Fine and ultrafine NMs may go through various routes in the atmosphere. Few NMs can either be expanded by compression of less volatile compounds or reduced in size by evaporation of adsorbed water and further volatiles, resulting in the dissimilarity in particle size dispersion but not the altogether concentration number [24]. Additionally, atmospheric NMs can accumulate, giving rise to enlarged element size while reducing the concentration quantity.

5.4.2 FATE AND BEHAVIOR IN WATER

NMs are introduced into the hydrological environment by various sources like as direct discharge of polluted water having NMs into the surface water, runoff or atmospheric deposition in soil and sediments. Moreover, colloidal transmit in porous media, recommend to facilitate the portability of ultrafine particles in soils based on several factors, such as (1) the physical or chemical characteristics of NPs that is configuration, range, outer coatings and steadiness;(2) the traits of soil or environment, is clay, sand, colloids, natural organic matter, irrigate chemistry or flow rates; and (3) interactivity of NPs with natural colloidal substance to facilitate surface coatings, accumulation or disintegration and adsorption to bigger particles.

According to Laux et al. [25], the introduction of nano-based materials into nature takes place by the discharge of their constituents at the time of use and in due course of their ultimate discard. So it is significant to chase and find out the kinetics and modification of the NMs in the surrounding world.

5.5 NMS' IMPACT ON HUMAN HEALTH

The potential for negative health consequences from exposure to engineered NMs may results from either unintentional production of the substances through fabricating, applying, discarding or recovering throughout environmental functions or through the intentional induction of these materials for personal care products, health applications or additional objectives. Basically, there are various pathways by which NMs come into contact with humans and incorporate into their bodies. According to Paschoalino et al. [26], additional pathways for the entry of NMs into receptor organisms take place via a respiratory tract (inhalation), skin (dermal) and the digestive system (ingestion).

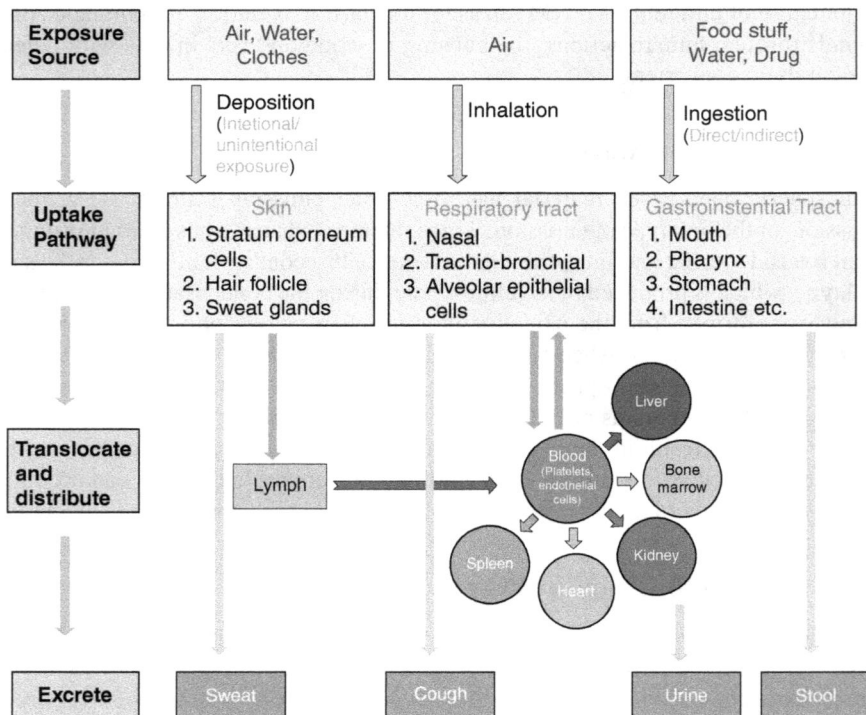

FIGURE 5.7 Impacts of NMs on human health.

5.5.1 EXPOSURE THROUGH INHALATION

As breathable air carries a distinct diversity of substances, namely bacteria, dust, or further pollutants, the lung has a specific clearance system to control such types of pollutants. Fine elements accumulate in the distinct segments of the lung wherever particles are swallowed up by macrophages in the alveolar section, further passing up to the bronchioles altogether with bronchi. Wide particles are straightly conveyed upward through the mucociliary clearance, also known as the mucociliary escalator. Subsequently, this mucus, carrying the alien substances, is eliminated from the lung by coughing, swallowing or spitting.

Furthermore, ultrafine particles are classified in the same range as tiny PM; they must be capable of passing through the air-blood barrier. The greater number of utilized dosages will be recognized by macrophages as a step out of the lungs. A small amount of the entirely inhaled dose passes with blood flow and ends up in secondary organelles, for example, the liver, heart, kidneys, bone marrow, spleen or heart. These finer NPs may cause respiratory, inflammatory and cardiovascular diseases.

5.5.2 EXPOSURE THROUGH SKIN

The exposure of human skin to NMs may take place either intentionally, like the application of cosmetics or other personal care products (creams, sunscreens, lotions) carrying silica, titanium oxides and ZnO-coated NPs, or unintentionally, via the clearly produced NPs during combustion, welding fume emissions, waxing of the skin, emissions from coal- and oil-fired power plants, natural gas manufacturing, transportation or the disposal of used NM-based products.

5.5.3 EXPOSURE THROUGH INGESTION

NM exposure may exist by way of direct or indirect ingestion of contaminated water, edibles, cosmetics and drugs, including food and others, from which NPs enter lymphatic cell tissues.

The absorption of NPs in the gastrointestinal pathways is influenced by a number of factors, including particle size, geometry, leg type and the surface charge of NPs, among others. NPs cleared from the respiratory tract through coughing and swallowing can now enter the intestinal tract for that reason. It is a vital target for NP exposure. Through an oxidative stress mechanism, the ingested titanium dioxide NMs can damage the cell membrane of the digestive gland.

5.6 NM DETECTION IN HUMAN BODY

Certain NMs are neither hazardous nor safe because certain NMs are in tune with bulk density and certain NPs similarly to smaller configuration and size, reactivity or retention time, along with spreading after conquering the biological barrier, subcellular as well as molecular interactions in the body may take part in a vast function to dictate the impact of toxicity. For toxicity evaluation, NM characterization is a beneficiary component [27–29]. The three key screening strategies for the toxicity potential of nano-based materials and ultrafine particles for humans and the environment are physicochemical characteristics, in vitro (cellular or noncellular) assays and in vivo assays.

5.6.1 Physicochemical Characterization

The extraordinary physicochemical characteristics of NMs are determined by their shape and size, particle distribution and high dispersion, crystalline conformation, chemical constitution, porosity and surface charge, which, along with their chemistry, may affect their toxicity and make it difficult to detect their destination and impact on living beings. NMs are found in various shapes and sizes, such as needles, spherical, hexagonal, polyhedral and tube-like structures. However, the configuration of NMs could have a potency on the kinetics of accumulation in addition to subsuming them in the body. Chemical constitution is one more principal variable that encompasses almost all material groups, namely, metals or metal-based oxides, among others.

The NPs' distribution and dispersion depend on the medium utilized to suspend them, perhaps acting on their biological function. In an aqueous medium, the toxicity screening of NMs requires stable scattering carrying precise particle configuration as well as particle distribution. Undergoing ambient surroundings, a few NPs may produce clustering and form dendrimers into chains or spherical structures. Surface modification can prevent agglomeration and fall into a distinct category with the concentration of surfactants through the coating of NPs. This prevention also switches their body dissemination, particularly in biological systems.

5.6.2 In Vitro Assays

In vitro testing methods can be carried out in a short time, it is cost-efficient and it gives an extremely thorough output of toxicologic testing and depiction [30]. It is a simpler assay and was first used for to evaluate cytotoxicity. In vitro assays comprise portal-entry toxicity or target-organ toxicity. The in vitro assay process is suitable for detecting NPs in a few affected human organs, as listed in Table 5.3.

These tests recommend a route of potential entry for cellular assessment through the routes discussed in the following subsections.

TABLE 5.3

List of Affected Organs Available for In Vitro Assays Modified from Reference 31

Affected Organ	Target Area	Endpoint	Probable Impact
Lung	Alveolar epithelial cell, endothelium cells, mucus or epithelium lining fluid.	Oxidative stress, genotoxicity, adhesive molecules	Toxicity, inflammation, carcinogenesis, cytotoxicity inflammation.

Affected Organ	Target Area	Endpoint	Probable Impact
Skin	Avascular epidermis	Cell viability, glucose utilization etc.	Cytotoxicity, inflammation, denature the protein or disclose the epitopes and alter the gene expression.
Liver	Hepatocytes, Kupffer cells	Signal transduction pathway, cytokine profile, oxidative stress	Particle interactions toxicity, inflammation.
Heart	Heart muscle	Oxidative stress, apoptosis, signal transduction pathway	Alter the cardiac function.
Kidney	Epithelial tubules, vasculature, renal tissue	LDH apoptosis, oxidative stress etc.	Function toxicity, inflammation.
Spleen	Lymphoid cells	Marker of lymphoid cell differentiation, functional and immune, responsivity	Consequence of immune response and immunopathology
Central and peripheral nervous system	Neuron cells	Apoptosis, metabolic level and efficacy on the nervous impulse or an ion balance	Toxicity, inflammation
Mucosa	Intestinal epithelium, oral cavity, nasal cavity, vagina	Cytokine profile, oxidative stress, signal transduction pathway	Cytotoxicity, inflammation
Blood	Platelets, megakaryocytes	Platelet activation, cytokine or chemokine release from leukocytes	Inflammation, immune response

5.6.2.1 Lung

It identifies a potential target for some aerial ultrafine particles together with in vitro models. The finer, airborne particles are deposited on the alveolar epithelial cells and encounter mucus or epithelium lining fluid. In continuity, they interface with macrophages and then go into the interstitium, where they assemble with fibroblasts along with endothelium cells or interface with the immune system. The endpoint concludes oxidative stress, genotoxicity, adhesive molecules and cell toxicity, inflammation, carcinogenesis, cytotoxicity and inflammation in the body.

5.6.2.2 Skin

Skin is the primary route of potential exposure to nanotoxicants. NPs could be transferred via a stratum sheet to be well sequestered within the epidermis to enhance the exposure time of NPs to feasible epidermal corneocytes. A variety of endpoints conclude cell viability, glucose utilization and others. In vitro tests are the best for conducting toxicity testing because they provide preliminary data as well as a relevant assessment of any risk.

5.6.2.3 Macrophages

Through numerous in vitro assays, it is possible to determine the effects of NPs on macrophages. During the release of LDH, cellular cytotoxicity may be estimated via a conventional method. When phagocytes, a lot of pathogenic particles, induce the liberation of cytokines such as TNF-alpha (interleukin-6) and nuclear transcription factors (nuclear factor kappa B), activation of macrophages along with an oxidative burst could be investigated [32]. In response to particles, nitric oxide and superoxide peroxynitrite, which is a tremendously harmful species, are also generated [33]. However, since the cytoskeleton is a central key to cell functions, it could be easily investigated when targeted by NPs.

5.6.2.4 Epithelium

It is the primary obstacle that confronts particles. In In Vitro studies, bronchial and alveolar epithelial cells are considered target cells. The NPs' translocation across epithelial cells may be a principal indication of toxicity. Moreover, for detecting NPs, the endpoints include toxicity measurements (e.g., LDH release and initiation of inflammation regarding transcription factors), oxidative and nitrosative stress, which are for cell destruction induced by pathogenic finer particles [34, 35]. Both can be measured during the regulation of superoxide mutase and glutathione peroxidase [36]. If we take cancer as an endpoint, the direct measurement of gene toxicity could be measured by the COMET assay (single-cell gel electrophoresis) and 8-hydroxydeoxyguanosine estimation.

5.6.2.5 Endothelial Cells

An ultrafine Perhaps has a huge effect on the endothelium cells. Cultured endothelium cells are completely applicable for the determination of NPs. Fibroblasts are liable to be affected by any particles in the NP–fibroblast interactions consist of two methods of reaction that can be initiated, such as pro-inflammatory effects or fibrogenic responses. If finer particles interconnect along with lymphocytes or modulating dendritic cells via an interstitial space, the immunopathological effects can be investigated. In in vitro assays, the endpoints and appropriate tests can be structured when macrophages or dendritic cells present NP effects via antigen presentation. In vivo, co-cultured epithelial cells and macrophages, or epithelial cells and endothelial cells, demonstrate an in vivo situation. The same configurations as in vivo are available in vitro, such as whole lung tissue slices (multiple pulmonary cell types) for methodology and culturing.

5.6.2.6 Mucosa

Mucosa lines certain organs and cavities all over the body. It is a moist tissue. It includes the nasal cavity, lungs, vagina and gastrointestinal tract or oral cavity. It is a prime way for the entrance of NPs. Intestinal epithelium cells may be examined using various procedures, such as immortalized cell lines as well as tissue constructs. Smith et al. [37] and Urayama et al. [38] used intestinal epithelial lining cells to evaluate the stimulation of several signal tracks, ROS/RNS release and cytokine.

5.6.2.7 Blood

In vitro tests using blood fractionalized by-products are used to evaluate the effect on blood circulation. The endpoints for evaluation of NPs are the initiation of platelets, the interactivity of red blood cells and leukocyte release of ROS/RNS, cytokine or chemokine [39, 40].

5.6.2.8 Spleen

The spleen is the wider site of lymphoid development as well as immune processing. The accumulation of NPs in the spleen could have major consequences for the immune response along with immune pathology. In an in vitro assay, the spleen can be isolated and studied for the effect of NPs on it. The exposure endpoints could conclude markers of lymphoid cell differentiation, functional and immune responsivity aspects.

5.6.2.9 Liver

Due to NMs, liver injury is characterized on the basis of histologic lesions (inflammation and necrosis) or at the molecular level. Endpoints include the signal transduction pathway, cytokine profile, oxidative stress and the potential impact of NMs on particle interactions, toxicity and inflammation. The most common mechanism is through the cytochrome P450 metabolic pathways. The isolated perfused liver is a complicated model system, and precision slicing of the liver is another modern model system; both are preferable for the investigation of fine nano-based materials.

5.6.2.10 Nervous System

5.6.2.10.1 Central Nervous System

It entails investigating the consequences of NPs on neurons and their functions. Apoptosis, metabolic level, and efficacy on nervous impulse or ion balance are the endpoints. There are various cells that could be studied for the effects of NPs, such as microglial cells, glial cells and astroctyes. For the peripheral nervous system, both sympathetic and parasympathetic neurons can be cultured and used to study the effects of NPs on the neuronal metabolism mechanism, electrical activity, viability and ionic homeostasis.

5.6.2.10.2 Heart

NMs can enter the heart muscle via microcirculation and alter cardiac function. The endpoint is oxidative stress, apoptosis and the signal transduction pathway. Cardiomyocytes can be cultured, and it is possible to examine the mechanism effects.

5.6.2.10.3 Kidney

The adverse effects of NPs could be assessed through in vitro assays. Permeability assays are used in the measurement of transmigration and perforation via the nephron tubules. In the existing cells, culture techniques NPs' effects can be evaluated on epithelial tubules and vasculature. There are various models available that are useful for endpoint investigation of NP–kidney interactions such as renal tissue slices used to evaluation of oxidation stress, translocation, toxicity and signal transduction responses [41, 42]. A low-cost substitute to kidney slice replica, immobilized cell lines perhaps facilitate mechanical statistics regarding the cellular toxicity effects of NMs.

5.6.2.11 Noncellular Assays

5.6.2.11.1 Production of Free Radicals
Most probably, in a cell-free system, all pathogens have the capacity for the generation of particles' free radicals. And this capability can cause oxidative stress, resulting in the contribution to the initiation of inflammation, cell injury and genotoxicity [35, 43, 44]. It can be assayed in vitro through plasmid DNA scission, electron paramagnetic spin resonance and others.

5.6.2.11.2 Adsorptive Properties
Different types of proteins require different NPs because NPs have a large surface area to adsorb proteins [45, 46], and this might alter how they can be handled by other cells or macrophages; it will become a central part of the study.

5.6.3 In Vivo Assays

In vivo assays included pulmonary exposures (deposition, translocation and biopersistence studies, various alternative animal models) injection, dermal and oral exposure.

5.6.3.1 Pulmonary Exposure
NPs can enter the respiratory tract through distinct pathways to further organs or tissues. These reactions can be started via NPs interacting with subcellular structures (endocytosis) through discrete target cells. It is necessary to be stated special attention to acknowledging such effects because vigorous life forms are very fine initially, but after sometime, severe effects are reflected in affected organs or organisms, such as the effects of ultrafine particles in asthmatic patients.

5.6.3.2 Deposition, Translocation and Biopersistence Studies
Through the inhalation process, exposure to airborne NPs leads to the deposition of numerous chambers of the respiratory tract. Corresponding to the probability that the airborne NPs may accumulate in the various compartment of the respiratory tract through breathing depends on three categories: aerodynamic or thermodynamic characteristics of nanoparticles, inhalation pattern or the three-dimensional configurations, as well as the anatomy of the respiratory tract [47].

Once NPs are deposited, the insoluble part of the NPs undergoes a clearance mechanism specific to the domain of the respiratory airway. Above the epithelium cells of the conducting air tract, mucociliary clearance gives quick transportation to the larynx for forwarding carry into the gastrointestinal tract and excretion. In the alveolar epithelium area, there is no speedy transportation occurring; in such an order, phagocytic activity may take place following a moderate clearance to the larynx. Due to their finite capacity, they become prominent along with paracellular transfer mechanism over compact coupling, undergoing inflamed circumstances; these mechanisms result in NPs' transmigration toward the interstitium and lymphatic drainage; all together, feasible transmigration could take place in endothelial cells into capillaries [47, 48]. Both direct effects on secondary organs through translocated NPs and indirect effects may also occur at respiratory tract retention sites along with adjoining biological systems such as cell, fluid, polypeptides and extracellular matrices.

Because of the high diversity of NPs and the underlying mechanism of translocation, accumulation or mediator-response studies of biopersistence in the secondary target organs and the NP's kinetics should be a high priority.

5.6.3.3 Animal Models

Due to enhancing susceptibility, the normal animal models do not apply because of the absence of NP effects. And that vulnerability may be from various elements such as age, modified organ operation, disease or genetic diversification and so on. Adaptive animal models consist of exposure of senescent, knockout or transgenic animals and animals exposed along with endangered organ complexes such as hypertension, diabetic models, virulent models and others. It is most important that the models are using should be relevant to the human disease states.

5.6.3.4 Oral Exposure

It is possible that during their life cycle, NMs may arrive in water reservoirs or be inadequately consumed. Single gavages at the dose of NP exposures represent the worst case of human vulnerability. For an oral divulgence assessment of morphological and chemical characteristics of the trial, materials should be should be screened regarding their transport in trial animals. The most common animals, rats or mice, are recommended for the animal model system, with no preference of gender for such studies; after four days of post-exposure, the feces are collected. The number of NMs are abolished versus those preserved must be examined. Moreover, GALT, mesenteric lymph nodes and the liver are used when inspecting for nanoscale particles. Whether subsuming is zero, then no need for evaluation of the systemic effects of oral exposure. However, if the absorption of NMs is evident, histology and functional assays can evaluate the assessment of systemic toxicity.

5.6.3.5 Injection

Certain nano-based materials are used for drug delivery in such instances, after the injection potential toxic effects of those NMs could be analyzed. Both rats and mice are suggested without gender preference. Whether viable labeled NPs must be introduced, its dissemination to a number of organs (such as liver, spleen, respiratory system, kidney, etc.) and eradication in the excreta must be observed for a period of time (week) and post-exposure assessment can be done with histology and functional assays.

5.6.3.6 Dermal Exposure

In addition to rats, pigs (often guinea pigs) and rabbits are employed to evaluate toxic effects and irritation. Rats are more common as they are small entities and most probably have existing datasets in the particular discipline of toxic kinetics. However, domestic pigs are utilized due to the skin is physical, biochemical, or anatomical like humans' skin.

After the exposure to NMs, the skin biopsies were taken for TEM and light microscopy to identify cellular changes and morphological alterations, respectively. There are various tests that can be performed for evaluating the toxic effects of NMs, such as analytic chemistry, hematology, and assessment of regional lymph glands' immune toxicity batteries [49, 50].

5.7 ADVANCED TECHNOLOGY

5.7.1 Computational Toxicology

Computational toxicology is a computer model that utilizes statistical as well as molecular biology applications along with a combination of various new screening technologies. It improves the prioritization of required data and attempts to evaluate the integrated hazards along with intentionally manufactured NMs for the safety of nature. This computational toxicology consists of four areas such as computational chemistry (molecular modeling, force fields, quantum chemistry etc.), molecular biology (genomics, proteomics, and metabonomics), computational biology or bioinformatics and system biology. It is formulated to enhance the capability to prioritize, screen and evaluate materials by increasing the potential of toxicity predictions.

5.8 CONCLUSION AND RECOMMENDATIONS

Nanoscale assembling materials are economically accessible and widely applied in the modern world as cleaning products, personal care or cosmetics, drugs, equipment and more. There are many reasons that anthropogenic nanoscale particles intentionally or unintentionally enter in the environment (air, water, or soil) or human body (inhalation, ingestion, penetration via skin), and they may alter the dispersion or toxicity of the environmental system and human health. Yet still no well-founded interpretation may presently be drawn out regarding the exposure together with release from products of NPs because there is basic insufficiency in knowledge regarding the refining and preparation of NM-based products. Scientists still have to know how they can correctly forecast chronic responses in animals and humans. Therefore, to attain well-founded answers regarding the healthiness of nano-based materials, some recommendations are as follows: (1) Standardized assays should be used to compare the properties and bioactivity of the NPs; (2) criteria should be established for safe and responsible uses of NMs; (3) standard assay test methods for biologically relevant nanoscale analysis should be developed; (4) the most relevant standard substance for in vitro assay methods should be formulated and specified; (5) reference materials as a source should be available to the researchers; and (6) validated predictive models are needed to develop guidelines for risk management. No doubt about the fact that all the suggestions are in need of great research, either using new, along with developed, standardized assay test methods or utilizing an existing one in the area to differentiate the properties and bioactivities of new NPs with benchmark particles.

REFERENCES

1. Bleeker, E.A., de Jong, W.H., Geertsma, R.E., Groenewold, M., Heugens, E.H., Koers-Jacquemijns, M., van de Meent, D., Popma, J.R., Rietveld, A.G., Wijnhoven, S.W., Cassee, F.R., &Oomen, A.G. (2013). Considerations on the EU definition of nanomaterials: Science to support policy making. *Regulatory Toxicology Pharmacology: RTP*, *65*(1), 119–125.
2. International Council of Chemical Associations (2010). *Core elements of a regulatory definitions of manufactured nanomaterials*. Available at: www.icca-chem.org/ICCADocs/Oct-2010_ICCA-Core-Elements-of-a-Regulatory-Definition-of-Manufactured-Nanomaterials.pdf?epslanguage=en. Accessed 10 January 2014.
3. International Organization for Standardization (2008). *Technical specification: Nanotechnologies-terminology and definitions for nano-objects-nanoparticle, nanofibre and nanoplate*. ISO/TS 80004-2:2008.Available at: https://www.iso.org/standard/44278.html.
4. International Organization for Standardization (2010). *Nanotechnologies-vocabulary-part1: Core terms*. ISO/TS 80004-1:2010.Available at: https://www.iso.org/obp/ui/#iso:std:iso:ts:80004:-1:ed-1:en.
5. EPA 100/B-07/001 (February 2007). Available at: https://archive.epa.gov/osa/pdfs/web/pdf/epa-nanotechnology-whitepaper-0207.pdf. Accessed 14 July 2020.
6. Baig, N., Kammakakam, I., & Falath, W.S. (2021). Nanomaterials: A review of synthesis properties, recent progress, and challenges. *Materials Advances*, *2*(6).
7. Geim, A.K., & Novoselov, K.S. (2007). The rise of graphene. *Nature Materials*, *6*, 183–191.
8. Duc Vu Quyen, N., Quang Khieu, D., Tuyen, T.N., Xuan Tin, D., & Thi Hoang Diem, B. (2019). Carbon nanotubes: Synthesis via chemical vapour deposition without hydrogen, surface modification, and application. *Journal of Chemistry*, *10*, 1–14.
9. Gao, S.J., Shi, Z., Zhang, W.B., Zhang, F., & Jin, J. (2014). Photoinduced superwetting single-walled carbon nanotube/TiO2 ultrathin network films for ultrafast separation of oil-in-water emulsions. *ACS Nano*, *8*(6), 6344–6352.
10. Park, H.-A., Liu, S., Salvador, P.A., Rohrer, G.S., & Islam, M.F. (2016). High visible-light photochemical activity of titania decorated on single-wall carbon nanotubes aerogels. *RSC Advances*, *6*, 22285–22294.
11. Radjenovic, J., & Sedlak, D.L. (2015). Challenges and opportunities for electrochemical processes as next-generation technologies for the treatment of contaminated water. *Environmental Science & Technology*, *49*(19), 11292–11302.
12. Rahaman, M.S., Vecitis, C.D., & Elimelech, M. (2012). Electrochemical carbon-nanotube filter performance toward virus removal and inactivation in the presence of natural organic matter. *Environmental Science & Technology*, *46*, 1556–1564.
13. Baby, R., Saifullah, B., & Hussein, M.Z. (2019). Carbon nanomaterials for the treatment of heavy metal-contaminated water and environmental remediation. *Nanoscale Research Letters*, *14*(1), 341.
14. Shirmardi, M., Mesdaghinia, A., Mahvi, A.H., Nasseri, S., & Nabizadeh, R. (2012). Kinetics and equilibrium studies on adsorption of acid red 18 (Azo-Dye) using multiwall carbon nanotubes (MWCNTs) from aqueous solution. *E-Journal of Chemistry*, *9*(4), 2371–2383.
15. Gupta, V.K., Kumar, R., Nayak, A., Saleh, T.A., & Barakat, M.A. (2013). Adsorptive removal of dyes from aqueous solution onto carbon nanotubes: A review. *Advances in Colloid and Interface Science*, *193*, 24–34.
16. Ai, L., & Jiang, J. (2012). Removal of methylene blue from aqueous solution with self-assembled cylindrical graphene–carbon nanotubes. *Chemical Engineering Journal*, *192*, 156–163.

17. Rajabi, M., Mahanpoor, K., & Moradi, O. (2017). Removal of dye molecules from aqueous solution by carbon nanotubes and carbon nanotube functional groups: Critical review. *RSC Advances*, *7*(74), 47083–47090.
18. Li, J.F., Rupa, E.J., Hurh, J., Huo, Y., Chen, L., Han, Y., & Yang, D.C. (2019). Cordycepsmilitaris fungus mediated zinc oxide nanoparticles for the photocatalytic degradation of methylene blue dye. *Optik*, *183*, 691–697.
19. Srivastava, S., Mehndiratta, P., Jain, A., & Gupta, N. (2013). Environmental pollution and nanotechnology. *Environment and Pollution*, *2*(2), 49.
20. Ivanković, T., & Hrenović, J. (2010). Surfactants in the environment. *Arhivzahigijenur adaitoksikologiju*, *61*(1), 95–109.
21. Prediger, P., Cheminski, T., de Figueiredo Neves, T., Nunes, W.B., Sabino, L., Picone, C.S.F., & Correia, C.R.D. (2018). Graphene oxide nanomaterials for the removal of non-ionic surfactant from water. *Journal of Environmental Chemical Engineering*, *6*, 1536–1545.
22. Cheminski, T., de Figueiredo Neves, T., Silva, P.M., Guimaraes, C.H., & Prediger, P. (2019). Insertion of phenyl ethyleneglycol units on graphene oxide as stabilizers and its application for surfactant removal. *Journal of Environmental Chemical Engineering*, *7*(2), 102976.
23. Abd El-Lateef, H.M., Ali, M.M.K., & Saleh, M.M. (2018). Adsorption and removal of cationic and anionic surfactants using zero-valent iron nanoparticles. *Journal of Molecular Liquids*, *268*, 497–505.
24. Zhang, Q., Worsnop, D.R., Canagaratna, M.R., & Jimenez, J.L. (2005). Hydrocarbon-like and oxygenated organic aerosols in Pittsburgh: Insights into sources and processes of organic aerosols. *Atmospheric Chemistry and Physics*, *5*(12), 3289–3311.
25. Peter, L., Christian, R., Andy, B.M., Joseph, B. D., Josephine, B., Cristina, C., Otto, C., Irina, L.E., Thomas, G., Gunnar, J., Harald, J., Heiko, K., Jutta, T., Ahmed, T., Andreas, S., Adrienne, S.J.A.M., Robert, Y.A., & Andreas, L. (2017). Biokinetics of nanomaterials: The role of biopersistence. *Science Direct*, *6*, 69–80.
26. Paschoalino, M. P., Marcone, G. P., & Jardim, W. F. (2010). Osnanomateriais e a questãoambiental. *Química Nova*, *33*(2), 421–430.
27. Oberdörster, G., Oberdörster, E., & Oberdörster, J. (2005). Nanotoxicology: An emerging discipline evolving from studies of ultrafine particles. *Environmental Health Perspectives*, *113*(7), 823–839.
28. Murdock, R.C., Braydich-Stolle, L., & Schrand, A.M. (2008). Characterization of nanomaterial dispersion in solution prior to in vitro exposure using dynamic light scattering technique. *Toxicological Sciences*, *101*(2), 239–253.
29. Warheit, D.B., & Donner, E.M. (2010). Rationale of genotoxicity testing of nanomaterials: Regulatory requirements and appropriateness of available OECD test guidelines. *Nanotoxicilogy*, *4*, 409–413.
30. Fadeel, B., Fornara, A., Toprak, M.S., & Bhattacharya, K. (2015). Keeping it real: The importance of material characterization in nanotoxicology. *Biochemical and Biophysical Research Communications*, *468*(3), 498–503.
31. Maynard, A., Donaldson, K., Castranova, V., Oberdorster, G., Julie, F., Kevin, A., Janet, M.C., Karn, B., Kreyling, W.G., Lai, D., Olin, S., Monteiro-Riviere, N.A., Warheit, D., & Yang, H. (2005). Principles for characterization the potential human health effects from exposure to nanomaterials: Elements of screening strategy. *Particle and Fibre Toxicology*, *2*(1), 8.
32. Brown, D.M., Roberts, N.K., & Donaldson, K. (1998). Effect of coating with lung lining fluid on the ability of fibres to produce a respiratory burst in rat alveolar macrophages. *Toxicology in Vitro*, 15–24.
33. Chao, C.C., Park, S.H., & Aust, A.E. (1996). Participation of nitric oxide and iron in the oxidation of DNA in asbestos-treated human lung epithelial cells. *Archives of Biochemistry and Biophysics*, *326*(1), 152–157.

34. Fubini, B., Aust, A.E., Bolton, R.E., Borm, P.J., Bruch, J., Ciapetti, G., & Muhle, H. (1998). Non-animal tests for evaluating the toxicity of solid xenobiotics: The report and recommendations of ECVAM workshop 30. *Alternatives to Laboratory Animals, 26*(5), 579–615.

35. Gilmour, P.S., Rahman, I., Donaldson, K., & MacNee, W. (2003). Histone acetylation regulates epithelial IL-8 release mediated by oxidative stress from environmental particles. *American Journal of Physiology-Lung Cellular and Molecular Physiology, 284*(3), L533–L540.

36. Janssen, Y.M., Van Houten, B., Borm, P.J., & Mossman, B.T. (1993). Cell and tissue responses to oxidative damage. *Laboratory Investigation; a Journal of Technical Methods and Pathology, 69*(3), 261–274.

37. Smith, J.M., Johanesen, P.A., Wendt, M.K., Binion, D.G., & Dwinell, M.B. (2005). CXCL12 activation of CXCR4 regulates mucosal host defense through stimulation of epithelial cell migration and promotion of intestinal barrier integrity. *American Journal of Physiology-Gastrointestinal and Liver Physiology, 288*(2), G316–G326.

38. Urayama, S., Musch, M.W., Retsky, J., Madonna, M.B., Straus, D., & Chang, E.B. (1998). Dexamethasone protection of rat intestinal epithelial cells against oxidant injury is mediated by induction of heat shock protein 72. *The Journal of Clinical Investigation, 102*(10), 1860–1865.

39. Nemmar, A., Hoylaerts, M.F., Hoet, P.H., Dinsdale, D., Smith, T., Xu, H., & Nemery, B. (2002). Ultrafine particles affect experimental thrombosis in an in vivo hamster model. *American Journal of Respiratory and Critical Care Medicine, 166*(7), 998–1004.

40. Oberdörster, G., Sharp, Z., Atudorei, V., Elder, A., Gelein, R., Lunts, A., Kreyling, W.,& Cox, C. (2002). Extrapulmonary translocation of ultrafine carbon particles following whole-body inhalation exposure of rats. *Journal of Toxicology and Environmental Health, Part A, 65*, 1531–1543.

41. Vickers, A.E., Rose, K., Fisher, R., Saulnier, M., Sahota, P., & Bentley, P. (2004). Kidney slices of human and rat to characterize cisplatin-induced injury on cellular pathways and morphology. *Toxicologic Pathology, 32*(5), 577–590.

42. Ricardo, S.D., Bertram, J.F., & Ryan, G.B. (1994). Reactive oxygen species in puromycinaminonucleoside nephrosis: In vitro studies. *Kidney International, 45*(4), 1057–1069.

43. McNeilly, J.D., Heal, M.R., Beverland, I.J., Howe, A., Gibson, M.D., Hibbs, L.R.,& Donaldson, K. (2004). Soluble transition metals cause the pro-inflammatory effects of welding fumes in vitro. *Toxicology and Applied Pharmacology, 196*(1), 95–107.

44. Nel, A.E., Diaz-Sanchez, D., & Li, N. (2001). The role of particulate pollutants in pulmonary inflammation and asthma: Evidence for the involvement of organic chemicals and oxidative stress. *Current Opinion in Pulmonary Medicine, 7*(1), 20–26.

45. Brown, D.M., Wilson, M.R., MacNee, W., Stone, V., & Donaldson, K. (2001). Size-dependent proinflammatroy effects of ultrafine polystyrene particles: A role for surface area and oxidative stress in the enhanced activity of ultrafines. *Toxicology and Applied Pharmacology, 175*, 191–199.

46. Kim, H., Liu, X., Kobayashi, T., Kohyama, T., Wen, F.Q., Romberger, D.J., & Rennard, S.I. (2003). Ultrafine carbon black particles inhibit human lung fibroblast-mediated collagen gel contraction. *American Journal of Respiratory Cell and Molecular Biology, 28*(1), 111–121.

47. Rothen-Rutishauser, B.M., Kiama, S.G., & Gehr, P. (2005). A three-dimensional cellular model of the human respiratory tract to study the interaction with particles. *American Journal of Respiratory Cell and Molecular Biology, 32*(4), 281–289.

48. Kiama, S.G., Cochand, L., Karlsson, L., Nicod, L.P., & Gehr, P. (2001). Evaluation of phagocytic activity in human monocyte-derived dendritic cells. *Journal of Aerosol Medicine, 14*(3), 289–299.

49. US EPA (1998). *Health effects test guidelines: OPPTS 870.2500 acute dermal irritation*. EPA 712-C-98-196. Available at: https://ntp.niehs.nih.gov/iccvam/suppdocs/fed docs/epa/epa_870_2500.pdf.

50. US EPA (1998). *Health effects test guidelines OPPTS 870.1200 acute dermal toxicity*. Available at: https://ntp.niehs.nih.gov/iccvam/suppdocs/feddocs/epa/epa-870-1200.pdf.

6 Fundamentals of Functional Materials
Applications for Clean Environment

Vidya Spriha Kujur, Mrinal Poddar, Tarun Kumar Dhiman, Avinash Kumar Singh, and Rahul Kumar

CONTENTS

DOI: 10.1201/9781003323464-6

6.1   INTRODUCTION

According to data from the World Water Reserve, water covers 70% of the total globe, but only 3% is fresh water, and out of this 3%, only 0.4% is accessible to human beings in the form of groundwater, present in rivers and other reservoirs, among others. This 0.4% of water serves billions of people every day. Each day the consumption of surface water is about 321 billion gallons per day, and that of groundwater is 77 billion gallons per day. As the population is growing simultaneously with new inventions and industrialization, there has been a significant threat to freshwater as it is being polluted every day with the excessive discharge of chemicals and household polluted waters into the water bodies and soil, which is contaminating these freshwater sources. Millions of people have scarce access to clean fresh water. These issues have raised the demand for reusability of the freshwater is growing every day, as there has been a shortage of water in every corner of the globe. Hence, several private firms and researchers from around the globe have been trying to get a solution for cleaning these freshwaters and making them as fresh as drinking water. Many have achieved success but not at the level required [Weblink 1]. Wastewater treatment contains three major components: the detection of organic and inorganic components and the detoxification of impure water. Here we explain the use of nanomaterials for all three components in detail with the latest example from the research being conducted all over the world. We explain zero dimensional (0D), one dimensional (1D), two dimensional (2D) and three dimensional (3D) nanostructures for the same. We explain the various pathways and mechanisms for detecting pollutants in the environment. We also explain the removal of pollutants using various nanomaterials based on the adsorption method. Thereafter, we explain the degradation of pollutants using the photocatalytic process.

There are various conventional methods that have been widely used until now for getting clean drinking water, such as filtration (bio-sand and ceramic filters, charcoal and activated-carbon bed), adsorption, distillation, ion exchange, electrochemical technology, chemical treatment, Ultraviolet (UV) radiation, precipitation and others [Weblink 2]. The effectiveness of these filters in removing organic contaminants is low. The turbidity and the growth of fungus and bacteria on the filter media require regular washing. With advancement, several plants started using reverse osmosis (RO) [Weblink 3] and multistage flash distillation (MSF) technology, which uses a thin film of polyamide composites. These purifiers are less economical, and the overall process involved in the system needs a huge amount of energy to function. Moreover, the films also get depleted over time as they are unable to handle a large amount of pressure and the volume purified by these purifiers is also not significant. With the development in nanotechnology, the hope of getting a brighter future in this area has increased; due to the outstanding and versatile characteristics of nanostructured materials, these are being tested for their widespread use in the water purification domain. With their in-depth potential, many fields in nanoscience and technology are still to be discovered. This chapter provides succinct insight into the areas of such diverse aspects of nano-technology, which gives potential applications of nanostructured materials in the field of water purification. It could be in the form of individual nanoparticles or could be used as the membranes to filter out the pollutants from the water (1).

We have already learned that the nanomaterials are the elemental or the com-posite materials at the nanoscale, that is, 10^{-9} meters, which also covers the study of atoms and molecules and their modification using several chemical and physical synthesis processes for achieving the purpose of desired chemical and physical prop-erties in them at the nanoscale. These materials existed and have been synthesized from ancient times for different unique purposes. However, the properties that these materials exhibit because of their unique and specific size and the structures that they develop at that particular size range were not known to humans. This particular nano range provides an increased surface-to-volume ratio to the typical material and, hence, increases its basic chemical and physical properties and provides it a different character than the bulk element or the composite. To date, carbonaceous materials are being used successfully to purify the polluted water to make it consumable, but there are certain limitations to their application as not all of the pollutants are successfully removable from water. Polluted water today contains more significant amounts of organic, inorganic and toxic pollutants from industrial production and household dis-charges. Apart from the naturally occurring pollutants, the excessive industrialization and urbanization have led to the causes of the different disease. This also causes dam-age to the natural reservoir, leading to environmental depletion and damage to the aquatic environment. Recent advancements in the research have shown the impor-tance of nanostructured materials in the purification of water successfully at lab-scale experiments and have suggested many nano-absorbent, nanocatalysts and nanostruc-tured membranes for an environment-friendly and economical way of wastewater treatment. The purpose that the nanomaterials need to serve in this application are

1. to selectively remove pollutants without affecting other essential minerals, permeability, concentration polarization and fouling;
2. being chemically and physically stable toward harmful chemicals; and
3. being economically and environmentally productive.

Other different nanoparticles such as the core–shell nanostructure of chitosan and gum arabic are being studied for removing heavy metal ions that are present in water. Gum arabic is a natural polysaccharide known for its properties of the absorption of metals, and chitosan possesses several carboxyl groups that act as the binding agents for the gum arabic, which may lead to better adsorption that has the potential in photocatalysis application (2).

Nanomaterials are smart materials and have extraordinary potential uses in the water purification field, but apart from their bright side, they also have darker aspects.

The smaller size of the nanoparticles makes them able to penetrable any living cell, which could be led to the change in the system of the cells; it could also bring changes in the DNA of the living body. These can pollute the environment as well, as they are extractable once they enter the soil or the water bodies and can harm flora and fauna, which could be very dangerous for life on earth. Therefore, they need to be used judiciously, and such nanostructures that could be extracted thoroughly after the purification is completed from the water bodies without harming the natural habitat or the living beings should be brought to light. Nanomaterials are being studied extensively due to their extraordinary unique properties that are not present in their bulk counterparts, like surface and size effect, quantum effect and macro-quantum tunneling effects. These properties lead to exceptional adsorption and reactivity of these nanoparticles and their subsequent composites, both of which help remove pollutants and/or reduce them into less harmful components of inorganic pollutants. Nanomaterials have been used to form different composites that can work as filtration units. These filtration units can filter out larger accumulated particles and/or bind with ions and eventually remove them from the sample solution (3).

6.2 VARIOUS TYPES OF NANOSTRUCTURES BASED ON THE DIMENSION

Nanoparticles have large surfaces area because of their small sizes and increased surface-to-volume ratio. The large surface area–to–volume ratio leads to more active sites and low intraparticle resistances. Apart from this, at such small-scale properties, such as a large number of nanopores, higher selectivity, high permeability and excellent mechanical and thermal stability, also exist (4). The size of the pore also affects the adsorption as the surface area is related to the surface area inversely (5). Along with the nanoscale properties of the material, the adsorption phenomenon gets enhanced by the hydrophobicity, working temperature and the working time of the filtration system. Almost every nanomembrane filter uses the same principle. In some systems, nanomaterials with a high adsorption capability, such as silver nanomaterials, are also used in the filter layer to adsorb the different specific pollutants before the membrane filtration process to speed up the process and obtain better results from the nanomembrane (6).

6.2.1 0D Nanostructures

Materials wherein all the dimensions are confined within the nanoscale fall in the category of 0D nanostructured materials. 0D nanostructures, such as homogeneous quantum dots (QDs), core–shell QDs and onion-shaped and hollow-shaped spheres. QDs are colloidal semiconductor nanocrystals with all three dimensions confined between in nanometerspace (7). QDs are unique in a sense that they display properties that lie between those of the bulk materials and the atoms/molecules. This results in unusually high surface-to-volume ratios. QDs are semiconductors in the nano range with significantly different electronic properties that correspond to their bulk substance. A QD is 0D relative to the bulk material. This confines the electrons in a QD to a minimal space. According to Pauli's exclusion principle, quantization of the energy levels take place; wherever the radius of nanomaterials is smaller than that of the Bohr exciton radius, the quantized and discrete energy levels of QDs make them closer to atoms rather than bulk materials in their properties (8;9). QDs are significantly used for optical applications because they exhibit bright and pure colors in addition to their ability to emit a rainbow of colors, along with their high efficiency, long lifetime and high extinction coefficient (10;11). Since QDs are 0D, they have a sharper density of states (DOSs) than bulk materials. Their small size also helps in less traveling of an electron within the particle. Thus, electronic devices can operate faster (12;13).

6.2.2   1D NANOSTRUCTURES

One-dimensional nanostructure has at least one of its dimensions in a nanoscale. These can be characterized by having either their length, breadth or height be less than 100 nm in size. The intense interest in 1D nanostructures has arisen due to their wide range of potential applications. One-dimensional nanostructures are facile nanostructures for exploring their optical, electronic and magnetic properties, among others. They include nanorods, nanotubes, nanowires, nanobelts, nanoribbons and hierarchical nanostructures. Metal oxides are those that have the mOx formulae (M = any metal element). Various metal oxides can be synthesized with TiO_2, CeO_2 and other metal oxide nanomaterials that are used as catalysts. These are used to optimize materials for fast and 100% degradation of pollutants, mainly organic pollutants present in the water that comes under the ozonation process (14). A photocatalyst, like TiO_2 nanoparticles, has been effectively studied for treating contaminated water that is polluted with organic pollutants such as chlorinated alkanes, benzenes and PCBs. The study shows that there was a complete degradation of organic carbon in contaminated water just by including TiO_2 nanoparticles as a photocatalyst (15). The mechanism for degrading organic dye involves the following steps:

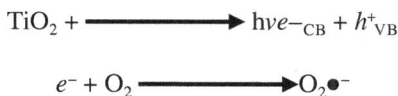

$$TiO_2 + \longrightarrow hve-_{CB} + h^+_{VB}$$

$$e^- + O_2 \longrightarrow O_2 \bullet^-$$

Initially, the formation of e^- and h^+ takes place, which act as a reductant and an oxidant, respectively. *Oxidation involves the following process:*

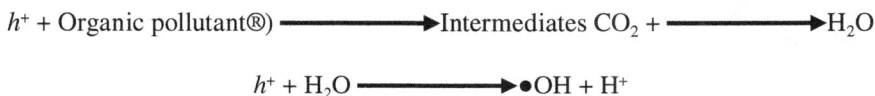

$$h^+ + \text{Organic pollutant®}) \longrightarrow \text{Intermediates } CO_2 + \longrightarrow H_2O$$

$$h^+ + H_2O \longrightarrow \bullet OH + H^+$$

Reduction involves the following process:

$$\bullet OH + \text{Org®c(R)} \longrightarrow \text{Intermediates} \longrightarrow CO_2 + H_2O$$

TiO_2 nanoparticles were also successfully used for the purification of microcystins in water in a "falling film" reactor (16). TiO_2 shows such an effective degradation of the organic compounds because of the regular production of enhanced hydroxyl radical production (17;18). Research into the TiO_2 nanoparticle is not limited just to a single compound, but its catalytic effects have also been studied by modifying and doping various noble metal TiO_2 nanoparticles. For example, doping Si onto TiO_2 was studied, and it was found that it was very effective in improving its efficiency, which was because there was an increase in the surface area and the crystallinity of the nanoparticles (19). Nobel metal–modified TiO_2 nanocrystals are used to degrade various organic dyes in the lab, like methylene blue, using the photocatalysis process under visible light, whereas nitrogen-doped and Fe(III-)-doped TiO_2 nanoparticles have shown good degradation results for degrading azo dyes and phenols. Apart from modification by doping, TiO_2 nanoparticles are also enhanced by deposition on porous Al_2O_3, which are used as catalysts to effectively remove total organic content (TOC) in water. Nanocomposites of other oxides such as mesoporous silica with TiO_2 are also used to successfully treat of aromatic pollutants present in water. Another study shows the degradation of 4-nitrophenol by using the nanocomposite of nanosized SO_4^{2-}/TiO_2. At the nanoscale, the structure of the nanoparticle also enhances its properties, and therefore, the nanotubes of TiO_2 have also been used in the water purification process, which, according to the study, was effective in degrading toluene better and were found very efficient than TiO_2 nanoparticles for the degradation of organic compounds in water. This material has not only been studied in the labs, but a few companies, such as Purifics Photo-Cat system, have developed a TiO_2-based commercial product that has shown high efficiency in removing organic compounds (20).

6.2.3 2D Nanostructures

Two-dimension nanostructured materials have two dimensions in the nanometer range. Two-dimensional nanostructures again have different characteristics from their bulk counterparts. They include junctions (continuous islands), branched structures, nanoplates, nanosheets, nanowalls and nanodisk-shaped nanostructures. Also, 2D nanostructures are of particular interest, not only for nanostructure synthesis but also for investigating and developing newer applications. These can also be used as templates for 2D structures of other materials. Among many types of nanostructures carbon nanotubes (CNTs) and graphene are the most studied nanostructures.

6.2.3.1 CNTs

CNTs are rolled sheets of graphene that are 2D in their structure. They have multiple unique properties, like high electrical conductivity and high tensile and mechanical strength. These properties of CNTs make them ideal candidates for research on 2D nanostructures. These can be found in two forms. Single-walled carbon nanotubes (SWNTs) and multiwalled carbon nanotubes (MWCNTs). Both forms can be synthesized via arc discharge method.

6.2.3.2 Graphene and Its Oxides

Graphene is the 2D hexagonal structure of a single layer of graphite. Graphene can also react with oxygen to form graphene oxide (GO) and reduced graphene oxide (RGO). Graphene can be synthesized via a wide variety of routes such as the arc discharge method, chemical vapor deposition (CVD) and microwave-assisted oxidation, among others.

6.2.4 3D NANOSTRUCTURES

Three-dimensional nanostructures have a comparatively large surface area and better properties when compared to bulk nanomaterials. These materials are characterized by having three arbitrary dimensions, nearly 100 nm in more than two dimensions. They possess a nanocrystalline structure, which means long-range crystalline order. It has been well studied and researched that the characteristics of nanostructured materials fully depend on its size, shape, dimension and morphology. Three-dimensional nanostructures have got worldwide attention for research due to their high surface area and large absorption sites for the adsorption of multiple molecules in a small space. Also, when these materials are synthesized with porosity, they could conduct better transport of atoms and molecules (21). Nanoflowers are one of the types of 3D nanomaterials that have a morphology like that of a flower at the nanoscale. They are a newly developed class of nanoparticles showing a very high surface area–to–volume ratio. These can be prepared by simple method of synthesis like wet chemical methods. Nanoflowers size ranges between 100–500 nm (22). There are several advantages of these nanoflowers:

1. Due to their high surface-to-volume ratio, the surface adsorption is enhanced, accelerating the reaction kinetics.
2. Nanoflowers have demonstrated better carrier immobility and charge transfer because of higher surface area.
3. Due to the large surface area, their efficiency of reaction on the surface is increased.

The following are some of nanostructured materials being used for this application and some being studied in the labs for better results at commercial scale:

1. Materials based on carbon are mostly composed of carbon and are present in different forms such as graphene, fullerenes and CNTs, which are capable of removing water pollutants from industrial discharge. They have the ability to initiate $\pi-\pi$ electrostatic interactions which facilitate them for the degradation (23). The CNTs poses large specific surface areas; they have small hollow and are layered in structure. This physical morphology makes them an excellent candidate as an adsorbent for a large number of organic pollutants and heavy metal removal. The modification of CNTs is easy, which can be done by chemical treatments and hence can increase adsorption capacity (24). These nanotubes enhance the flow rates through

their pores, which are small due to the presence of a smooth interior. Using nanotubes helps save energy because they promote flow rates that ultimately reduces required pressure to push water through the tubes. This can further be cleaned using autoclave or ultrasonication technique at 121°C for 30 minutes and can be reused with same filtering efficiency (25).

2. Example of metal-based nanomaterials are nanogold, nanosilver and QDs, among others, and metal oxides include zinc oxide, titanium dioxide, aluminum oxide and others.

3. Dendrimers are the polymers in the nanorange and are constructed from a large number of small units in a branched manner. Its surface has a number of chain ends that are capable of adapting or performing specific chemical functionalities, which makes them one of the catalytic candidates.

4. Nanoparticles and composites combine to other nanoparticles present in a solution. Other nanoparticles like nanoclays are incorporated to increase thermal and mechanical barriers and to enhance fire-retardant characteristics.

Based on this principle, nanomembranes are used in filtration. The different sizes of the molecules lead to the different principle of filtration and differentiate microfiltration (MF) and ultrafiltration (UF). The nanofiltration (NF) and RO display the effects of ionic charge of molecules and the surface of the membrane, hydrophobicity, and electrical polarity/ion-exchange membranes. The membranes such as MF, UF, NF and RO act on the hydraulic pressure.

6.3 VARIOUS PATHWAYS AND THE MECHANISM FOR THE DETECTION OF POLLUTANTS

The exponential growth of the chemical and agrochemical industries over many decades has resulted in the release of various new chemical compounds in the environment. According to an organisation for economic co-operation and development (OECD) report, the daily use of chemicals related to 70,000 to 1,00,000 mainly organic synthetic chemicals, and this number is also increasing continually (26). Organic pollutants comprise harmful herbicides and insecticides that are used for agricultural purposes and to control pests. Mostly, harmful organic chemicals are classified as persistent organic pollutants (POPs). These are the organic materials that have the ability to withstand a long duration of time in the environment because they resist all kinds of degradation, such as chemical, photolytic and biological degradation, among others (27;28). POPs are distinguished with high and low solubility in water, which lead to their bioaccumulation in the fatty tissue. They are highly toxic even at lower concentration. These are known to be semi-volatile in nature, which permits them to occur in vapor form or be adsorbed on the particles suspended in the atmosphere, hence facilitating long-range transport via the atmosphere. Table 6.1 shows the various POPs and their adverse effects.

There are various detection and removal mechanisms used to detect and remove all types of organic and inorganic pollutants, such as electrochemical, optical and

TABLE 6.1

Various POPs and Their Adverse Effects

S.N.	Name of POPs	Uses	Type of Substance	Adverse Effect	Banned Year
1	Aldrin	To kill termites, grasshoppers, Western corn rootworm	Insecticides	Exposure of Aldrin leads to headaches, nausea and vomiting, anorexia, muscle twitching, and myoclonic jerking	1990
2	Chlordane	termite-treatment	Pesticides	affect the human immune system and cancer disease	1988
3	Dieldrin	to control termites, textile pests, insect-borne diseases	Pesticides	Parkinson's disease, breast cancer	1974
4	Endrin	to control rodents	Insecticides	central nervous system	1991
5	Heptachlor	to kill soil insects and termites	Pesticides	Exposure of newborn animals to high doses of heptachlor causes a decrease in their body weight	1987
6	Hexachlorobenzene	as a seed treatment	Fungicide	metabolic disorder	1981
7	Mirex	to control fire ant	Insecticides	Lead to changes in liver caused by fat, causes convulsion and hyperexcitabilit	1976
8	Toxaphene	on cotton, cereal, grain, fruits, nuts, and vegetables	Insecticides	Exposure of toxaphene causes change in thyroid, liver, and kidneys' morphology	1990

(Continued)

TABLE 6.1 *(Continued)*
Various POPs and Their Adverse Effects

S.N.	Name of POPs	Uses	Type of Substance	Adverse Effect	Banned Year
9	Polychlorinated biphenyls (PCBs)	dielectric fluid, coolant fluids of electrical appliances, and carbonless copy paper	Organic Chlorine compound	It produces toxic effects leading to neurotoxicity and endocrine disruption	1978
10	DDT	Used to restrict malaria and typhus disease and in agricultural use	Organochlorine and insecticides	It causes cancer and its agricultural use threat to wildlife (birds)	1972
11	Dioxins	by-products of high-temperature processes	Organochlorine	It causes in human immune and enzyme disorders	1979

capacitive, among others (29;30;31;32;33). Most of these mechanisms involve some form of a mediator that impedes their real-world application. Thus, an optical measurement-based method, such as photoluminescence, offers a much better choice for studying the detection and identification of chemicals for a clean and safe environment (34;35). Among these, photoluminescence spectroscopy is one of the important tools used for detecting and determining several pollutants (36;37). This technique is simple compared to the other techniques and can be used as portable detection system. This can be explained in various pathways such as fluorescence resonance energy transfer (FRET), photo-induced electron transfer (PET) or inner filter effect, among others (38). Here we have explained the first two techniques in details.

6.3.1   FRET

FRET involves the phenomenon of energy transfer that takes place from an excited donor to ground-state acceptor. It is a technique to measure the distance between two chromophores, namely, a donor and acceptor pair. FRET is a distance-dependent phenomenon in which the distance between the donor and the acceptor must lie between more than 20 and below 80 Å. This technique is considered essential to understanding the biological system and application in optoelectronics and thin film device.

In the process of FRET (Figure 6.1), an electron is excited from its donor molecule to its electronically higher state so that it can transfer its energy to a nearby acceptor molecule through the nonradiative process. Moreover, there should be a long-range dipole–dipole interaction between donor and acceptor. Energy transfer leads to a decrease or quenching of the fluorescence intensity of the donor and is responsible for a decrease of the lifetime and increase the intensity of the fluorescence of the acceptor molecule (39) [Weblink 4]. The following is the relationship for the Forster radius (R_o):

$$R_o = 9.78 \times 10^3 \ (k^2 . \ n^{-4} . \ Q_D . \ J)^{1/6} \ A^0$$

k^2—Dipole angular orientation of each molecule
n—Solution's refractive index
Q_D—Fluorescence quantum yield of the donor in the absence of acceptor
J—Spectral overlap integral of the donor–acceptor pair

6.3.2 PET

PET is a technique in which excited-state electron transfer takes place from its excited donor to ground-state acceptor molecule. Here, two methods are discussed regarding the electron transfer prss:

1. In the first scheme as shown in Figure 6.2, when the neutral electron-rich donor (such as olefin compounds and amines) molecule undergoes excitation caused by photon energy absorption, and it goes to the higher state, confronts another electron acceptor molecule and is electron-deficient; then electron transfer takes place within these entities. When an electron is transferred from the donor to the acceptor, a radical is produced. In the donor's case, a cationic radical is produced, and in an acceptor's case, anion is produced by accepting an electron from the excited donor molecule.

   Generally, in the initial stage, an ionic pair of a donor–acceptor radical is produced and then diffuses into the free ions that are in the solution, which partially depends on the solvent's polarity. Solvent polarity will favor the

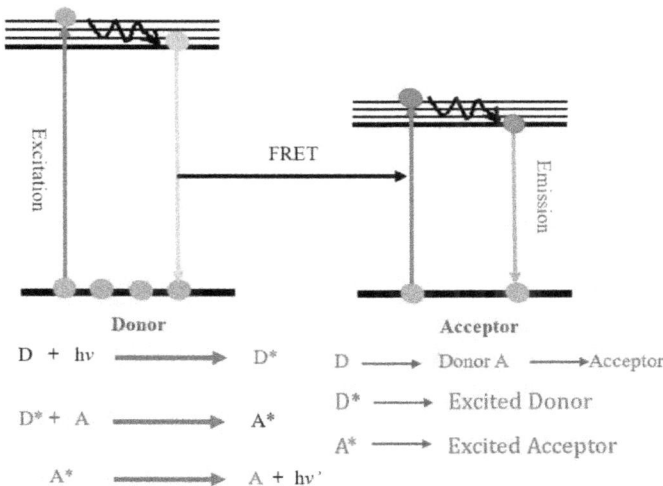

FIGURE 6.1 FRET process.

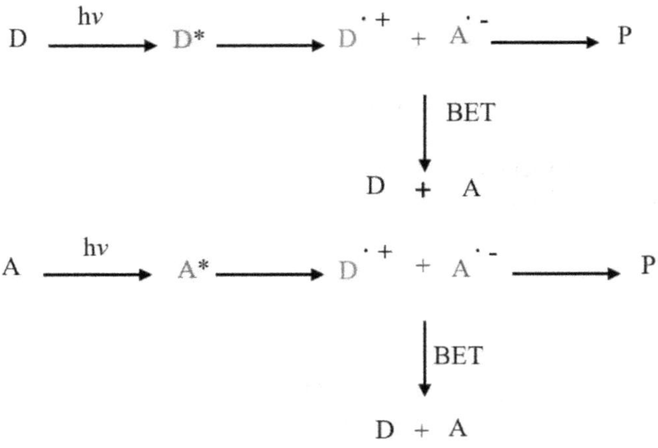

FIGURE 6.2   PET process.

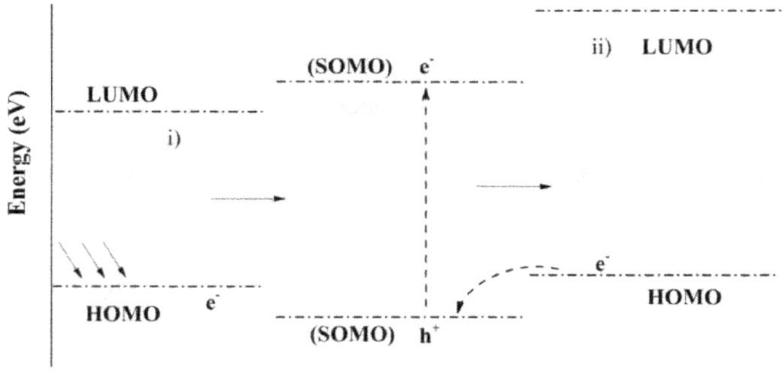

FIGURE 6.3   Oxidative photo-induced electron transfer process.

transfer of an electron rather than favor the transfer of energy due to the high polarity of solvents (such as dimethyl sulfoxide [DMSO], acetonitrile, etc.) and produce ionic radicals' electronic transfer initiated by photons. Once the ionic radicals are formed, they undergo the following two reactions:

a. They can go through chemical reaction and go-to formation of product, under which several chemical reactions are possible.

b. In another case, back electron transfer (BET) takes place in which the acceptor anion radical donates its electron to the donor cation radical and produces a donor and an acceptor in the process by wasting the energy of the photon. BET is always an energy-wasting process because it reproduces the starting point from the donor and the acceptor is reproduced.

2. In the second scheme as shown in Figure 6.3, the acceptor undergoes excitation by absorption of a photon instead of the donor and forms the

excited-state acceptor molecule. This excited acceptor molecule encounters an electron from the donor molecule, and in the process of donating an electron, the donor becomes the cation radical and by the process of absorbing, the electron acceptor becomes the anion radical. The donor–acceptor anion radical and cation radicals produced can go to the product by chemical reaction, or it can merely undergo the BET reaction.

6.4 REMOVAL OF POLLUTANTS USING VARIOUS NANOMATERIALS BASED ON ADSORPTION

Chemical pollutants are present in water sources naturally or through human activities (industrial wastes, mining by-products, excess fertilizers, etc.; (40). These pollutants are nonbiodegradable and harmful, even in a minute amount. Natural water sources have most of these pollutants in trace amounts or can have higher amounts depending on the external contamination sources. These pollutants include inorganic salts, heavy metal ions, metalloid ions, complexes of metals and metalloids and mineral acids. Pollution due to inorganic contaminants is because of the following:

1. The generated chemicals (anthropogenic pollutants) or their by-products not being able to decompose in the environment
2. Improper disposal of chemicals in the environment

The introduction of the high level of heavy metal in the environment started with the start of a global industrial revolution. Due to industrialization and improper use of agrochemicals (like fertilizers), the amount of heavy metal pollutants and inorganic pollutants has increased on the surface of the earth beyond the amount from the deposition from natural sources. Heavy metals accumulation may disrupt aquatic life and may contaminate the organisms that are later consumed as food. Direct contact with different inorganic contaminants through food or water may cause adverse effects to the liver, kidneys, the blood, the cardiovascular system, the gastrointestinal system, the integumentary system, bones or the nervous system depending on the type of inorganic contaminant, targeted host, concentration of accumulation and level of exposure. A different inorganic pollutant may enter or stay over the surface of the soil system due to deficient or excess management of agriculture practices, untreated discharge of industrial emission, leakage of landfills, wastes from households, wet depositions, dry deposition, industrial waste disposal and volcanic eruption. Nitrogen and phosphorous compounds can cause algal growth.

As we studied earlier, carbonaceous materials use the adsorption process to extract impurities. Similarly, there are several nanomaterials that use this process with the help of the functional groups attached to their surface sites and with the help of pores and its size. There are many factors that influence the adsorption rate and the spectrum of pollutants to be filtered. Some factors that have been studied are discussed (41;42).

The chemical adsorption between the adsorbent and adsorbate occurs due to the ionic bonds and the forces are similar to the covalent bond and follow Coulomb's law during the chemisorption. Moreover, during physisorption, the physical forces such, as Van der Waals forces, act between the adsorbate and adsorbent. However, these forces are weaker than the chemisorption. The various nano-adsorbents suggested are the nanomesh, nanoporous ceramics, nanotubes, nano-membranes, magnetic nanoparticles, nanofibrous alumina filters, nanoporous clays, polymers such as polypyrrole–CNT composites, cyclodextrin nanoporous polymers and others (43).

The materials based on carbon exhibit extraordinary physical as well as chemical properties due to their sp2 hybridized carbon bonds, which makes them feasible for the use in the electrical, thermal and mechanical applications (44). There are many allotropes of carbon that are highly useful in different areas. Therefore, it is being studied in the field of water detoxification. Nanoporous activated carbon (NAC) is widely being used for this purpose.

6.4.1 GRAPHENE AND GRAPHENE-BASED NANOSTRUCTURES

Graphene constitutes of the carbon atoms bond in a hexagonal structure whereas GO is an oxidized form of graphene. Graphene is hydrophobic, but GO also its production is effortless; hence, graphene oxide is used centrally for desalination. Graphene and GO, which are 1–2 atomic layers thick, are widely used as the membranes in several water purifiers, which serve the purpose of desalination and the separation of other organic compounds or the impurities in the water. Primarily, membranes based on graphene can be produced by inducing pores of nanoscale in a graphene sheet. Furthermore, the graphene sheet could be produced using the CVD over the copper substrate as a membrane and then etched by oxygen plasma to create the nanopores. The hydrophilic nature of the GO make it highly selective, and it has the potential for 100% desalination (45;46). These also cause antifouling and antimicrobial effects in purification; are tolerant of chlorine, acids and bases; and are more comfortable to clean at high temperature. The potential of graphene lies in further progress to develop low-cost membrane synthesis. GO, having strong hydrogen bonds and hydrophilicity, is being mainly used in the membrane filtration processes. RGO is also another structure reduced from GO that contains fewer bonding groups and a smaller number of sp2 aromatic domains. Both GO and RGO have 2D laminar carbons that have like structures, due to which they possess attributes such as surface defects, specific edges and structural wrinkles. These specifications make GO and RGO much more desirable than graphene alone (44;47;48).

Different nanomaterials like CNTs, zeolites and dendrimers show exceptional adsorption. These have found application in the removal of organic waste materials from wastewater (49). CNTs are of particular interest due to excellent results in water treatment. Also, CNTs can show very high adsorption of organic waste (50). NAC fibers, which are prepared with CNTs using the electrospinning technique, have shown large adsorption of organic waste than the activated carbon, which is in a granular form (51). When compared with activated carbon, both SCNTs and MWCNTs display higher adsorption capacities (52). For treating sorbents such as chlorophenols, herbicides, dichloro-diphenyl-trichloroethane (DDTs) and more, MWCNTs have been used and studied (53). Finally, the polymers that are functionalized with CNTs show effective results in removing organic pollutants present in the water (54).

GO is efficient in removing organic pollutants and metal ions from wastewater. Graphene-based nanocomposites have presented an outstanding ability in adsorbing various organic compounds and dyes (55), antibiotics (56), polycyclic aromatic hydrocarbons (PAHs) (57), phenolic compounds (58), pesticides (47) and more. The aqueous phase enhances the adsorption phenomenon. The strong adsorption of these materials is attributed due to the abundance of the pore size distribution, incredibly high specific surface area and different other surface properties, which make them interactable to a large number of molecules at less time with the organic compounds with the help of hydrophobic effects and ethylenediamine (EDA) interactions, hydrogen bonding interactions and others.

Carbon nanotubes (CNTs) are the tubular form of graphene sheets that can single layer, double layer or multilayer. They exhibit a large number of physical and chemical properties due to which they are being studied in the field of water purification too. They can form a strong π–πelectrostatic interaction. The adsorption energy and the charge transfer of hydrogen atoms (H_2) of the tubular structures at interstitial and grooved sites also enhance the surface activity of CNTs. There are several ways in which CNTs' activities could be harnessed for good results; for example, they could be aggregated into the aqueous phase to create a lot of interstitial spaces and grooves, ultimately leading to high energy adsorption, which could help in removing organic contaminants. They are also being developed as a nanomesh by developing nanotubes that are grown on the flexible porous medium that are capable of attaching at least onto the functional group. Moreover, these nanotubes are attached with different functional groups to make them effective in removing different contaminants. A study has reported MWCNTs coated with alumina to be efficient adsorbents for removing Pb(II) from industrial effluent. The absorption increases in pH between 3 and 7 and depends on the contact time, dosage of adsorbent and agitation speed (1). Among various pollutants, there are many heavy metal ions and metalloids that need attention due to their toxic nature and potential carcinogenicity. Some of them have been studied in Section 6.4.1.

6.4.2 HEAVY METALS AND METALLOIDS

Heavy metals exist naturally in the earth's crust and can easily be found in rocks, soils, sediments, water sources and more. They can also be found in microorganisms at certain standard concentrations. The standard concentration is increased due to anthropogenic release (59). These heavy metals fall under the group of metals and metalloids with a density equal to or higher than $4 \pm 1g/cm^3$. They cannot be destroyed or degraded and so stay and accumulate in the environment (60). Burning fossil fuels causes an increase in heavy metals in the environment. Many industries, like mines, thermal power plants, smelters, fertilizers, electronics, metallurgical and more, also contribute to the increase in heavy metal concentrations. These heavy metals in small amounts can be transported from a source location to different areas even out of country or continent (61). The presence of heavy metals at low concentrations is not harmful, but their persistence and accumulation can cause higher concentrations in that targeted location that can exhibit toxicity to plant, soil microorganisms and the food chain. Nickel, cobalt, chromium and copper show relatively high toxicity toward plants, whereas arsenic, cadmium, lead and mercury are more toxic to humans and animals.

6.4.2.1 Chromium

Chromium is present in two oxidation states. In its hexavalent state (Cr(VI)), it is toxic to humans, has high mobility and is soluble in water but not in the trivalent state, that is, Cr(III), form. Oxidizing compounds present in the soil reduce from Cr(VI) to Cr(III); this reduces its carcinogenicity. Some part of Cr(III) could be oxidized to form Cr(VI) when at a high pH; this oxidation is prompted by manganese oxides. Chromium is used in the chemical, metallurgical, leather, textile and refractory industries because of its aesthetics and durability (62). In humans, intake of Cr compounds causes skin irritation and ulcers inside the nose, and prolonged exposure can cause harmful effects to different body organs, like in kidneys, the nervous system and the cardiovascular system, which could also lead to cancer. It can affect seed emergence in plants, reduce their growth and decrease the production of dry matter. Many different types of nanoparticles and their composites have been used for removing or reducing hexavalent chromium in water sources (63).

6.4.2.2 Lead

Lead is a natural constituent in the earth's crust and is very toxic and carcinogenic. Lead is common in the mining industry, the smelting industry and electroplating (64). Lead has different effects on adults and children. In adults, the fingers become frail, blood pressure rises and there are abnormalities in the nervous system. In children, a reduction in body and mind growth can be seen. At high concentrations, lead can cause multiple organ failure, which leads to death. It can also cause a reduction in fertility level in men and abnormal child growth during pregnancy or may cause miscarriage. In plants, it causes a darkening of leaves, stunted foliage and an increase in the number of shoots. Many different types of nanoparticles and their composite have been used for removing or reducing lead in water sources.

6.4.2.3 Arsenic

Arsenic is inorganic and a highly toxic metalloid. Arsenic interferes with the Adenosine Triphosphate (ATP) generation pathway. At low exposure levels, it can cause nausea and vomiting; impair blood cell functioning, that is, red and white; and cause abnormalities in heartbeat. Long-term exposure leads to a darkening of the skin and soles, kidney and liver failure and can cause cancer of the bladder, lungs, skin and liver. In plants, it causes necrotic spots on old leaves of red-brown color, with the roots becoming yellow-brown, and a reduction in the growth of the plant. Many different types of nanoparticles and their composite have been used to remove or reduce arsenic from water sources.

6.4.2.4 Copper

Copper is an important nutrient for human beings; however, in large concentrations, it can lead to diseases such as anemia, liver cirrhosis and damage to the liver and kidneys (65;66). Intaking water with high concentrations of copper can cause anemia, kidney and liver damage, diarrhea, pain in abdomen, vomiting, headache and nausea in children (67). It causes chlorosis in plants, which creates chlorotic patches in old leaves, and can cause severe chlorosis in younger leaves as well at a high concentration; a yellow color in the leaves and stem; a purple coloration on the lower side of the midrib; and less branching and inhibited growth of the roots. Many different types of nanoparticles and their composite have been used for removing or reducing copper in water sources (68).

6.4.2.5   Mercury

Mercury is considered the most toxic heavy metal for humans when consumed. Minamata disease is caused by mercury poisoning, which causes a loss of physical control over bodily movements, muscle weakness, hearing and speech damage, hand and foot numbness and a loss of peripheral vision. At extreme poisoning, it can cause insanity, paralysis, coma and lead to death. Short-term time exposure can cause skin irritation, diarrhea, abnormal blood pressure, vomiting and damage to the lungs. Longtime exposure may cause organ failure and death. In plants, it can lead to extreme stunting of roots and seedlings, chlorosis, brown leaf tips, a reduction of growth and tumor-like emergences in deformed carrots. Many different types of nanoparticles and their composite have been used for removing or reducing mercury in water sources.

6.4.2.6   Nickel

Nickel pollutants are added to the environment through use in different industrial processes such as through the connectors, lead frames and tableware; in electroplating processes; plastics manufacturing processes; finishing processes for metals; nickel-cadmium batteries; in fertilizers; in mining industries; and metallurgical operations. It can cause different problems, such as damage to the kidneys and lungs and can cause nausea, vomiting, diarrhea, renal edema, pulmonary fibrosis and skin dermatitis. In plants, it causes chlorosis similar to iron deficiency, necrosis, growth inhibition and decreased area of the leaves. Many different types of nanoparticles and their composite have been used for removing or reducing nickel in water sources.

6.4.2.7   Cadmium

The chemical properties of cadmium are similar to zinc, and so it mimics the behavior of zinc's uptake and metabolic functions. Cadmium shows high toxicity even at low concentrations, it can quickly move through plants and its transport speed is also very high. Its mobility is higher than zinc and lead. Low-concentration exposure for a long time can damage kidney and lungs; also it can weaken bones. High-concentration exposure for a long time can lead to abnormalities in the biocycles of the body and can cause irritation of stomach, vomiting and diarrhea. It can cause renal tubular dysfunction that is associated with atrophic kidney and osteomalacia (Itai-Itai disease). It also shows carcinogenic properties. In plants, it causes moderate growth, leaves with brown margins, chlorosis and necrosis, leaf curving, reddish veins and petioles, as well as short brown roots, a reduction of growth and a purple coloring. Many different types of nanoparticles and their composite have been used for removing or reducing cadmium in water sources.

6.4.2.8   Zinc

Zinc is used in many areas due to its different properties. It possesses excellent antioxidant properties which can protect the body from premature aging of muscle and skin. However, its rising concentration in the environment, along with its nonbiodegradable property, is causing toxicity to the environment. Zinc is hugely used in industries such as galvanization, batteries, paint, smelting, fertilizers and pesticides, the burning of fossil fuels, polymer stabilizers and more. Zinc causes illnesses similar to those that lead causes and can be diagnosed as lead poisoning by mistake (69). Zinc metal is relatively nontoxic compared to other heavy metals. However, in excess quantities, it can cause problems with growth and reproduction (70). It can cause vomiting, diarrhea, blood in the urine, icterus, liver and kidney failure and anemia. Many different types of nanoparticles and their composite have been used for removing or reducing zinc in water sources.

6.4.2.9   Selenium

Selenium shows chemical properties similar to sulfur and can be found in sulfide ores where it can replace sulfide. It is a necessary element in cell function in organisms, as well as animals. It is a contaminant that is distributed through the oil refining, mining and power generation industries. At an appropriate amount, it helps with the proper functioning of thyroid glands. At high concentrations, it causes toxicity and imparts abnormality in organs. Short-term high-concentration exposure causes abnormalities in the respiratory system. High-concentration exposure causes hair loss, stomach pain and nail weakness. It can also cause neurological abnormalities. In plants, it causes chlorosis of veins, spotting of black color, bleaching and yellow color of young leaves and roots with spots of pink color. Many different types of nanoparticles and their composite have been used for removing or reducing selenium in water sources.

6.4.2.10   Antimony

Antimony is a chalcophilic, naturally available element. Sb(III) is highly toxic to humans with carcinogenic properties. They are consumed through contaminated water and food. Antimony is primarily introduced into the water sources through the mining of antimony. Due to its flexibility, it is mixed into alloys for many applications like lead storage batteries manufacturing, bearings, castings and more. It is also used in textiles, paper, and paints and in the production of explosives, antimony salts, ruby glass and medicines. It shows high mobility and a complex reaction with water that causes it to persist in the environment and cause adverse effects. It causes respiratory problems, pneumoconiosis, increased blood pressure, abdominal pain, diarrhea, vomiting, ulcers, antimony spots over the skin and more (71)

Similar to heavy metal and metalloids, there are other polluting compounds that also pollute the environment and are explained in Section 6.4.2.

6.4.3   Other Polluting Compounds

6.4.3.1   Ammonia

Ammonium is usually present in low concentrations in water bodies. Primary sources of ammonium accumulation are due to agricultural runoff, atmospheric deposition, domestic and industrial effluents and runoff from waste sites, construction sites and animal feedlots, among others. It also contributes to polluting surface and groundwater (72). Ammonium ion and unionized ammonia are interchanged upon changing pH and temperature. Ammonia is much more toxic than ammonium ions. Ammonium ions are still more than ammonia in water sources due to a close-to-neutral pH, which favors ammonium ions. Ammonia is a neutral molecule that can freely move in the epithelial membranes of different organisms. It damages gills; causes asphyxiation; stimulates glycolysis; hinders the Krebs cycle, which leads to progressive acidosis that causes a reduction in the oxygen-carrying capacity of the blood and a disruption in blood vessels; and affect functions of liver and kidney.

6.4.3.2 Hydrogen Sulfide

Hydrogen sulfide (H_2S) is present in decaying wastewater (sewage) as bacteria produce it from decaying plant and animal protein or from the direct reduction of sulfate (73). Hydrogen sulfide is a harmful compound; it can cause pulmonary edema, unconsciousness or even death in high concentrations. Long-term exposure can cause chronic respiratory system problems and central nervous system problems like fatigue, headache, poor memory and dizziness. In aquatic life, it can cause extreme growth of algae and the consumption of oxygen in the water through sedimentation, thereby killing bottom aquatic life (74).

6.4.3.3 Phosphorus Compounds

Phosphorus is present in the environment in the form of phosphates, mineral deposits and soil. Phosphorus is taken up in plants for the food chain and so is used in fertilizers to enhance their properties. It is also present in wastewater, due to human consumption, and so it can be taken up and used in fertilizers. Excess quantities of phosphorus can also speed up the eutrophication of rivers and lakes, which is caused by a reduction in dissolved oxygen from sources of water. This occurs due to an increase in minerals and organic nutrients. This excess of minerals and organic nutrients causes excess growth of algae in lakes. It causes problems with water quality, including raising the cost of purification; it also lowers the recreational and conservation values of impoundments, leading to a loss of livestock, and has a possible harmful effect of toxins of algal on the drinking water (75).

6.5 DEGRADATION OF THE POLLUTANTS USING PHOTOCATALYSIS PROCESS

Photocatalysis deals with reactions based on the interaction of nanoparticles with light energy. In the water purification field, it is generally used with metallic nanoparticles or oxides such as the TiO_2, ZnO_2, SiO_2 and others as they have broad and high photocatalytic activity for various pollutants (76;77;78). This mechanism is the most suitable for mass water purification. The principle of the photocatalysis process is initiated by the light energy absorbed by the nanoparticles, and the rest of the process is controlled by the nanoparticles themselves. This requires a meager consumption of energy and is economical for mass application (79;29). The photocatalysis process generally is used to degrade organic pollutants, toxic chemicals and others. In this process, the nanoparticles release the free O^-/OH-ions at its surface by absorbing the light energy that interacts with organic pollutants to break them down into smaller molecules without leaving any harmful effects in the water (80;81). This method is effective, and it could be helpful in mass cleaning as it possesses the potential an extraction and reusability possibility due to its magnetic properties. However, it needs to be further studied for its mass application and effects on the environment.

6.5.1 PHOTOCATALYSIS MECHANISM

Photon/light energy greater than the bandgap of the nanoparticle excites this nanocatalyst and helps generate the electron and hole pair due to which the pollutants get adsorbed on the surface of the nanostructure by lattice oxygen process. The mentioned reactions take place at the surface of the structure as shown in Figure 6.4.5 (42;82).

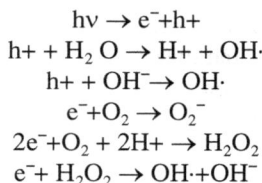

$$hv \rightarrow e^- + h+$$
$$h+ + H_2O \rightarrow H+ + OH\cdot$$
$$h+ + OH^- \rightarrow OH\cdot$$
$$e^- + O_2 \rightarrow O_2^-$$
$$2e^- + O_2 + 2H+ \rightarrow H_2O_2$$
$$e^- + H_2O_2 \rightarrow OH\cdot + OH^-$$

Finally, the organic pollutants break down to $\cdot OH + O_2 = CO_2 + H_2O$ +other reduced compounds. For the most part, there are two prospects: The electrons and holes generated by photon incident could recombine via radiative or nonradiative actions that go through a rapid vibrational process. Second, trapping of these e^- and h^+ could take place in the defect states which are created by impurities, which later either recombine or are transferred to the surface of the nanostructure for the surface reactions to take place such as photocatalysis (83). Here, these entities could react with adsorbed species that are bonded covalently or adsorbed physically over the nanostructure's surface. In the valence band (VB), the holes interact to chemisorbed or covalently absorbed H_2O molecules and form highly reactive free radicals like $OH^\cdot$. However, in the conduction band (CB), electrons interact with O_2 to form $OH^\cdot$ radicals that eventually react with organic dyes to completely remove them from water. The efficiency of the photo-catalyst depends on sufficient bandgap, the material's morphology, the large and exposed surface area, the stability of the material and reusability (84).

Photocatalysis mainly refers to the reaction that involves oxidation reaction and reduction reaction simultaneously over the catalyst's surfaces, which are arbitrated by the presence of h^+ entities or holes on VB and the presence of e^- entities or electrons on the CB that are generated when UV-Visible spectroscopy (UV-Vis) light radiation is absorbed by the catalyst. These are called the photo-generated pairs of electrons and holes that induce the formation of highly aggressive and reactive hydroxyl (OH^-) or superoxide radical species. These types of reactive entities are hefty enough that can oxidize and decompose the organic compounds or stinky

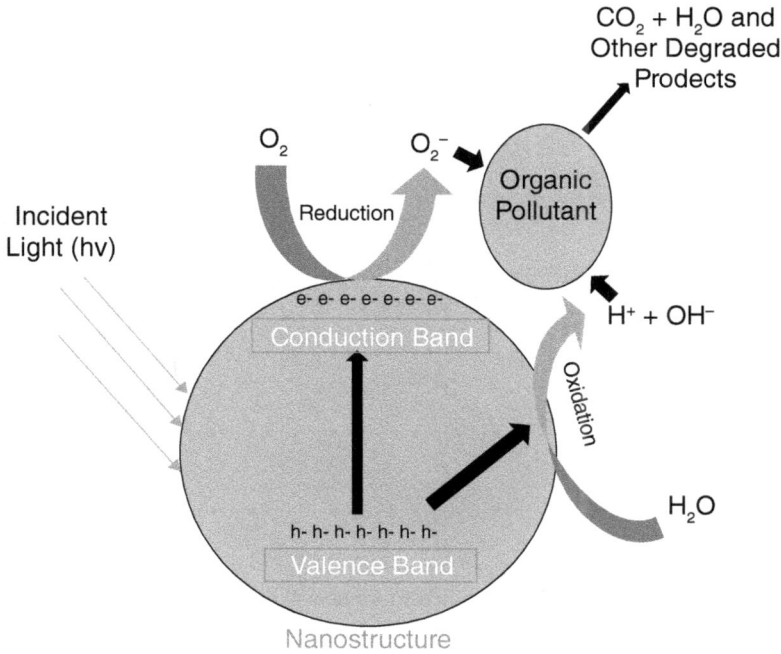

FIGURE 6.4 Mechanism of photocatalysis using the nanoparticles.

gases and can also kill bacteria (85). The photocatalysis process has been considered a very efficient technique that can be used for the mineralization of harmful organic pollutants and perilous inorganic materials and can also act as a disinfectant against microbes. This result is possible because of the OH⁻ ions that form during reaction processes behave as strong oxidizing agents (86;48).

Nanocomposites such as ferrites are used for the process, which uses a different path of photocatalysis known as Fenton catalysis as shown in Figure 6.5 (87). The Fenton process is an advanced oxidation process mostly used for the treatment of acidic wastewater, which was reported by Fenton in 1894. The advantage of this method is the use of iron that is found in abundance; also, it is nontoxic material used in the presence of hydrogen peroxide, which again is a nontoxic and environmentally friendly agent (2). Apart from the several advantages of the Fenton process, it also has a few disadvantages, such as it cannot completely mineralize organic compounds because generally less than half of organic carbons can be reduced to CO_2. Also, this method generates sludge because of the iron and flocculation of molecules.

This process involves H_2O_2 and ferrous salts in the reaction, which leads to the generation of hydroxyl radicals. This method is also called as photo-Fenton process. It starts with the oxidation of iron (II) to iron (III) present in the nanostructure due to the presence of hydrogen peroxide, which forms hydroxyl radicals and hydroxide. Later, iron molecules undergo reduction process to from iron (II) by the interaction of another hydrogen peroxide molecule, which helps in the formation

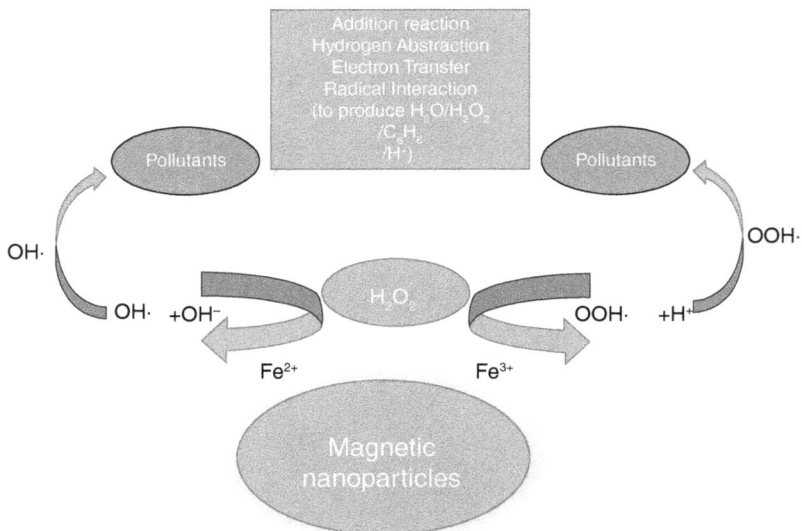

FIGURE 6.5   Mechanism of Fenton process for removal of water pollutants.

of hydroperoxyl and protons. Free radicals in this process then initiate the secondary reactions. The reactions involved in this process are swift and exothermic later, which reduces the pollutant compounds to water and carbon dioxide via oxidation. The pH of the Fenton reagent plays a vital role in the overall process. The reaction kinetics is enhanced by the maintained pH as the presence of both iron (II) and iron (III) is essential and their solubilities are directly related to it. For the reaction to take place rapidly, an acidic medium is needed (88;89).

6.6 CONCLUSION

Nanotechnology paves the way for developing science and technology in this, as well as the coming, centuries that can enable us to synthesize novel functional materials. As explained, various dimensional nanomaterials can be synthesized using a wide variety of routes. Zero-dimensional, 1D, 2D and 3D nanomaterials can be prepared in various shape and sizes, such spherical, cylindrical, nanoflowers, nanotubes, nanocubes and tetrapods, among others. Using these multidimensional nanomaterials, we can purify the raw wastewater containing inorganic and organic contaminants. Among the various inorganic contaminants, the toxicity of sodium, chromium, lead, arsenic copper, mercury, nickel, cadmium, zinc, selenium and antimony was explained in detail. Among other important toxic elements, the harmful effects of ammonia, phosphorous and hydrogen sulfide have also been explained. POPs, such as insecticides, pesticides and organochlorine-based compounds, are some of the most toxic compounds. Their detection and remediation using nanomaterials was also explained. Among different nanomaterials, carbonaceous nanomaterials like CNTs, NAC and GO were explained for their exceptional adsorption properties and application in water purification. Inorganic nanomaterials like TiO_2, ZnO, CeO_2 and others, and their nanocomposites with other nanomaterials were explained for use in detoxifying wastewater. Among various nanomaterials and techniques used, nano-adsorbents and photocatalysis are two of the most often used. The number of nano-materials confirms the potential of nanotechnology-based materials for future water treatment applications. The future of nanomaterials as commercially suitable materials for wastewater treatment depends on their large-scale production and application at commercial scale.

ABBREVIATIONS

Full Form	Abbreviation
Adenosine triphosphate	ATP
Back electron transfer	BET
Carbon nanotubes	CNTs
Chemical vapor deposition	CVD
Conduction band	CB
Dichloro-diphenyl-trichloroethane	DDT
Dimethyl sulfoxide	DMSO
Ethylenediamine	EDA
Fluorescence resonance energy transfer	FRET
Graphene oxide	GO
Microfiltration	MF
Multiwalled carbon nanotubes	MWCNT
Nanofiltration	NF
Nanoporousactivated carbon	NAC
One-dimensional	1D
Organisation for Economic Co-operation and Development	OECD
Persistent organic pollutants	POPs
Photo-induced electron transfer	PET
Polychlorinated biphenyls	PCBs
Polycyclic aromatic hydrocarbons	PAHs
Quantum dots	QDs
Reduced graphene oxide	RGO
Reverse osmosis	RO
Single-walled carbon nanotubes	SWNTs
Three-dimensional	3D
Two-dimensional	2D
Ultrafiltration	UF
UV-Visible spectroscopy	UV-Vis
Valence band	VB
Zero-dimensional	0D

REFERENCES

1. Gangadhar, G., Maheshwari, U., & Gupta, S. (2012). Application of nanomaterials for the removal of pollutants from effluent streams. *Nanoscience & Nanotechnology-Asia*, 2, 140–150.
2. Giri, A. S., & Golder, A. K. (2014). Fenton, photo-fenton, H2O2 photolysis, and TiO2 photocatalysis for dipyrone oxidation: Drug removal, mineralization, biodegradability, and degradation mechanism. *Industrial & Engineering Chemistry Research*, 53, 1351–1358.
3. Das, S., Chakraborty, J., Chatterjee, S., & Kumar, H. (2018). Prospects of biosynthesized nanomaterials for the remediation of organic and inorganic environmental contaminants. *Environmental Science: Nano*, 5, 2784–2808.
4. Morris, J., & Willis, J. (2007). *US environmental protection agency nanotechnology white paper*. US Environmental Protection Agency.
5. Bhawana, P., & Fulekar, M. (2012). Nanotechnology: Remediation technologies to clean up the environmental pollutants. *Research Journal of Chemical Sciences*, 2231, 606X.
6. Naidu, L., Saravanan, S., Goel, M., Periasamy, S., & Stroeve, P. (2016). A novel technique for detoxification of phenol from wastewater: Nanoparticle assisted nano filtration (NANF). *Journal of Environmental Health Science and Engineering*, 14, 9.
7. Petryayeva, E., Algar, W. R., & Medintz, I. L. (2013). Quantum dots in bioanalysis: A review of applications across various platforms for fluorescence spectroscopy and imaging. *Applied Spectroscopy*, 67, 215–252.
8. Reimann, S. M., & Manninen, M. (2002). Electronic structure of quantum dots. *Reviews of Modern Physics*, 74, 1283.
9. Bawendi, M. G., Steigerwald, M. L., & Brus, L. E. (1990). The quantum mechanics of larger semiconductor clusters ("quantum dots"). *Annual Review of Physical Chemistry*, 41, 477–496.
10. Yoffe, A. D. (2001). Semiconductor quantum dots and related systems: Electronic, optical, luminescence and related properties of low dimensional systems. *Advances in Physics*, 50, 1–208.
11. Sargent, E. H. (2012). Colloidal quantum dot solar cells. *Nature Photonics*, 6, 133.
12. Nirmal, M., & Brus, L. (1999). Luminescence photophysics in semiconductor nanocrystals. *Accounts of Chemical Research*, 32, 407–414.
13. Zhao, Y., & Burda, C. (2012). Development of plasmonic semiconductor nanomaterials with copper chalcogenides for a future with sustainable energy materials. *Energy & Environmental Science*, 5, 5564–5576.
14. Orge, C. A., Órfão, J. J., Pereira, M. F., De Farias, A. M. D., Neto, R. C. R., & Fraga, M. A. (2011). Ozonation of model organic compounds catalysed by nanostructured cerium oxides. *Applied Catalysis B: Environmental*, 103, 190–199.
15. Chitose, N., Ueta, S., Seino, S., & Yamamoto, T. A. (2003). Radiolysis of aqueous phenol solutions with nanoparticles. 1. Phenol degradation and TOC removal in solutions containing TiO2 induced by UV, γ-ray and electron beams. *Chemosphere*, 50, 1007–1013.
16. Shephard, G. S., Stockenström, S., De Villiers, D., Engelbrecht, W. J., & Wessels, G. F. (2002). Degradation of microcystin toxins in a falling film photocatalytic reactor with immobilized titanium dioxide catalyst. *Water Research*, 36, 140–146.
17. Han, X., Kuang, Q., Jin, M., Xie, Z., & Zheng, L. (2009). Synthesis of titania nanosheets with a high percentage of exposed (001) facets and related photocatalytic properties. *Journal of the American Chemical Society*, 131, 3152–3153.
18. Liu, S., Yu, J., & Jaroniec, M. (2011). Anatase TiO2 with dominant high-energy {001} facets: Synthesis, properties, and applications. *Chemistry of Materials*, 23, 4085–4093.

19. Iwamoto, S., Iwamoto, S., Inoue, M., Yoshida, H., Tanaka, T., & Kagawa, K. (2005). XANES and XPS study of silica-modified titanias prepared by the glycothermal method. *Chemistry of Materials*, 17, 650–655.

20. Bae, E., & Choi, W. (2003). Highly enhanced photoreductive degradation of perchlorinated compounds on dye-sensitized metal/TiO2 under visible light. *Environmental Science & Technology*, 37, 147–152.

21. Wang, J. N., Su, L. F. & Wu, Z. P. (2008). Growth of highly compressed and regular coiled carbon nanotubes by a spray-pyrolysis method. *Crystal Growth and Design*, 8, 1741–1747.

22. Ye, R., Zhu, C., Song, Y., Lu, Q., Ge, X., Yang, X., Zhu, M. J., Du, D., Li, H., & Lin, Y. (2016). Bioinspired synthesis of all-in-one organic–inorganic hybrid nanoflowers combined with a handheld pH meter for on-site detection of food pathogen. *Small*, 12, 3094–3100.

23. Hedderman, T. G., Keogh, S. M., Chambers, G., & Byrne, H. J. (2006). In-depth study into the interaction of single walled carbon nanotubes with anthracene and p-terphenyl. *The Journal of Physical Chemistry B*, 110, 3895–3901.

24. Zargar, R. A., Arora, M., Alshahrani, T., & Shkir, M. (2021a). Screen printed novel ZnO/MWCNTs nanocomposite thick films. *Ceramic International*, 47, 6084–6093.

25. Zargar, R. A., Hussan, M. M., Kumar, K., Nagal, V., Bashir, A., Alshahrani, B., Alshahrani, T., & Shkir, M. (2021b). Development and characterization of $(ZnO)_{0.90}(CNT)_{0.10}$ thick film for photovoltaic application. *Optik*, 248, 167975.

26. Schwarzenbach, R. P., Gschwend, P. M., & Imboden, D. M. (2016). *Environmental organic chemistry*. John Wiley & Sons.

27. Nasrabadi, T., Bidhendi, G. N., Karbassi, A., Grathwohl, P., & Mehrdadi, N. (2011). Impact of major organophosphate pesticides used in agriculture to surface water and sediment quality (Southern Caspian Sea basin, Haraz River). *Environmental Earth Sciences*, 63, 873–883.

28. Gilliom, R. J., Barbash, J. E., Kolpin, D. W., & Larson, S. J. (1999). Peer reviewed: Testing water quality for pesticide pollution. *Environmental Science & Technology*, 33, 164A–169A.

29. Dhiman, T. K., & Singh, S. (2019). Enhanced catalytic and photocatalytic degradation of organic pollutant rhodamine-B by LaMnO3 nanoparticles synthesized by non-aqueous sol-gel route. *Physica Status Solidi (a)*, 216(11), 1900012.

30. Kumar, R., Lakshmi, G. B. V. S., Dhiman, T. K., Singh, K., & Solanki, P. R. (2021). Highly sensitive amoxicillin immunosensor based on aqueous vanadium disulphide quantum dots. *Journal of Electroanalytical Chemistry*, 892, 115266.

31. Dhiman, T. K., Ahlawat, A., & Solanki, P. R. (2021). ZnOnps based photocatalytic reactor for degradation of multiple organic pollutants driven by solar light-based UV irradiation. *SPAST Abstracts*, 1(01). Retrieved from https://spast.org/techrep/article/view/1747.

32. Singh, A. K., Dhiman, T. K., Lakshmi, G. B. V. S., Raj, R., Jha, S. K., & Solanki, P. R. (2022). Rapid and label-free detection of Aflatoxin-B1 via microfluidic electrochemical biosensor based on manganese (III) oxide (Mn3O4) synthesized by co-precipitation route at room temperature. *Nanotechnology*, 33(28), 285501.

33. Verma, D., Dhiman, T. K., Mukherjee , M. D., & Solanki , P. R. (2021). The Development of Green Synthesized AgNps Anchored RGO nanocomposite based amperometric sensor for BPA detection. *SPAST Abstracts*, 1(01). Retrieved from https://spast.org/techrep/article/view/180.

34. Garimella, L. B., Dhiman, T. K., Kumar, R., Singh, A. K., & Solanki, P. R. (2020). One-step synthesized ZnO np-based optical sensors for detection of aldicarb via a photoinduced electron transfer route. *ACS Omega*, 5(6), 2552–2560.

35. Singh, A. K., Ahlawat, A., Dhiman, T. K., Lakshmi, G., & Solanki, P. R. (2021). Degradation of Methyl Parathion using Manganese oxide (MnO2) nanoparticles through photocatalysis. *SPAST Abstracts*, 1(01). Retrieved from https://spast.org/techrep/article/view/1027.

36. Sajwan, R. K., Pandey, S., Kumar, R., Dhiman, T. K., Eremin, S. A., & Solanki, P. R. (2021). Enhanced fluorescence of mercaptopropionic acid-capped zinc sulfide quantum dots with moxifloxacin in food and water samples via reductive photoinduced electron transfer. *Environmental Science: Nano*, 8(9), 2693–2705.

37. Singh, A. K., Sri, S., Garimella, L. B., Dhiman, T. K., Sen, S., & Solanki, P. R. (2022). Graphene quantum dot-based optical sensing platform for aflatoxin B1 detection via the resonance energy transfer phenomenon. *ACS Applied Bio Materials*, 5(3), 1179–1186.

38. Hashmi, S. Z. H., Dhiman, T. K., Chaudhary, N., Singh, A. K., Kumar, R., Sharma, J. G., Kumar, A., & Solanki, P. R. (2021). Levofloxacin detection using L-cysteine capped MgS quantum dots via the photoinduced electron transfer process. *Frontiers in Nanotechnology*, 2. https://doi.org/10.3389/fnano.2021.616186.

39. Hussain, S. A. (2009). *An introduction to fluorescence resonance energy transfer (FRET).* arXiv preprint arXiv:0908.1 815. https://arxiv.org/abs/0908.1815.

40. Speight, J. G. (2016). *Environmental organic chemistry for engineers.* Butterworth-Heinemann.

41. Hu, A., & Apblett, A. (2014). *Nanotechnology for water treatment and purification.* Springer.

42. Kaur, B., Chand, S., Singh, K., & Malik, A. K. (2019). Detoxification of dye contaminated water by Mn-doped ZnS nanostructures. *Bulletin of Materials Science*, 42, 61.

43. Yan, X., Shi, B., Lu, J., Feng, C., Wang, D. & Tang, H. (2008). Adsorption and desorption of atrazine on carbon nanotubes. *Journal of Colloid and Interface Science*, 321, 30–38.

44. Boretti, A., Al-Zubaidy, S., Vaclavikova, M., Al-Abri, M., Castelletto, S., & Mikhalovsky, S. 2018. Outlook for graphene-based desalination membranes. *NPJ Clean Water*, 1, 5.

45. Buelke, C., Alshami, A., Casler, J., Lewis, J., AL-Sayaghi, M., & Hickner, M. A. (2018). Graphene oxide membranes for enhancing water purification in terrestrial and space-born applications: State of the art. *Desalination*, 448, 113–132.

46. Shan, S., Zhao, Y., Tang, H., & Cui, F. (2017). *A mini-review of carbonaceous nano-materials for removal of contaminants from wastewater.* IOP Conference Series: Earth and Environmental Science. IOP Publishing, 012003.

47. Li, F., Yuasa, A., Ebie, K., Azuma, Y., Hagishita, T., & Matsui, Y. (2002). Factors affecting the adsorption capacity of dissolved organic matter onto activated carbon: Modified isotherm analysis. *Water Research*, 36, 4592–4604.

48. Yang, K., Wang, J., Chen, X., Zhao, Q., Ghaffar, A., & Chen, B. (2018). Application of graphene-based materials in water purification: From the nanoscale to specific devices. *Environmental Science: Nano*, 5, 1264–1297.

49. Savage, N., & Diallo, M. S. (2005). Nanomaterials and water purification: Opportunities and challenges. *Journal of Nanoparticle Research*, 7, 331–342.

50. Singh, N., & Gupta, S. K. (2016). Adsorption of heavy metals: a review. *International Journal of Innovative Research in Science, Engineering and Technology*, 5(2), 2267–2281.

51. Mangun, C. L., Yue, Z., Economy, J., Maloney, S., Kemme, P., & Cropek, D. (2001). Adsorption of organic contaminants from water using tailored ACFs. *Chemistry of Materials*, 13, 2356–2360.

52. Cai, Y.-Q., Cai, Y.-E., Mou, S.-F., & Lu, Y.-Q. (2005). Multi-walled carbon nanotubes as a solid-phase extraction adsorbent for the determination of chlorophenols in environmental water samples. *Journal of Chromatography A*, 1081, 245–247.

53. Zhou, M., Nemade, P. R., Lu, X., Zeng, X., Hatakeyama, E. S., Noble, R. D., & Gin, D. L. (2007). New type of membrane material for water desalination based on a cross-linked bicontinuous cubic lyotropic liquid crystal assembly. *Journal of the American Chemical Society*, 129, 9574–9575.
54. Salipira, K., Mamba, B., Krause, R., Malefetse, T., & Durbach, S. (2007). Carbon nanotubes and cyclodextrin polymers for removing organic pollutants from water. *Environmental Chemistry Letters*, 5, 13–17.
55. Addison, R. F., & Brodie, P. F. (1987). Transfer of organochlorine residues from blubber through the circulatory system to milk in the lactating grey seal Halichoerus grypus. *Canadian Journal of Fisheries and Aquatic Sciences*, 44(4), 782–786.
56. Shen, Q., Jiang, L., Zhang, H., Min, Q., Hou, W., & Zhu, J.-J. (2008). Three-dimensional dendritic PT nanostructures: Sonoelectrochemical synthesis and electrochemical applications. *The Journal of Physical Chemistry C*, 112, 16385–16392.
57. Teng, X., Liang, X., Maksimuk, S., & Yang, H. (2006). Synthesis of porous platinum nanoparticles. *Small*, 2, 249–253.
58. Charnock, C., & Kjønnø, O. (2000). Assimilable organic carbon and biodegradable dissolved organic carbon in Norwegian raw and drinking waters. *Water Research*, 34, 2629–2642.
59. Mohammed, A. S., Kapri, A., & Goel, R. (2011). *Heavy metal pollution: Source, impact, and remedies. Biomanagement of metal-contaminated soils*. Springer.
60. Duffus, J. H. (2002). "Heavy metals" a meaningless term? (IUPAC Technical Report). *Pure and Applied Chemistry*, 74, 793–807.
61. Pacyna, J. M., & Pacyna, E. G. (2001). An assessment of global and regional emissions of trace metals to the atmosphere from anthropogenic sources worldwide. *Environmental Reviews*, 9, 269–298.
62. Sharma, M., Joshi, M., Nigam, S., Shree, S., Avasthi, D. K., Adelung, R., Srivastava, S. K., & Mishra, Y. K. (2019). ZnO tetrapods and activated carbon-based hybrid composite: Adsorbents for enhanced decontamination of hexavalent chromium from aqueous solution. *Chemical Engineering Journal*, 358, 540–551.
63. Kumar, R., Singh, A. K., Dhiman, T. K., Lakshmi, G. B. V. S., Solanki, P. R., & Singh, K. (2021). Transition metal dichalcogenide quantum dots based optical detection platform for Cu2+ ions in water. *SPAST Abstracts*, 1(01).
64. Moyo, M., Nyamhere, G., Sebata, E., & Guyo, U. (2016). Kinetic and equilibrium modelling of lead sorption from aqueous solution by activated carbon from goat dung. *Desalination and Water Treatment*, 57, 765–775.
65. Dhiman, T. K., Poddar, M., Lakshmi, G. B. V. S., Kumar, R., & Solanki, P. R. (2021). Non-enzymatic and rapid detection of glucose on PVA-CuO thin film using Arduino UNO based capacitance measurement unit. *Biomedical Microdevices*, 23(3), 1–11.
66. Dhiman, T. K., Lakshmi, G. B. V. S., Kumar, R., Asokan, K., & Solanki, P. R. (2020). Non-enzymatic detection of glucose using a capacitive nanobiosensor based on PVA capped CuO synthesized via co-precipitation route. *IEEE Sensors Journal*, 20(18), 10415–10423.
67. Madsen, H., Poulsen, L., & Grandjean, P. (1990). Risk of high copper content in drinking water. *Ugeskrift for Laeger*, 152, 1806–1809.
68. Kumar, R., Lakshmi, G. B. V. S., Singh, K., & Solanki, P. R. (2019). A novel approach towards optical detection and detoxification of Cr (VI) to Cr (III) using L-Cys-VS2QDs. *Journal of Environmental Chemical Engineering*, 7(4), 103202.
69. McCluggage, D. (1991). Heavy metal poisoning. *NCS Magazine*. The Bird Hospital, Co.
70. Nolan, K. R. (1983). Copper toxicity syndrome. *Journal of Orthomolecular Psychiatry*, 12, 270–282.
71. Zeng-Ping, N., & Tang-Fu, X. (2007). Supergene geochemical behavior and environmental risk of antimony. *Earth and Environment*, 35, 176–182.

72. Camargo, J. A., & Alonso, Á. (2006). Ecological and toxicological effects of inorganic nitrogen pollution in aquatic ecosystems: A global assessment. *Environment International*, 32, 831–849.
73. World Health Organization. (1981). *Environmental health criteria 19, hydrogen sulfide*. International Programme on Chemical Safety, 40.
74. Nagaraj, A., Munusamy, M. A., Al-Arfaj, A. A., & Rajan, M. (2018). Functional ionic liquid-capped graphene quantum dots for chromium removal from chromium contaminated water. *Journal of Chemical & Engineering Data*, 64, 651–667.
75. Remmen, K., Niewersch, C., Wintgens, T., Yüce, S., & Wessling, M. (2017). Effect of high salt concentration on phosphorus recovery from sewage sludge and dewatering properties. *Journal of Water Process Engineering*, 19, 277–282.
76. Zargar, R. A., Arora, M., & Bhat, R. A. (2018). Study of nanosized copper-doped ZnO dilute magnetic semiconductor thick films for spintronic device applications. *Applied Physics A*, 124, 36.
77. Dhiman, T. K., Lakshmi, G. B. V. S., Dave, K., Roychoudhury, A., Dalal, N., Jha, S. K., Kumar, A., Han, K. H., & Solanki, P. R. (2021b). Rapid and label-free electrochemical detection of fumonisin-B1 using microfluidic biosensing platform based on Ag-CeO2 nanocomposite. *Journal of the Electrochemical Society*, 168(7), 077510.
78. Kujur, V. S., Gaur, R., Gupta, V., & Singh, S. (2022). Significantly enhanced UV-light-driven photocatalytic performance of ferroelectric (K0.5Na0.5) NbO3: Effect of corona-poling and particle size. *Journal of Physics and Chemistry of Solids*, 110751.
79. Zargar, R. A., Boora, N., Hassan, M. M., Khan, A., & Hafiz, A. K. (2020). Screen printed TiO$_2$ film: A candidate for photovoltaic applications. *Materials Research Express*, 7, 065904.
80. Harshulika, Ahlawat, A., Dhiman, T. K., & Solanki, P. R. (2021). Photocatalytic degradation of gentamycin using TiO2 nanoparticle driven by UV light irradiation. *SPAST Abstracts*, 1(01). Retrieved from https://spast.org/techrep/article/view/1939.
81. Singh, A. K., Lakshmi, G. B. V. S., Dhiman, T. K., Kaushik, A., & Solanki, P. R. (2021). Bio-active free direct optical sensing of aflatoxin B1 and ochratoxin A using a manganese oxide nano-system. *Frontiers in Nanotechnology*, 2, 621681.
82. Singh, P., & Ikram, S. (2016). Role of nanomaterials and their applications as photocatalyst and sensors: A review. *Nano Research and Applications*, 2.
83. Ahlawat, A., Dhiman, T. K., Solanki, P. R., & Rana, P. S. (2023). Facile synthesis of carbon dots via pyrolysis and their application in photocatalytic degradation of rhodamine B (RhB). *Environmental Science and Pollution Research*, 1–8. https://doi.org/10.1007/s11356-023-25604-6.
84. Ahlawat, A., Dhiman, T. K., Solanki, P. R., & Rana, P. S. (2022). Enhanced photocatalytic degradation of p-Nitrophenol and phenol red through synergistic effects of a CeO2-TiO2 nanocomposite. *Catalysis Research*, 2(4), 1–13.
85. Poddar, M., Lakshmi, G. B. V. S., Sharma, M., Chaudhary, N., Nigam, S., Joshi, M., & Solanki, P. R. (2022). Environmental friendly Polyacrylonitrile nanofiber mats encapsulated and coated with green algae mediated Titanium oxide nanoparticles for efficient oil spill adsorption. *Marine Pollution Bulletin*, 182, 113971.
86. Khin, M. M., Nair, A. S., Babu, V. J., Murugan, R., & Ramakrishna, S. (2012). A review on nanomaterials for environmental remediation. *Energy & Environmental Science*, 5, 8075–8109.
87. Kujur, V. S., & Singh, S. (2020). Structural, magnetic, optical and photocatalytic properties of GaFeO3 nanoparticles synthesized via non-aqueous solvent-based sol–gel route. *Journal of Materials Science: Materials in Electronics*, 31(20), 17633–17646.
88. Chacón, J. M., Leal, M. T., Sánchez, M., & Bandala, E. R. (2006). Solar photocatalytic degradation of azo-dyes by photo-Fenton process. *Dyes and Pigments*, 69, 144–150.

89. Solís-López, M., Durán-Moreno, A., Rigas, F., Morales, A. A., Navarrete, M., & Ramírez-Zamora, R. M. (2014). Assessment of copper slag as a sustainable Fenton-type Photocatalyst for water disinfection. In *Water reclamation and sustainability* (pp. 199–227). Elsevier.

WEBLINKS

Weblink1: https://worldwaterreserve.com/water-crisis/water-scarcity-facts/
Weblink2: http://water-purifiers.com/different-drinking-water-purification-techniques/
Weblink3: www.drinkmorewater.com/water-purification-methods
Weblink4: https://chem.libretexts.org/Bookshelves/Physical_and_Theoretical_Chemistry_Text book_Maps/Supplemental_Modules_(Physical_and_Theoretical_Chemistry)/Funda mentals/Fluorescence_Resonance_Energy_Transfer

7 Metal Oxide with Activated Carbon for Wastewater Treatment

*Farha Shahabuddin, Ab Qayoom Mir,
Badri Vishal Meena, Afsar Ali, Saleem
Ahmad Yatoo, and Rayees Ahmad Zargar*

CONTENTS

7.1 INTRODUCTION

Globally, water pollution and energy disaster are the main common challenges at present, which are mainly responsible for rapid industrial growth and the cause of the human pollution race. Currently, researchers have been focusing on environmentally friendly and sustainable technologies for the purification of water.[1–3] The consumption of fossil fuels enhances the emission of greenhouse gases and the pollution of soil and water in the environment from different sources like transportation, manufacturing, construction, industrial site, households, and deforestation. Hence, to minimize environmental pollution and create a clean environment for future generations, we need to aim at limiting the use of fossil fuels and industrial wastes and maximizing the use of environmentally friendly materials. Mainly, water pollution is created in the environment due to the presence of contaminants that are used in our daily routines, the burning of fossil fuels, bathing, and construction that are enhanced by human activities. Along with sustainable abundance, clean energy sources like wind, hydropower, and solar are the better option for minimizing water pollution than petroleum-based fuels. However, the use of these energy sources for wide applications is limited due to technological confinement. Solar energy acts as a pillar of renewable energy and a source of clean and sustainable energy, which has shown a high tendency toward eco-friendly photocatalyzed green fuel generation, that can satisfy future human energy needs. Herein, several physical factors such as daytime, season, geographical position, and intensity of sunlight are major limitations.[3] The transformation of solar energy into chemical energy is a convenient way to minimize energy requirements at the world level and is responsible for the shift in environmental pollution. Photocatalysts play a pivotal role in this type of conversion.[4, 5] Photocatalysis has attracted great attention to minimizing hazards of pollutants and the formation of sustainable green energy sources. Generally, photochemical materials are based on metal-free organic compounds; semiconductors, mainly semiconductor systems, have been focused on photocatalytic activity. Fujishima and Honda in 1972 discovered photoactive water oxidation by applying TiO_2.[6] Several water purification methods, such as sedimentation, photocatalytic, membrane filtration, precipitation, ion exchange, adsorption, assimilation, and degradation by microorganisms, have a pivotal role in converting wastewater into fresh and clean water that can be utilized by living organisms.

Wastewater treatment is highly desirable/essential for purifying contaminated water into clean water, which could be utilizable to living organisms and human beings. Meanwhile, the huge growth in the industrial sector, such as paper, pulp, pharmaceutical, dye, and metal refineries, directly contaminates the clean and safe water.[7] Apart from this, wastes generated in metal refineries are more toxic due to the presence of a maximum portion of heavy metals that are directly released into our environment and affect aquatic ecosystems and the health of biotic life. Unfortunately, the inflation of various heavy metals including of mercury, arsenic, cadmium, lead, copper, and nickel have shown a very toxic nature at negligible concentrations.[8, 9] Various photocatalytic cells such as photocatalytic reactors, immobilized TiO_2, and mesoporous zinc oxide (ZnO) have been utilized for degrading organic pollutants along with pharmaceutical drug molecules.[9–11] Herein, different types of photocatalytic systems have been used for water treatment, air purification, and the decomposition of polyaromatic based complex organic pollutants in wastewater.

Generally, hydroxyl radicals and radical anions play a crucial role in the oxidation of organic pollutants at the surface of MONs and AC nanocomposites. In addition, these

have broad applications at various industrial levels such as in the paper, pulp, pharmaceutical, pesticide, and dye industries. Several types of MONs/AC nanocomposites act
as photocatalysts, in which hole generation takes place in the VB; the reducing species accepts electrons in the CB to generate strong oxidizing agents such as hydroxyl
radicals. Meanwhile, electrons in the CB are received by electron-accepting species to
generate radical anions.[4, 12] However, a broad range of metal oxides have been constructed for water treatment with low catalytic efficiency. In this regard, researchers are
searching for noble catalytic material with a large surface area and high efficiency.[13,
14] The ongoing research is mainly making a hub point for the development of MONs/
AC nanocomposites for treating polluted water. This new modulation in the adsorbent
could enhance its surface area along with the efficiency for pollutant removal manifold.

7.2 METHODS OF POLLUTANT REMOVAL FROM WASTEWATER

For the past few decades, environmental pollution is the dominant issue in our atmosphere, which is responsible for serious problems such as contamination of water and
deterioration of air along with decreasing fertility of the earth. Among them, water pollution is the one of most dominant environmental concerns for the whole world. Several
water pollution factors, such as algal formation, eutrophication, and groundwater pollution, have a direct negative impact on freshwater resources. The presence of various
types of pollutants in the water body has a direct impact on our aquatic ecosystem and
living organisms. So, it is mandatory to get rid of environmental pollutants by using
eco-friendly methods. Herein, the following pollutant removal methods have been
extensively used, including flocculation, precipitation, assimilation, adsorption, biodegradation, and complexation, and are playing a vital role in water purification.[8, 15]

7.2.1 ASSIMILATION

Mainly, assimilation is related to the microbial aggregate layer. Nutrients, such as
nitrogen, carbon, and phosphorous, are extracted by photosynthetic microorganisms for their microbial growth. This type of microorganism tends to absorb nitrate,
nitrite, and ammonium as the primary source of nitrogen.[16] However, nitrogen,
carbon, and phosphorus are extracted by microorganisms in the two steps of the
anaerobic and aerobic processes. Nitrogen and phosphorous absorption take place in
the aeration period (first step). The weak absorption of phosphorous also takes place
during agitation (second step). The organic materials that are adsorbed by microorganisms act as the main nutrient sources for their growth. A maximum part of these
adsorbed organic materials is utilized for maintaining the growth and formation of
the active part of the cell.[17, 18]

7.2.2 ADSORPTION

Generally, adsorption is one type of mechanism, in which heavy metal ions are
adsorbed on the surface of adsorbents at a variable operating conditions such as
temperature, adsorbent amount, adsorption time, and pH. Certain types of microorganisms also adsorb metal ions from low concentrations of thrash metal ions

in the wastewater body. Therefore, several examples of biomass have a great tendency toward the adsorption of heavy metals at the crouched level. Interestingly, a few groups of aggregated microorganisms tend to adsorb a variety of pollutants in the water body, which have been generally referred to as bio-adsorption. A few examples of microorganisms included in this category are such as bacteria, algae, and fungi. All these microorganisms contain specific features, such as flocculation and adhesion, which make them able to adsorb heavy metals, toxic materials, and organic dyes from wastewater. This method has shown flagship properties such as high removal capacity, easy implementation, and low operating costs.[8, 19–22]

7.2.3 BIODEGRADATION

For the last few decades, microorganisms have been used in our environment to treat wastewater. Biodegradation is the natural process in which biological agents such as fungi, bacteria, and green plants have been applied to decomposed toxic stuff in wastewater.[23, 24] However, biological wastewater treatment plants play a pivotal role in transforming toxic compounds into neutral products and helping remove carbon-based materials. Along with this, microbial aggregation can be influenced by environmental factors, operation conditions, process configuration, biotic interactions, stochastic process, and dispersal limitations.[25–27] Herein, the efficiency of microbial communities shows a variable degree of pollutant degradation for industrial and municipal wastewater treatment plants because both contain different components of organic loading, which alter bacteria's growth and metabolism. Generally, biodegradability is a parameter for determining the effective amount of nutrients along with the presence of toxic substances in the pollutant water. The biodegradability of any polluted water body can be measured by the ratio of BOD to COD. [28–32]

7.2.4 FLOCCULATION

Different types of impurities such as fine suspended solids, dissolved solids, and inorganic and organic small particles are major pollutants that are responsible for wastewater. These small solid suspended particles contain a surface charge that is not easier to separate from the water body. So the separation of this type of colloidal particle is very tactical and challenging for industries and water treatment plants. Flocculation is a very crucial step in industrial wastewater treatment. Various promising and advanced methodologies have been developed for separating small colloidal particles from a water body, in which flocculation/coagulation is one of the most suitable methods for the separation of suspended and colloidal small solid particles from wastewater.[33, 34] Coagulation/flocculation is a simple, cheap, and very efficient method for wastewater treatment such as textile, paper, pulp, refinery wastewater, and oil mill effluent. In this process, generally variable types of coagulant/flocculants such as inorganic flocculants (salt of multivalent metal), organic polymeric flocculants, and biopolymer flocculants are playing a pivotal role because of their tendency to flocculate effectively at low concentrations. Along with this, biopolymer-based coagulant utilization is increasing drastically due to its biodegradability and environmentally friendly nature. Based on the mechanism, the initially added coagulant/flocculant neutralizes the charged colloidal particles after that they aggregate or agglomerate into large particles (flocs) that settle and help in water purification.[35, 36]

7.2.5 COMPLEXATION

Complexation is one of the foremost processes for removing heavy metals in contaminated water. The simple principle behind metal ion separation is the coordination of heavy metal ions with macromolecular ligand species, which increases the aggregated molecular weight. As a result, this compound settles to the bottom, and then the water is purified. However, the complexation method has shown high separation efficiency due to the low energy requirement for selective binding of metal ions and water-soluble polymeric ligand species.[37, 38] Herein, polymeric ligands with carboxylic acid and amine groups have been extensively used for this complexation method. Several examples of macropolymeric ligands such as chitosan, carboxyl methylcellulose, polyvinylethyleimine, polyvinyl alcohol, and polyacrylic acid have been used to complex heavy metal ions.[39–41]

7.3 SYNTHESIS OF MONS/AC NANOCOMPOSITES

Generally, synthesis methods for MONs/AC nanocomposites have been classified based on a different mode of application. Several types of synthesis methods have been explained for the formation of MONs adsorbed within AC.

7.3.1 MICROMECHANICAL EXFOLIATION

First, this method has been utilized by Novoselov and Gaim for generating graphene with a few layers by using adhesive material. Also, it is helpful only for creating TMOs with a few layer. All physical properties, such as the stoichiometric ratio of atoms, along with an order of molecules have been unaltered after the formation of TMOs. However, it is only applicable to synthesizing layered nanomaterials.[42] It is used in variable operating conditions such as an electric field or by using epoxy resin.

7.3.2 ULTRASONIC EXFOLIATION

This process is more convenient and more effective for constructing single- to few-layered nanomaterials than the micromechanical process. However, the selection of an apt solvent and appropriate time for sonication play a vital role in this method. The selection of good solvents is very essential to avoid reaggregation of exfoliate nanomaterials and used as a dispersing medium for delaminating Vander Waal solids. This method is highly desirable due to providing pure single-layer nanomaterials, which are highly dictated to electronic gadgets. WO_2 and MoS_2 nanomaterials can be constructed by this process.[43–45]

7.3.3 Ion-Exchange Exfoliation

The previously discussed two synthetic methods are useful only for Van der Waal materials, in contrast, both methods are not appropriate for ionic solids due to revealing strong ionic attraction at the atomic level. This method is less applicable to the synthesis of TMOs. Seong Ju et al. have developed $LiCoO_2$ nanomaterial by the ion-exchange exfoliation method.[46]

7.3.4 Lithium-Intercalated Exfoliation

This method has a huge attraction toward the synthesis of a single-layer nanomaterial. The lithium-intercalated exfoliation method can be used to construct single nanosheets containing nanomaterial that is not an easy task by another process. The amount of single-layer nanomaterials can be altered by lithium-based TMOs. In this method, the yield of a single-layer formation is controlled by the formation of lithium-incorporated TMOs. This synthesis method required a huge temperature, so it is necessary to perform reaction very carefully to avoid the formation of nanoparticles of a lithium precursor. Along with this, the generation of intercalated ions is mainly caused by lowering the interaction between layers in the plane by charge movements (Figure 7.1).[42] MoS_2 nanoparticles have been developed by this method.

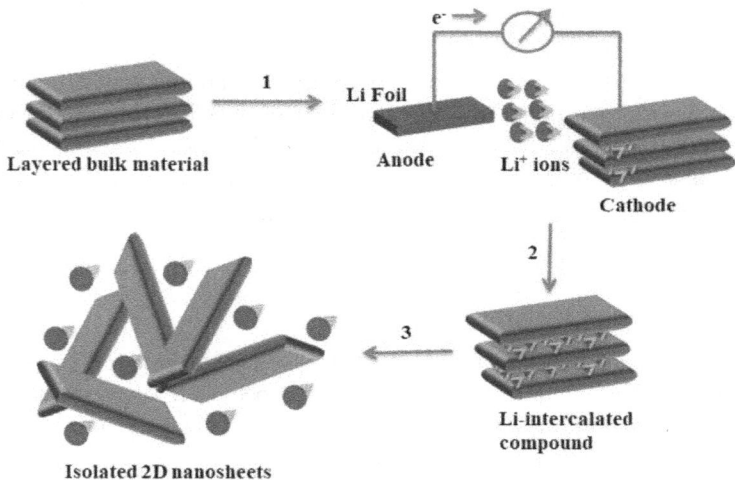

FIGURE 7.1 Lithium-intercalated method for construction of nanomaterials. [47]

7.3.5 COPRECIPITATION METHOD

The coprecipitation process is a very emerging method for the synthesis of nanomaterials. It is very effective in the formation of simple, cost-effective, uniform, and controllable particle sizes. Various examples of MONs, including ZnO/Cu, ZnO-CuO and graphene oxide/Fe_3O_4, have been constructed by coprecipitation methods. Also, a few examples of binary nanocomposites TMOs, such as $CuWO_4$, $CuWO_4$/NiO, NiO-ZnO, and $CeO_2.CuO.ZnO$, have been synthesized by this method. These all examples are revealing that the coprecipitation method is a very efficient, simple, and easy method for deploying MONs.[19, 48–52]

7.3.6 SOL-GEL METHOD

The sol-gel process is one type of wet-chemical method. In this process, gel formation takes place after sol formation by drying the solvent. However, a variable number of applications, such as sensing, antibacterial, and catalysis, are mainly affected by the pore size, structure, constitution, and dimensions of nanomaterials. The formation of nanomaterials can be controlled by changes in the nature of the solvent, the pH of the reaction medium, and solvent viscosity. The sol-gel method has broad attention for the synthesis of nanomaterials because of their low temperatures required for reaction, high purity of products, and defined particle size. The sol-gel method is a very emerging technique for the synthesis of TMO nanocomposites such as NiO/TiO_2, ZnO/CuO, WO_3/TiO_2, ZnO/reduced graphene oxide, CdO/ZnO, ZnO/SnO_2, and ZnO/SiO_2.[14, 53–60]

7.3.7 MICROWAVE-ASSISTED METHOD

This process has shown few advantages compared to other methods such as low reaction time (min), great energy efficiency, and desirable shape and size of TMOs. A few examples of TMOs such as SnO_2, α-Ni(OH)$_2$, $K_{0.17}MnO_2$, and CuSe have been synthesized by this strategy. However, α-Ni(OH)$_2$ can be constructed by applying Ni(NO$_3$)$_2$.6H$_2$O, urea, deionized water, and ethylene glycol as precursors (Figure 7.2). [61, 62]

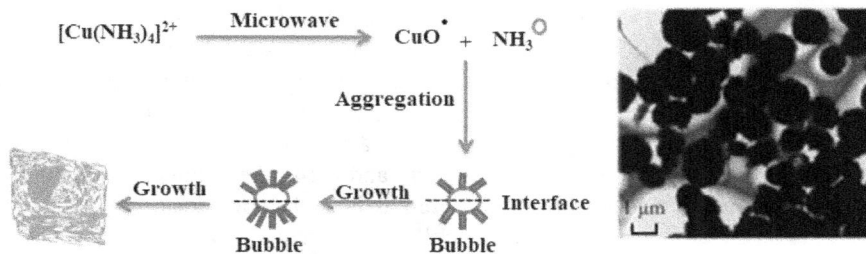

FIGURE 7.2 Microwave-assisted method for the development of TMO nanomaterials.[63]

7.3.8 CVD

This method loaded surface substrate within gas molecules in a quartz reactor, which operates at variable pressure, temperature, and gas flow rate.[64–66] Generally, atypical CVD setup is assembled by engaging a quartz reactor, a mass flow chamber, thermocouples for temperature adjustment, a gas delivery operator, a vacuum chamber, and an energy device. We can use variable metal oxide substrates such as Ni, Cu, and Fe, along with CH_4 and C_2H_2, are used as carbon sources in the CVD method. However, the CVD method has a good capability to develop excellent properties containing TMOs with a huge surface area, uniform thickness, a scalable size, and high photocatalytic features. In the CVD process, the required material can be doped on the surface of the substrate by deploying optimum temperature. The CVD process is a vital method for the synthesis of desirable physical and chemical properties of high-purity TMOs/AC nanocomposites.[67, 68]

7.4 APPLICATION OF MONS/AC NANOCOMPOSITES

Interestingly, the applications of TMOs/AC nanocomposites are increasing drastically in the area of nanotechnology because it contains distinctive physiochemical properties. TMOs/AC nanocomposites have auspicious applicability for photocatalysis, the oxidation of organic pollutants, water purification, and biomedicine.[69, 70] The AC or activated charcoal is one type of carbon source that has revealed large porosity along with huge surface area, making them highly fascinating regarding their catalytic properties and chemical reactions. Interestingly, a very low amount of AC (approximately 1 g) has shown a huge surface area (approximately 500 m^2/g) due to its potency of large porosity. Commercial AC has shown broad-scale applicability for removing heavy metals, such as mercury, cadmium, nickel, and copper; clearing industrial polluted water; and oxidizing organic dyes and pesticides along with various types of toxic and hazardous stuff. Also, AC provided a clue to researchers of its applicability to the separation, purification, and catalytic activities of gases.[71] The high cost and regeneration are the main drawbacks. Thus, there is a need to search for synthetic techniques for low-cost AC. However, carbon-based materials such as biochar, carbon aerogel, AC, graphene, CNTs, and carbon nanofibers have a wide attraction to researchers for their application as adsorbents.

The performance of AC depends on its chemical nature. The efficiency and activity of AC can be enhanced to many folds by altering variable physical and chemical parameters. Herein, lignocellulose biomass has been used as a precursor for the development of AC via the pyrolysis method. By the way, detaching moisture and volatile compounds from the bio-precursor should be necessary for the enhancement of AC's porosity at an extended level.[73] The main motive for synthesis methods of AC is a huge surface area with large porosity and high pore volume. AC possesses a heterogeneous pore structure. Also, there has been a tremendous increase in the literature on the use of nanoparticles for pollutant degradation owing to their surface area. Nanosized metal oxides adsorbed with adsorption material have a huge surface area, fast kinetics, high capacity, and specific affinity for variable types of contaminants.[74, 75] In the quest to find a more affordable, efficient adsorbent researchers

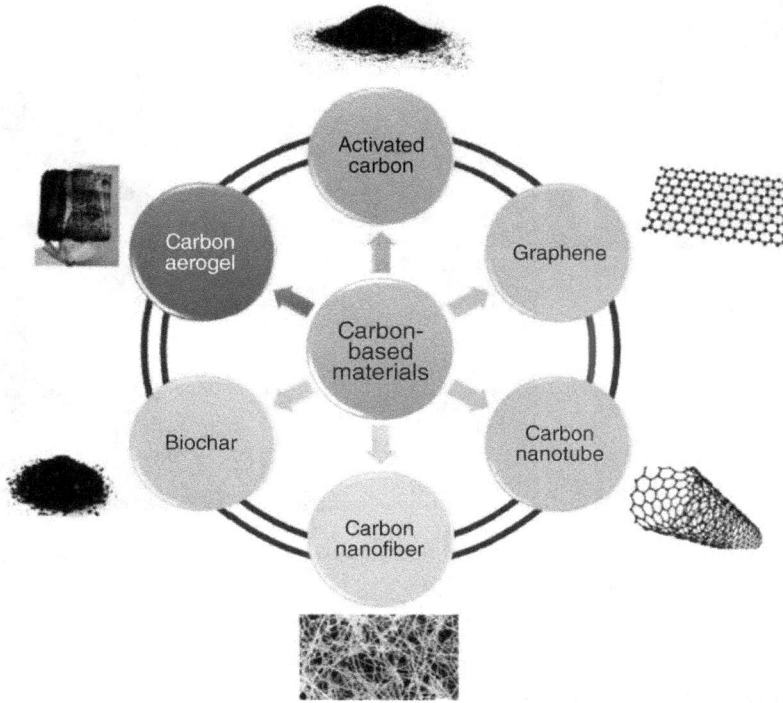

FIGURE 7.3   Carbon-based adsorbent for wastewater treatment. Adapted with permission.[72]

have used a mix of both AC/nanocomposite for pollutant removal from wastewater. The composites of nanoparticles with AC induce the surface area of the adsorbent along with the firmness of the surface. With this modification, the adsorption capacity of the composite increases tremendously compared to traditional only AC or nanoparticles. Herein, MONs/AC nanocomposites can be synthesized by dispersing inorganic oxides within the AC matrix. MONs/AC nanocomposites have been utilized for various types of chemical reactions, the decomposition of organic pollutants, and use for catalysis. TMOs are predominantly useful for this purpose due to their low production cost, more selectivity, and effortless regeneration. As one example, iron-based Fe_3O_4/AC nanocomposites have been used highly effectively as a Fenton-like catalyst for oxidizing organic contaminants. In this case, AC provides a huge surface area and porosity; as a result, its catalytic efficiency has been enhanced.[76, 77] One form of Fe_3O_4, amorphous ferric oxide (FeOOH) is a highly porous oxide in nature, which has a huge affinity for removing arsenic from wastewater.[76, 78] Fe_3O_4/AC has shown the highest affinity for arsenic removal compared to only AC. Also, several iron oxide/AC nanocomposites have been developed by surface functionalization on AC using $FeSO_4$ and $FeCl_2$.[71, 79]

Along with this, TiO_2/AC nanocomposites have been reported purifying wastewater, whereby AC adsorbed the pollutant, which cooperated with the

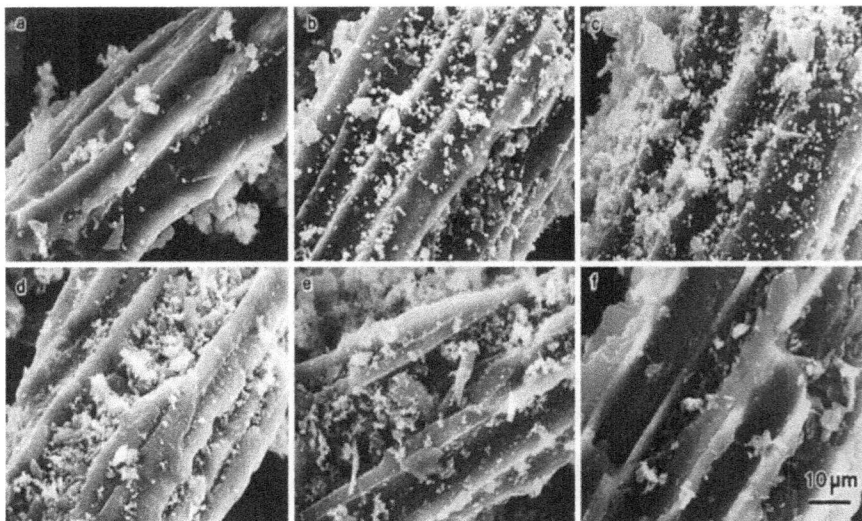

FIGURE 7.4 SEM micrograph of TiO$_2$/AC composites prepared for water purification. Adapted with permission of reference 68.[80]

photocatalytic activity of TiO$_2$.[80, 81] Meanwhile, magnetically active TiO$_2$/ AC nanocomposites also have been developed by impregnating of Fe$_3$O$_4$, making them more suitable for the easy separation of suspended photocatalyst from the reaction medium that has been used for phenol degradation under UV light illumination.[4, 82] Herein, SnO$_2$/AC nanocomposites have been synthesized by the wet impregnation method by heating of resulting product at 700–800°C. Interestingly, SnO$_2$/AC nanocomposites have revealed high porosity and uniform pore size compared to parent AC or SnCl$_2$.[83] Research has worked on enhancing the physical properties, such as the thermophysical, mechanical, and electrical properties, of ceramic by incorporating AC in different ratios.[71] However, Al$_2$O$_3$/AC nanocomposites have excellent properties for protecting the wall ceramic monolith via the formation of a uniform layer of AC. Also, Al$_2$O$_3$/AC has shown catalytic applications such as catastrophic filtration and preserving nature. [84]

The catalytic activity of CeO$_2$ has been enhanced by doping AC with CeO$_2$, which provides a huge surface area along with electron movement between CeO$_2$ and AC. These CeO$_2$/AC nanocomposites revealed excellent catalytic features for the ozonation of oxalic acid and aniline.[86, 87] In another study, also this has been found that CeO$_2$/AC nanocomposite showed magnificent photocatalytic properties for adsorption and degradation of 4-chlorophenol in an aqueous medium compared to pure CeO$_2$.[87] The efficiency of cerium-based nanocomposites can be altered by the amount of Ce^{3+} cations. Mainly, CeO$_2$/AC nanocomposites can be prepared by the impregnation process by using cerium nitrate as a precursor.[88, 89] Herein,

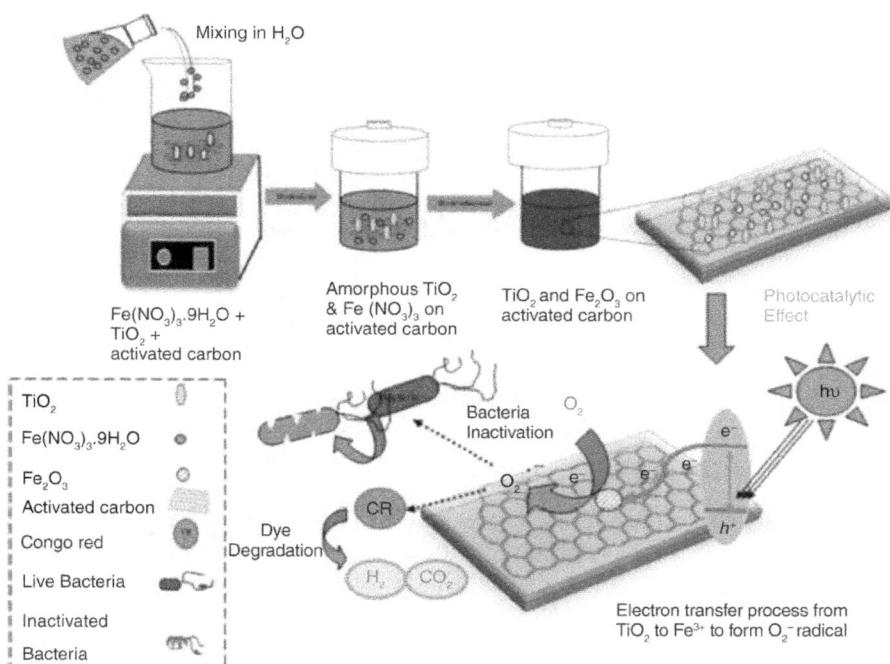

FIGURE 7.5 Schematic representation for MON/AC nanocomposite formation and used for photocatalytic degradation of Congo red dyes. Adapted with permission of ref.[85]

Geneche et al. have developed Fe_3O_4/AC in the variable ratio of Fe_3O_4 and AC via coprecipitation methods and applied for Cr(VI) removal from wastewater. The ratio 2:1 w/w ratio of AC and Fe_3O_4 nanoparticles has shown the highest Cr(VI) removal efficiency. However, AC and magnetic property of nanocomposites play a crucial role in magnetic separation and regeneration.[90] Interestingly, Sillanpaa et al. have constructed Fe_3O_4/AC nanocomposites by coprecipitation method for removing Cr(VI), Cu(II), and Cd(II) cations from wastewater (Table 7.1). The highest removal efficiency has been observed at pH-2 for Cr(VI) and pH-6 for Cu(II) and Cd(II) at room temperature.[91] Another study has been performed by Subramanian et at.in which they have developed $Fe_3O_4/$nanosphere for removal of As(III) and As(V) from water bodies. A carbon-based nanosphere consists of a high surface area, high pore volume, and large hierarchical mesopore/macropore structures. Based on the composition of nanocomposites, 37% weight of iron oxide has represented the highest adsorption efficiency for As(III) and As(V), which is very high than pure iron oxide–based adsorbents.[92] In one study, Fe_3O_4/AC has been deployed as an adsorbent for the removal of BB dye from polluted water. This study also revealed that synthesized nanocomposites have shown the best adsorbent efficiency of BB in an aqueous medium.[93–95]

TABLE 7.1

MON/AC Nanocomposite with Specific Applications

MON/AC Nanocomposite	Application	References
Fe_3O_4/AC, TiO_2/AC, ZnO/AC	Decomposition of organic molecules, hydroxylation of benzene, and acylation of alcohols	94 and 95
WO_3/AC	Oxidation of methanol and ethanol, hydrogenation of ethylene, combustion of toluene	82
SnO_2/AC	Low temperature of CO	83
Fe_3O_4/AC	Cr(VI), Cu(II) and Cd(II) removal from wastewater	91
Al_2O_3/AC	Hydroprocessing methods	84
Fe_3O_4/AC	As(III) and As(V) removal	92
Fe_3O_4/AC	Removal of BB	93

7.5 PHOTOCATALYTIC DEGRADATION

Nanotechnology is playing a pivotal role in the domain of photocatalytic chemistry. Nanomaterials have outstanding features including optimized light irradiation, lighter weight, low recombination energy, and high surface area, which has attracted researchers' attention for utilizing nanomaterials for photocatalytic application. Along with this, nanomaterials are more promising than micromaterials due to containing high electromagnetic wave interaction, and low ionization potential. Nanomaterial-based photocatalysts have shown broad contributions toward environmental clearance and energy application such as CO_2 reduction into hydrocarbon fuels, hydrogen evolution by water oxidation, wastewater treatment, air purification, and organic pollutant degradation. Several methods, such as chemical, biological, and electrochemical methods, for degrading organic molecules have been reported, but they are less effective. However, photocatalysts have spotlight attraction compared to other methods such as advanced oxidation method, UV treatment, ozonation, UV with H_2O_2 treatment, and the Fenton process for organic molecules degradation due to the formation of less toxic intermediates and effective degradation.[15, 96–98]

7.6 DYES DEGRADATION

Water pollution caused by industrial dye effluents rising environmental problems for human health and aquatic life. Today, dyes are widely used in the industries like textile, pharmaceuticals, leather, paper, and more, which are hazardous, toxic, and carcinogenic to our ecosystem and humankind. Mainly, dye effluent contains variable aromatic and polyaromatic dyestuffs, and their substitutes have been shown resistant to several water treatment methods such as coagulation, bio-treatment, and AC adsorption. However, AOPs are showing characteristic dye degradation activity and efficiency in an aqueous medium at room temperature. However, researchers have explored variable AOPs methods such as ozonation, sonochemical oxidation, persulfate oxidation, Fenton oxidation method, UV/H_2O_2, zeolite, and its derivatives, BiOBr/graphene oxide nanocomposite, photocatalyst bismuth oxyhalides/reduced graphene oxide nanocomposite (Table 7.2).[51, 62–71]

TABLE 7.2

Percentage of Degradation of Dye by Metal Oxide with AC

Metal Oxide with AC	Dye	Optimal Condition	% of Degradation/ (mg/g)	Ref.
TiO$_2$-AC	Ruthenium N-3 dye	180 min	99%	59
ZCP-AC	Methylene blue	20 min	98.12%	60
Fe-MAC	Methylene blue	75 min	99%	61
CeO2-AC	Methylene blue	60 min	98%	62
ZnO-Ac	Methylene blue	60 min	98%	62

During the manufacturing of dye, many liters of freshwater have been used that need to be treated before discharge into the river and more.[99–102]

7.7 PHARMACEUTICALS/PESTICIDES REMOVAL

Over the last few decades, the level of organic molecules such as pharmaceutical, dyes, and pesticide wastes has been increasing from nanograms to micrograms per liter of polluted water. Generally, wastewater generated from the hospital, municipal, wastewater, farmhouses, and veterinary wastes have been included in this category. Herein, the worldwide utilization of medication, fertilization, and pesticides has had a direct effect on human life and environmental pollution. Various types of residual drugs can be generated metabolism disruption and drug resistance in the human body and aquatic life.[103–105] However, a trace amount of drugs and pesticides has persisted in water resources after treatment by filtration, membrane distillation, and reverse osmosis, which is indicated the low efficiency of these methods. Apart from this, various drug molecules such as paraben, diazepam, carbamazepine, and diclofenac are biologically resisted. However, a broad range of techniques such as adsorption, air stripping, biological methods, and reverse osmosis have been applied for the deployment of pharmaceutical drug pollutants from water bodies. Based on the effectiveness, the adsorption method is more promising due to a lower treatment price, the simplicity of the process, low maintenance, and higher efficiency.[106–108]

Z. Fallah et al. reported metal oxide for the removal of organic pollutants such as drug molecules and pesticides.[109] Al-Jabari et al. applied iron oxide for the absorption of levofloxacin (80% in 240 min). Abdelillah Ali Elhussein et al. prepared a CeO$_2$-based metal-organic framework that showed 95.78 mg g^{-1} absorption of 2,4-dichlorophenoxyacetic acid in the batch system. Yin et al. reported a core–shell-based Fe-FeOOH structure has been used to remove ibuprofen (nearly 98% in 30 min).[109] Roghaye Behnam et al. synthesized an aluminum oxide and magnesium oxide nanomaterial for the adsorption of diazinon (0.025 g/ul in less than 10 min) and methyl parathion (0.05 g/ul in 10 min), respectively.[110]

7.8 TOXIC METAL REMOVAL

Today, heavy metal pollution caused by metal refineries has been one of the most serious and toxic to our environment. Herein, detaching heavy metals from the surroundings is the primary issue for researchers. Even low exposures to heavy metals are hugely toxic to humans.[111, 112] The availability of heavy metals such as mercury, cadmium, copper, and nickel in the aqueous system has continuous effects on humans and aquatic life when present at more-than-acceptable levels. However, heavy metals cannot be destroyed or degraded; they can be only adsorbed from wastewater. Heavy metal toxicity can occur through sources such as drinking water contamination, increasing concentration of metal nanoparticles near the source of emission, or circulation via the food chain.[113, 114] Industries release the untreated wastewater into the river that contains carcinogenic and harmful toxic metals such as chromium, zinc, arsenic, and others. Continuously, the applications of heavy metals are increasing, which has increased its concentration in natural water sources. Various technologies, including adsorption, precipitation, and ion exchange, have been applied for detaching heavy metals from contaminated water. However, the adsorption method is playing a pivotal role in removing heavy metals.[8, 115] Few examples of adsorbents including silica, AC, and graphene have been utilized for heavy metals rid of wastewater.[116, 117] Hoang Thu Ha et al. reported about 32.57 mg g^{-1} removal efficiency of arsenic by using Fe_3O_4-AC material.[118] In the other literature, Wang Weilong et al. synthesized Fe/Mn (1:15) mixed oxide for removing hexavalent chromium up to 26.6 mg g^{-1}.[59]

7.9 MECHANISM OF DYE DEGRADATION

The photocatalysis on the surface of the photocatalyst takes place because of the reaction between organic pollutants and strong oxidizing agents, which are generated by the light illumination or the oxidizing species. Mainly, metal oxides (CuO, Fe_2O_3, WO_4, ZnO, ZnO, TIO_2, WO_3, SnO_2, Ta_2O_5, CeO_2, CuO, ZrO_2, Fe_2O_3) and SnO_2 act as photocatalysts used for photocatalysis under light illumination for degradation of organic and inorganic contaminants. However, the selectivity of photocatalysts for the photocatalytic mechanism mainly stands on the reduction and oxidation step. For organic dye degradation, TMO nanomaterials are more efficient and reactive because TMOs possess good properties such as light absorption, indirect bandgap energy, the presence of highly active surface area, and the negligible recombination of charge carriers.[119]

Upon light illumination, absorption of photons takes place at the photocatalyst and generated charge carriers (hole and electron pairs) followed by transfer of photogenerated carriers in the photocatalyst.[120, 121] Generally, a suitable hybrid photocatalyst would be induced photocatalytic activity by lowering the interactivity of electron–hole pairs.[122]

Yu et al. synthesized a $WS_2/WO_3/H_2O$-hybrid photocatalyst that showed photocatalytic decomposition of methyl orange (approx. 90–95%) under a visible range of solar illumination. Peitao Liu et al. synthesized α-Fe_2O_3@N-doped MoS_2-hybrid photocatalyst for detaching of Rhodamine B in the visible range of solar irradiation and showed a 26.4 times higher decomposition tendency than MoS_2 nanomaterials.[123] Similarly, Yang et al. constructed $MoS_2/ZnO/P$-hybrid photocatalysts that revealed a high oxidation tendency (95%) for methylene blue and Rhodamine B in presence of sunlight illumination.[124] Another strategy was also reported for enhancing the adsorption efficiency of TMO nanomaterials that can be incorporated

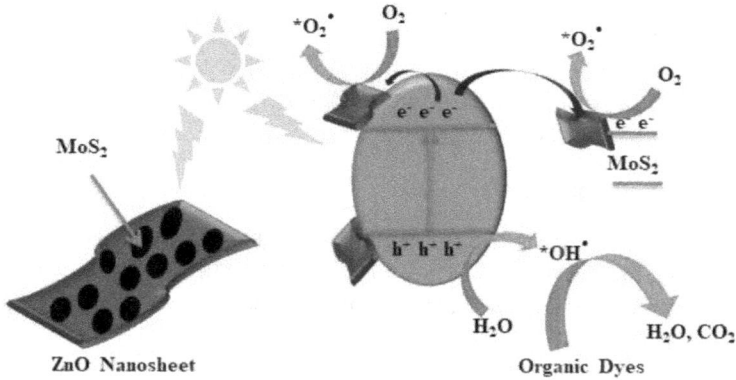

FIGURE 7.6 Photocatalysis mechanism of ZnO/MoS$_2$ nanomaterial. Adopted with permission of reference 125 [125].

by supporting MONs on the surface of AC. Along with this, adsorbent materials should contain specific properties such as high surface areas with apt pore sizes and excellent conductivity for facile charge transfer. However, various types of TiO$_2$/AC nanocomposite have been reported for specifical properties, in which a group of researchers have utilized carbon-based nanosphere, CNTs, and MWCNTs as adsorbents.

7.10 CONCLUSION

Recently, TMO/AC nanocomposites have been getting huge awareness because their distinctive features included low density, thermal stability, large reactive surface area, and high porosity. However, a different form of AC provided high conductivity, excellent mechanical strength, and thermal stability which are preferable for advanced composite applications. The presence of AC, along with nanoparticles, makes nanocomposites free from eco-toxic divalent ions. Mainly, this chapter explored the photocatalytic mechanisms of photocatalyst, classification, and synthetic approach and improving the physical and chemical properties of TMOs, along with AC and its application.

ABBREVIATIONS

AC	Activated carbon
AOPs	Advanced oxidation processes
BB	Bismarck brown dye
BOD	Biological Oxygen Demand
CB	Conduction band
CNTs	Carbon nanotubes
COD	Chemical Oxygen Demand
CVD	Chemical vapor deposition

MONs   Metal oxide nanocomposites
MWCNTs   Multiwalled carbon nanotubes
TMOs   Transition metal oxides
UV   Ultraviolet
VB   Valence band

ACKNOWLEDGMENTS

The author FS would like to thank AMU Aligarh for support and AA would like to acknowledge the support from DST, India (NPDF Fellow-PDF/2021/000449) for this book chapter contribution. Moreover, the author would also like to thank IIT Delhi for its facilities.

REFERENCES

1. Walter, M. G.; Warren, E. L.; McKone, J. R.; Boettcher, S. W.; Mi, Q.; Santori, E. A.; Lewis, N. S. Solar Water Splitting Cells. *Chemical Reviews.***2010**, *110* (11), 6446–6473. https://doi.org/10.1021/cr1002326.
2. Balat, M.; Balat, H. Biogas as a Renewable Energy Source—A Review. *Energy Sources, Part A: Recovery, Utilization, and Environmental Effects* **2009**, *31* (14), 1280–1293. https://doi.org/10.1080/15567030802089565.
3. Haque, F.; Daeneke, T.; Kalantar-zadeh, K.; Ou, J. Z. Two-Dimensional Transition Metal Oxide and Chalcogenide-Based Photocatalysts. *Nano-Micro Lett.***2017**, *10* (2), 23. https://doi.org/10.1007/s40820-017-0176-y.
4. Naikwade, A. G.; Jagadale, M. B.; Kale, D. P.; Gophane, A. D.; Garadkar, K. M.; Rashinkar, G. S. Photocatalytic Degradation of Methyl Orange by Magnetically Retrievable Supported Ionic Liquid Phase Photocatalyst. *ACS Omega* **2020**, *5* (1), 131–144. https://doi.org/10.1021/acsomega.9b02040.
5. Yoon, T. P. Visible Light Photocatalysis as a Greener Approach to Photochemical Synthesis. *Nature Chemistry* **2010**, *2*, 6.
6. Hernández-Alonso, M. D.; Fresno, F.; Suárez, S.; Coronado, J. M. Development of Alternative Photocatalysts to TiO2: Challenges and Opportunities. *Energy & Environmental Science* **2009**, *2* (12), 1231–1257. https://doi.org/10.1039/B907933E.
7. Le Van, K.; Luong Thi, T. T. Activated Carbon Derived from Rice Husk by NaOH Activation and Its Application in Supercapacitor. *Progress in Natural Science: Materials International* **2014**, *24* (3), 191–198. https://doi.org/10.1016/j.pnsc.2014.05.012.
8. Karnib, M.; Kabbani, A.; Holail, H.; Olama, Z. Heavy Metals Removal Using Activated Carbon, Silica and Silica Activated Carbon Composite. *Energy Procedia* **2014**, *50*, 113–120. https://doi.org/10.1016/j.egypro.2014.06.014.
9. Olabemiwo, F. A.; Tawabini, B. S.; Patel, F.; Oyehan, T. A.; Khaled, M.; Laoui, T. Cadmium Removal from Contaminated Water Using Polyelectrolyte-Coated Industrial Waste Fly Ash. *Bioinorganic Chemistry and Applications* **2017**, *2017*, 1–13. https://doi.org/10.1155/2017/7298351.
10. Singh, P.; Mondal, K.; Sharma, A. Reusable Electrospun Mesoporous ZnO Nanofiber Mats for Photocatalytic Degradation of Polycyclic Aromatic Hydrocarbon Dyes in Wastewater. *Journal of Colloid and Interface Science* **2013**, *394*, 208–215. https://doi.org/10.1016/j.jcis.2012.12.006.
11. Locardi, F.; Sanguineti, E.; Fasoli, M.; Martini, M.; Costa, G. A.; Ferretti, M.; Caratto, V. Photocatalytic Activity of TiO2 Nanopowders Supported on a New Persistent Luminescence Phosphor. *Catalysis Communications* **2016**, *74*, 24–27. https://doi.org/10.1016/j.catcom.2015.10.037.

12. Akpan, U. G.; Hameed, B. H. Parameters Affecting the Photocatalytic Degradation of Dyes Using TiO2-Based Photocatalysts: A Review. *Journal of Hazardous Materials* **2009**, *170* (2–3), 520–529. https://doi.org/10.1016/j.jhazmat.2009.05.039.
13. Liu, Y.; Yu, Y.-X.; Zhang, W.-D. MoS_2/CdS Heterojunction with High Photoelectrochemical Activity for H $_2$ Evolution under Visible Light: The Role of MoS_2. *The Journal of Physical Chemistry C* **2013**, *117* (25), 12949–12957. https://doi.org/10.1021/jp4009652.
14. Perera, S. D.; Mariano, R. G.; Vu, K.; Nour, N.; Seitz, O.; Chabal, Y.; Balkus, K. J. Hydrothermal Synthesis of Graphene-TiO2 Nanotube Composites with Enhanced Photocatalytic Activity. *ACS Catalysis* **2012**, *2* (6), 949–956. https://doi.org/10.1021/cs200621c.
15. Shukor, S. A. A.; Hamzah, R.; Bakar, M. A.; Noriman, N. Z.; Al-Rashdi, A. A.; Razlan, Z. M.; Shahriman, A. B.; Zunaidi, I.; Khairunizam, W. Metal Oxide and Activated Carbon as Photocatalyst for Waste Water Treatment. *IOP Conference Series: Materials Science and Engineering* **2019**, *557* (1), 012066. https://doi.org/10.1088/1757-899X/557/1/012066.
16. Allen, A. E.; Booth, M. G.; Frischer, M. E.; Verity, P. G.; Zehr, J. P.; Zani, S. Diversity and Detection of Nitrate Assimilation Genes in Marine Bacteria. *Applied and Environmental Microbiology* **2001**, *67* (11), 5343–5348. https://doi.org/10.1128/AEM.67.11.5343-5348.2001.
17. Allen, A. E.; Howard-Jones, M. H.; Booth, M. G.; Frischer, M. E.; Verity, P. G.; Bronk, D. A.; Sanderson, M. P. Importance of Heterotrophic Bacterial Assimilation of Ammonium and Nitrate in the Barents Sea During Summer. *Journal of Marine Systems* **2002**, *38* (1–2), 93–108. https://doi.org/10.1016/S0924-7963(02)00171-9.
18. Jiang, X.; Dang, H.; Jiao, N. Ubiquity and Diversity of Heterotrophic Bacterial NasA Genes in Diverse Marine Environments. *PLoS One* **2015**, *10* (2), e0117473. https://doi.org/10.1371/journal.pone.0117473.
19. Kumar, A. S. K.; Jiang, S.-J.; Tseng, W.-L. Effective Adsorption of Chromium(VI)/Cr(III) from Aqueous Solution Using Ionic Liquid Functionalized Multiwalled Carbon Nanotubes as a Super Sorbent. *Journal of Materials Chemistry A* **2015**, *3* (13), 7044–7057. https://doi.org/10.1039/C4TA06948J.
20. Duan, C.; Ma, T.; Wang, J.; Zhou, Y. Removal of Heavy Metals from Aqueous Solution Using Carbon-Based Adsorbents: A Review. *Journal of Water Process Engineering* **2020**, *37*, 101339. https://doi.org/10.1016/j.jwpe.2020.101339.
21. Ngah, W. S. W.; Fatinathan, S. Adsorption of Cu(II) Ions in Aqueous Solution Using Chitosan Beads, Chitosan–GLA Beads and Chitosan–Alginate Beads. *Chemical Engineering Journal* **2008**, *143* (1–3), 62–72. https://doi.org/10.1016/j.cej.2007.12.006.
22. Anson, F. C.; Shi, C.; Steiger, B. Novel Multinuclear Catalysts for the Electroreduction of Dioxygen Directly to Water. *Accounts of Chemical Research* **1997**, *30* (11), 437–444. https://doi.org/10.1021/ar960264j.
23. Goodell, B.; Qian, Y.; Jellison, J.; Richard, M. Decolorization and Degradation of Dyes with Mediated Fenton Reaction. *Water Environment Research* **2004**, *76* (7), 2703–2707.
24. Aneyo, I. A.; Doherty, F. V.; Adebesin, O. A.; Hammed, M. O. Biodegradation of Pollutants in Waste Water from Pharmaceutical, Textile and Local Dye Effluent in Lagos, Nigeria. *Journal of Health and Pollution* **2016**, *6* (12), 34–42. https://doi.org/10.5696/2156-9614-6.12.34.
25. Langenheder, S.; Székely, A. J. Species Sorting and Neutral Processes Are Both Important during the Initial Assembly of Bacterial Communities. *ISME Journal* **2011**, *5* (7), 1086–1094. https://doi.org/10.1038/ismej.2010.207.
26. Griffin, J. S.; Wells, G. F. Regional Synchrony in Full-Scale Activated Sludge Bioreactors Due to Deterministic Microbial Community Assembly. *ISME Journal* **2017**, *11* (2), 500–511. https://doi.org/10.1038/ismej.2016.121.

27. De Lucas, A.; Rodríguez, L.; Villaseñor, J.; Fernández, F. J. Denitrification Potential of Industrial Wastewaters. *Water Research* **2005**, *39* (15), 3715–3726. https://doi.org/10.1016/j.watres.2005.06.024.

28. Wang, X.; Wen, X.; Yan, H.; Ding, K.; Zhao, F.; Hu, M. Bacterial Community Dynamics in a Functionally Stable Pilot-Scale Wastewater Treatment Plant. *Bioresource Technology* **2011**, *102* (3), 2352–2357. https://doi.org/10.1016/j.biortech.2010.10.095.

29. Ibarbalz, F. M.; Figuerola, E. L. M.; Erijman, L. Industrial Activated Sludge Exhibit Unique Bacterial Community Composition at High Taxonomic Ranks. *Water Research* **2013**, *47* (11), 3854–3864. https://doi.org/10.1016/j.watres.2013.04.010.

30. Amat, A. M.; Arques, A.; García-Ripoll, A.; Santos-Juanes, L.; Vicente, R.; Oller, I.; Maldonado, M. I.; Malato, S. A Reliable Monitoring of the Biocompatibility of an Effluent along an Oxidative Pre-Treatment by Sequential Bioassays and Chemical Analyses. *Water Research* **2009**, *43* (3), 784–792. https://doi.org/10.1016/j.watres.2008.11.017.

31. Chen, Y.; Lan, S.; Wang, L.; Dong, S.; Zhou, H.; Tan, Z.; Li, X. A Review: Driving Factors and Regulation Strategies of Microbial Community Structure and Dynamics in Wastewater Treatment Systems. *Chemosphere* **2017**, *174*, 173–182. https://doi.org/10.1016/j.chemosphere.2017.01.129.

32. Pholchan, M. K.; Baptista, J. de C.; Davenport, R. J.; Curtis, T. P. Systematic Study of the Effect of Operating Variables on Reactor Performance and Microbial Diversity in Laboratory-Scale Activated Sludge Reactors. *Water Research* **2010**, *44* (5), 1341–1352. https://doi.org/10.1016/j.watres.2009.11.005.

33. Ahmad, A. L.; Ismail, S.; Bhatia, S. Optimization of Coagulation–Flocculation Process for Palm Oil Mill Effluent Using Response Surface Methodology. *Environmental Science & Technology* **2005**, *39* (8), 2828–2834. https://doi.org/10.1021/es0498080.

34. Ahmad, A.; Wong, S.; Teng, T.; Zuhairi, A. Improvement of Alum and PACl Coagulation by Polyacrylamides (PAMs) for the Treatment of Pulp and Paper Mill Wastewater. *Chemical Engineering Journal* **2008**, *137* (3), 510–517. https://doi.org/10.1016/j.cej.2007.03.088.

35. Divakaran, R.; Sivasankara Pillai, V. N. Flocculation of Kaolinite Suspensions in Water by Chitosan. *Water Research* **2001**, *35* (16), 3904–3908. https://doi.org/10.1016/S0043-1354(01)00131-2.

36. Nasser, M. S.; James, A. E. The Effect of Polyacrylamide Charge Density and Molecular Weight on the Flocculation and Sedimentation Behaviour of Kaolinite Suspensions. *Separation and Purification Technology* **2006**, *52* (2), 241–252. https://doi.org/10.1016/j.seppur.2006.04.005.

37. Trivunac, K.; Stevanovic, S. Removal of Heavy Metal Ions from Water by Complexation-Assisted Ultrafiltration. *Chemosphere* **2006**, *64* (3), 486–491. https://doi.org/10.1016/j.chemosphere.2005.11.073.

38. Aliane, A.; Bounatiro, N.; Cherif, A. T.; Akretche, D. E. Removal of Chromium from Aqueous Solution by Complexation—Ultrafiltration Using a Water-Soluble Macroligand. *Water Research* **2001**, *35* (9), 2320–2326. https://doi.org/10.1016/S0043-1354(00)00501-7.

39. Baek, K.; Kim, B.-K.; Cho, H.-J.; Yang, J.-W. Removal Characteristics of Anionic Metals by Micellar-Enhanced Ultrafiltration. *Journal of Hazardous Materials* **2003**, *99* (3), 303–311. https://doi.org/10.1016/S0304-3894(03)00063-3.

40. Bodzek, M.; Korus, I.; Loska, K. Application of the Hybrid Complexation-Ultrafiltration Process for Removal of Metal Ions from Galvanic Wastewater. *Desalination* **1999**, *121* (2), 117–121. https://doi.org/10.1016/S0011-9164(99)00012-0.

41. Luisa Cervera, M.; Carmen Arnal, M.; de la Guardia, M. Removal of Heavy Metals by Using Adsorption on Alumina or Chitosan. *Anal Bioanal Chem* **2003**, *375* (6), 820–825. https://doi.org/10.1007/s00216-003-1796-2.

42. Chhowalla, M.; Shin, H. S.; Eda, G.; Li, L.-J.; Loh, K. P.; Zhang, H. The Chemistry of Two-Dimensional Layered Transition Metal Dichalcogenide Nanosheets. *Nature Chem* **2013**, *5* (4), 263–275. https://doi.org/10.1038/nchem.1589.
43. Coleman, J. N.; Lotya, M.; O'Neill, A.; Bergin, S. D.; King, P. J.; Khan, U.; Young, K.; Gaucher, A.; De, S.; Smith, R. J.; Shvets, I. V.; Arora, S. K.; Stanton, G.; Kim, H.-Y.; Lee, K.; Kim, G. T.; Duesberg, G. S.; Hallam, T.; Boland, J. J.; Wang, J. J.; Donegan, J. F.; Grunlan, J. C.; Moriarty, G.; Shmeliov, A.; Nicholls, R. J.; Perkins, J. M.; Grieveson, E. M.; Theuwissen, K.; McComb, D. W.; Nellist, P. D.; Nicolosi, V. Two-Dimensional Nanosheets Produced by Liquid Exfoliation of Layered Materials. *Science* **2011**, *331* (6017), 568–571. https://doi.org/10.1126/science.1194975.
44. Smith, R. J.; King, P. J.; Lotya, M.; Wirtz, C.; Khan, U.; De, S.; O'Neill, A.; Duesberg, G. S.; Grunlan, J. C.; Moriarty, G.; Chen, J.; Wang, J.; Minett, A. I.; Nicolosi, V.; Coleman, J. N. Large-Scale Exfoliation of Inorganic Layered Compounds in Aqueous Surfactant Solutions. *Advanced Materials* **2011**, *23* (34), 3944–3948. https://doi.org/10.1002/adma.201102584.
45. Wang, H.; Yang, X.; Shao, W.; Chen, S.; Xie, J.; Zhang, X.; Wang, J.; Xie, Y. Ultrathin Black Phosphorus Nanosheets for Efficient Singlet Oxygen Generation. *Journal of the American Chemical Society* **2015**, *137* (35), 11376–11382. https://doi.org/10.1021/jacs.5b06025.
46. Kim, T. W.; Oh, E.-J.; Jee, A.-Y.; Lim, S. T.; Park, D. H.; Lee, M.; Hyun, S.-H.; Choy, J.-H.; Hwang, S.-J. Soft-Chemical Exfoliation Route to Layered Cobalt Oxide Monolayers and Its Application for Film Deposition and Nanoparticle Synthesis. *Chemistry—A European Journal* **2009**, *15* (41), 10752–10761. https://doi.org/10.1002/chem.200901590.
47. Eda, G.; Yamaguchi, H.; Voiry, D.; Fujita, T.; Chen, M.; Chhowalla, M. Photoluminescence from Chemically Exfoliated MoS$_2$. *Nano Letters* **2011**, *11* (12), 5111–5116. https://doi.org/10.1021/nl201874w.
48. Egizbek, K.; Kozlovskiy, A. L.; Ludzik, K.; Zdorovets, M. V.; Korolkov, I. V.; Marciniak, B.; M, J.; Chudoba, D.; Nazarova, A.; Kontek, R. Stability and Cytotoxicity Study of NiFe2O4 Nanocomposites Synthesized by Co-Precipitation and Subsequent Thermal Annealing. *Ceramics International* **2020**, *46* (10), 16548–16555. https://doi.org/10.1016/j.ceramint.2020.03.222.
49. Jan, T.; Azmat, S.; Mansoor, Q.; Waqas, H. M.; Adil, M.; Ilyas, S. Z.; Ahmad, I.; Ismail, M. Superior Antibacterial Activity of ZnO-CuO Nanocomposite Synthesized by a Chemical Co-Precipitation Approach. *Microbial Pathogenesis* **2019**, *134*, 103579. https://doi.org/10.1016/j.micpath.2019.103579.
50. Thambidurai, S.; Gowthaman, P.; Venkatachalam, M.; Suresh, S. Enhanced Bactericidal Performance of Nickel Oxide-Zinc Oxide Nanocomposites Synthesized by Facile Chemical Co-Precipitation Method. *Journal of Alloys and Compounds* **2020**, *830*, 154642. https://doi.org/10.1016/j.jallcom.2020.154642.
51. Shahrbabak, M. S. N.; Sharifianjazi, F.; Rahban, D.; Salimi, A. A Comparative Investigation on Bioactivity and Antibacterial Properties of Sol-Gel Derived 58S Bioactive Glass Substituted by Ag and Zn. *Silicon* **2019**, *11* (6), 2741–2751. https://doi.org/10.1007/s12633-018-0063-2.
52. Witoon, T.; Permsirivanich, T.; Donphai, W.; Jaree, A.; Chareonpanich, M. CO2 Hydrogenation to Methanol over Cu/ZnO Nanocatalysts Prepared via a Chitosan-Assisted Co-Precipitation Method. *Fuel Processing Technology* **2013**, *116*, 72–78. https://doi.org/10.1016/j.fuproc.2013.04.024.
53. Taufik, A.; Albert, A.; Saleh, R. Sol-Gel Synthesis of Ternary CuO/TiO 2/ZnO Nanocomposites for Enhanced Photocatalytic Performance under UV and Visible Light Irradiation. *Journal of Photochemistry and Photobiology A: Chemistry* **2017**, *344*, 149–162. https://doi.org/10.1016/j.jphotochem.2017.05.012.

54. Ahmed, M. A. Synthesis and Structural Features of Mesoporous NiO/TiO2 Nanocomposites Prepared by Sol–Gel Method for Photodegradation of Methylene Blue Dye. *Journal of Photochemistry and Photobiology A: Chemistry* **2012**, *238*, 63–70. https://doi.org/10.1016/j.jphotochem.2012.04.010.

55. Ovchinnikov, O.; Aslanov, S.; Smirnov, M.; Perepelitsa, A.; Kondratenko, T.; Selyukov, A.; Grevtseva, I. Colloidal Ag $_2$ S/SiO $_2$ Core/Shell Quantum Dots with IR Luminescence. *Optical Materials Express* **2021**, *11* (1), 89. https://doi.org/10.1364/OME.411432.

56. Amin, S. A.; Pazouki, M.; Hosseinnia, A. Synthesis of TiO2–Ag Nanocomposite with Sol–Gel Method and Investigation of Its Antibacterial Activity against E. Coli. *Powder Technology* **2009**, *196* (3), 241–245. https://doi.org/10.1016/j.powtec.2009.07.021.

57. Ganchimeg, P.; Tan, W. T.; Yusof, N. A.; Goh, J. K. Voltammetric Oxidation of Ascorbic Acid Mediated by Multi-Walled Carbon Nanotubes/Titanium Dioxide Composite Modified Glassy Carbon Electrode. *Journal of Applied Sciences* **2011**, *11* (5), 848–854. https://doi.org/10.3923/jas.2011.848.854.

58. Hosseinzadeh Sanatkar, T.; Khorshidi, A.; Sohouli, E.; Janczak, J. Synthesis, Crystal Structure, and Characterization of Two Cu(II) and Ni(II) Complexes of a Tetradentate N2O2 Schiff Base Ligand and Their Application in Fabrication of a Hydrazine Electrochemical Sensor. *Inorganica Chimica Acta* **2020**, *506*, 119537. https://doi.org/10.1016/j.ica.2020.119537.

59. Ghorbani, M.; Abdizadeh, H.; Golobostanfard, M. R. Hierarchical Porous ZnO Films Synthesized by Sol–Gel Method Using Triethylenetetramine Stabilizer. *SN Applied Sciences* **2019**, *1* (3), 267. https://doi.org/10.1007/s42452-019-0274-1.

60. Ansari, F.; Sobhani, A.; Salavati-Niasari, M. Simple Sol-Gel Synthesis and Characterization of New CoTiO3/CoFe2O4 Nanocomposite by Using Liquid Glucose, Maltose and Starch as Fuel, Capping and Reducing Agents. *Journal of Colloid and Interface Science* **2018**, *514*, 723–732. https://doi.org/10.1016/j.jcis.2017.12.083.

61. Zhu, J.; Li, Q.; Bi, W.; Bai, L.; Zhang, X.; Zhou, J.; Xie, Y. Ultra-Rapid Microwave-Assisted Synthesis of Layered Ultrathin Birnessite K0.17MnO2 Nanosheets for Efficient Energy Storage. *Journal of Materials Chemistry A* **2013**, *1* (28), 8154. https://doi.org/10.1039/c3ta11194f.

62. Jang, J.; Jeong, S.; Seo, J.; Kim, M.-C.; Sim, E.; Oh, Y.; Nam, S.; Park, B.; Cheon, J. Ultrathin Zirconium Disulfide Nanodiscs. *Journal of the American Chemical Society* **2011**, *133* (20), 7636–7639. https://doi.org/10.1021/ja200400n.

63. Zhang, Y.; Liu, S.; Liu, W.; Liang, T.; Yang, X.; Xu, M.; Chen, H. Two-Dimensional MoS $_2$ -Assisted Immediate Aggregation of Poly-3-Hexylthiophene with High Mobility. *Physical Chemistry Chemical Physics* **2015**, *17* (41), 27565–27572. https://doi.org/10.1039/C5CP05011A.

64. An, H.; Lee, W.-J.; Jung, J. Graphene Synthesis on Fe Foil Using Thermal CVD. *Current Applied Physics* **2011**, *11* (4), S81–S85. https://doi.org/10.1016/j.cap.2011.03.077.

65. Sun, L.; Yuan, G.; Gao, L.; Yang, J.; Chhowalla, M.; Gharahcheshmeh, M. H.; Gleason, K. K.; Choi, Y. S.; Hong, B. H.; Liu, Z. Chemical Vapour Deposition. *Nat Rev Methods Primers* **2021**, *1* (1), 5. https://doi.org/10.1038/s43586-020-00005-y.

66. Yavari, F.; Chen, Z.; Thomas, A. V.; Ren, W.; Cheng, H.-M.; Koratkar, N. High Sensitivity Gas Detection Using a Macroscopic Three-Dimensional Graphene Foam Network. *Scientific Reports* **2011**, *1* (1), 166. https://doi.org/10.1038/srep00166.

67. Liu, H.; You, C. Y.; Li, J.; Galligan, P. R.; You, J.; Liu, Z.; Cai, Y.; Luo, Z. Synthesis of Hexagonal Boron Nitrides by Chemical Vapor Deposition and Their Use as Single Photon Emitters. *Nano Materials Science* **2021**, *3* (3), 291–312. https://doi.org/10.1016/j.nanoms.2021.03.002.

68. Li, X.; Cai, W.; An, J.; Kim, S.; Nah, J.; Yang, D.; Piner, R.; Velamakanni, A.; Jung, I.; Tutuc, E.; Banerjee, S. K.; Colombo, L.; Ruoff, R. S. Large-Area Synthesis of High-Quality and Uniform Graphene Films on Copper Foils. *Science* **2009**, *324* (5932), 1312–1314. https://doi.org/10.1126/science.1171245.

69. Kappera, R.; Voiry, D.; Yalcin, S. E.; Branch, B.; Gupta, G.; Mohite, A. D.; Chhowalla, M. Phase-Engineered Low-Resistance Contacts for Ultrathin MoS2 Transistors. *Nature Materials* **2014**, *13* (12), 1128–1134. https://doi.org/10.1038/nmat4080.
70. Acerce, M.; Voiry, D.; Chhowalla, M. Metallic 1T Phase MoS2 Nanosheets as Supercapacitor Electrode Materials. *Nature Nanotechnology* **2015**, *10* (4), 313–318. https://doi.org/10.1038/nnano.2015.40.
71. Bagheri, S.; Julkapli, N. M. Effect of Hybridization on the Value-Added Activated Carbon Materials. *International Journal of Industrial Chemistry* **2016**, *7* (3), 249–264. https://doi.org/10.1007/s40090-016-0089-5.
72. Adeola, A. O.; Abiodun, B. A.; Adenuga, D. O.; Nomngongo, P. N. Adsorptive and Photocatalytic Remediation of Hazardous Organic Chemical Pollutants in Aqueous Medium: A Review. *Journal of Contaminant Hydrology* **2022**, *248*, 104019. https://doi.org/10.1016/j.jconhyd.2022.104019.
73. Liu, K.-G.; Rouhani, F.; Gao, X.-M.; Abbasi-Azad, M.; Li, J.-Z.; Hu, X.; Wang, W.; Hu, M. L.; Morsali, A. Bilateral Photocatalytic Mechanism of Dye Degradation by a Target Designed Ferrocene-Functionalized Cluster Under Natural Sunlight. *Catalysis Science & Technology* **2020**, *10*.
74. Qu, X.; Alvarez, P. J. J.; Li, Q. Applications of Nanotechnology in Water and Wastewater Treatment. *Water Research* **2013**, *47* (12), 3931–3946. https://doi.org/10.1016/j.watres.2012.09.058.
75. Gong, K.; Du, F.; Xia, Z.; Durstock, M.; Dai, L. Nitrogen-Doped Carbon Nanotube Arrays with High Electrocatalytic Activity for Oxygen Reduction. *Science* **2009**, *323* (5915), 760–764. https://doi.org/10.1126/science.1168049.
76. Du, X.; Wang, C.; Chen, M.; Jiao, Y.; Wang, J. Electrochemical Performances of Nanoparticle Fe₃O₄/Activated Carbon Supercapacitor Using KOH Electrolyte Solution. *The Journal of Physical Chemistry C* **2009**, *113* (6), 2643–2646. https://doi.org/10.1021/jp8088269.
77. Ranjithkumar, V.; Hazeen, A. N.; Thamilselvan, M.; Vairam, S. Magnetic Activated Carbon-Fe₃O₄ Nanocomposites—Synthesis and Applications in the Removal of Acid Yellow Dye 17 from Water. *Journal of Nanoscience and Nanotechnology* **2014**, *14* (7), 4949–4959. https://doi.org/10.1166/jnn.2014.9068.
78. Barala, S. K.; Arora, M.; Puri, C.; Saini, K. K.; Kotnala, R. K.; Saini, P. K. Ferrofluid/Activated Carbon Composites for Water Purification and EMI Shielding Applications. *MHD* **2013**, *49* (3–4), 277–281. https://doi.org/10.22364/mhd.49.3-4.5.
79. Yang, N.; Zhu, S.; Zhang, D.; Xu, S. Synthesis and Properties of Magnetic Fe3O4-Activated Carbon Nanocomposite Particles for Dye Removal. *Materials Letters* **2008**, *62* (4–5), 645–647. https://doi.org/10.1016/j.matlet.2007.06.049.
80. Zhang, Y.; Tang, Z.-R.; Fu, X.; Xu, Y.-J. TiO₂ –Graphene Nanocomposites for Gas-Phase Photocatalytic Degradation of Volatile Aromatic Pollutant: Is TiO₂ –Graphene Truly Different from Other TiO₂ –Carbon Composite Materials? *ACS Nano* **2010**, *4* (12), 7303–7314. https://doi.org/10.1021/nn1024219.
81. Yang, M.-Q.; Zhang, N.; Pagliaro, M.; Xu, Y.-J. Artificial Photosynthesis over Graphene–Semiconductor Composites. Are We Getting Better? *Chemical Society Reviews* **2014**, *43* (24), 8240–8254. https://doi.org/10.1039/C4CS00213J.
82. Peñas-Garzón, M.; Gómez-Avilés, A.; Bedia, J.; Rodriguez, J.; Belver, C. Effect of Activating Agent on the Properties of TiO2/Activated Carbon Heterostructures for Solar Photocatalytic Degradation of Acetaminophen. *Materials* **2019**, *12* (3), 378. https://doi.org/10.3390/ma12030378.
83. Che Othman, F. E.; Yusof, N.; Ismail, A. F.; Jaafar, J.; Wan Salleh, W. N.; Aziz, F. Preparation and Characterization of Polyacrylonitrile-Based Activated Carbon Nanofibers/Graphene (GACNFs) Composite Synthesized by Electrospinning. *AIP Advances* **2020**, *10* (5), 055117. https://doi.org/10.1063/5.0008012.

84. Kim, J.-H.; Lee, C.-H.; Kim, W.-S.; Lee, J.-S.; Kim, J.-T.; Suh, J.-K.; Lee, J.-M. Adsorption Equilibria of Water Vapor on Alumina, Zeolite 13X, and a Zeolite X/ Activated Carbon Composite. *Journal of Chemical & Engineering Data* **2003**, *48* (1), 137–141. https://doi.org/10.1021/je0201267.

85. Baruah, M.; Ezung, S. L.; Supong, A.; Bhomick, P. C.; Kumar, S.; Sinha, D. Synthesis, Characterization of Novel Fe-Doped TiO2 Activated Carbon Nanocomposite towards Photocatalytic Degradation of Congo Red, E. Coli, and S. Aureus. *Korean Journal of Chemical Engineering* **2021**, *38* (6), 1277–1290. https://doi.org/10.1007/ s11814-021-0830-4.

86. Dutta, M.; Basu, J. K. Fixed-Bed Column Study for the Adsorptive Removal of Acid Fuchsin Using Carbon–Alumina Composite Pellet. *International Journal of Environmental Science and Technology* **2014**, *11* (1), 87–96. https://doi.org/10.1007/ s13762-013-0386-x.

87. Faria, P. C. C.; Órfão, J. J. M.; Pereira, M. F. R. Mineralization of Substituted Aromatic Compounds by Ozonation Catalyzed by Cerium Oxide and a Cerium Oxide-Activated Carbon Composite. *Catalysis Letters* **2009**, *127* (1–2), 195–203. https://doi.org/10.1007/ s10562-008-9670-7.

88. Takahashi, M.; Mori, T.; Ye, F.; Vinu, A.; Kobayashi, H.; Drennan, J. Design of High-Quality Pt?CeO$_2$ Composite Anodes Supported by Carbon Black for Direct Methanol Fuel Cell Application. *Journal of the American Ceramic Society* **2007**, *90* (4), 1291–1294. https://doi.org/10.1111/j.1551-2916.2006.01483.x.

89. Gonçalves, A.; Silvestre-Albero, J.; Ramos-Fernández, E. V.; Serrano-Ruiz, J. Carlos.; Órfão, J. J. M.; Sepúlveda-Escribano, A.; Pereira, M. F. R. Highly Dispersed Ceria on Activated Carbon for the Catalyzed Ozonation of Organic Pollutants. *Applied Catalysis B: Environmental* **2012**, *113–114*, 308–317. https://doi.org/10.1016/j.apcatb.2011.11.052.

90. Kaur, J.; Kaur, M.; Ubhi, M. K.; Kaur, N.; Greneche, J.-M. Composition Optimization of Activated Carbon-Iron Oxide Nanocomposite for Effective Removal of Cr(VI) Ions. *Materials Chemistry and Physics* **2021**, *258*, 124002. https://doi.org/10.1016/j. matchemphys.2020.124002.

91. Jain, M.; Yadav, M.; Kohout, T.; Lahtinen, M.; Garg, V. K.; Sillanpää, M. Development of Iron Oxide/Activated Carbon Nanoparticle Composite for the Removal of Cr(VI), Cu(II) and Cd(II) Ions from Aqueous Solution. *Water Resources and Industry* **2018**, *20*, 54–74. https://doi.org/10.1016/j.wri.2018.10.001.

92. Su, H.; Ye, Z.; Hmidi, N.; Subramanian, R. Carbon Nanosphere–Iron Oxide Nanocomposites as High-Capacity Adsorbents for Arsenic Removal. *RSC Advances* **2017**, *7* (57), 36138–36148. https://doi.org/10.1039/C7RA06187K.

93. Gholamvaisi, D.; Azizian, S.; Cheraghi, M. Preparation of Magnetic-Activated Carbon Nanocomposite and Its Application for Dye Removal from Aqueous Solution. *Journal of Dispersion Science and Technology* **2014**, *35* (9), 1264–1269. https://doi.org/10.1080/01 932691.2013.843465.

94. Ho, M. Y.; Khiew, P. S. Heat-Treated Fe$_3$O$_4$—Activated Carbon Nanocomposite for High Performance Electrochemical Capacitor. *AMR* **2014**, *894*, 349–354. https://doi. org/10.4028/www.scientific.net/AMR.894.349.

95. Hao, W.; Björkman, E.; Yun, Y.; Lilliestråle, M.; Hedin, N. Iron Oxide Nanoparticles Embedded in Activated Carbons Prepared from Hydrothermally Treated Waste Biomass. *ChemSusChem* **2014**, *7* (3), 875–882. https://doi.org/10.1002/cssc.201300912.

96. DasinehKhiavi, N.; Katal, R.; KholghiEshkalak, S.; Masudy-Panah, S.; Ramakrishna, S.; Jiangyong, H. Visible Light Driven HeterojunctionPhotocatalyst of CuO–Cu2O Thin Films for Photocatalytic Degradation of Organic Pollutants. *Nanomaterials* **2019**, *9* (7), 1011. https://doi.org/10.3390/nano9071011.

97. Usman, M.; Ahmed, A.; Yu, B.; Peng, Q.; Shen, Y.; Cong, H. Photocatalytic Potential of Bio-Engineered Copper Nanoparticles Synthesized from FicusCarica Extract for the

Degradation of Toxic Organic Dye from Waste Water: Growth Mechanism and Study of Parameter Affecting the Degradation Performance. *Materials Research Bulletin* **2019**, *120*, 110583. https://doi.org/10.1016/j.materresbull.2019.110583.

98. Gnanaprakasam, A.; Sivakumar, V. M.; Thirumarimurugan, M. Influencing Parameters in the Photocatalytic Degradation of Organic Effluent via Nanometal Oxide Catalyst: A Review. *Indian Journal of Materials Science* **2015**, *2015*, 1–16. https://doi.org/10.1155/2015/601827.

99. Zain, N. M.; Lim, C. M.; Usman, A.; Keasberry, N.; Thotagamuge, R.; Mahadi, A. H. Synergistic Effect of TiO2 Size on Activated Carbon Composites for Ruthenium N-3 Dye Adsorption and Photocatalytic Degradation in Wastewater Treatment. *Environmental Nanotechnology, Monitoring & Management* **2021**, *16*, 100567. https://doi.org/10.1016/j.enmm.2021.100567.

100. Steplin Paul Selvin, S.; Ganesh Kumar, A.; Sarala, L.; Rajaram, R.; Sathiyan, A.; Princy Merlin, J.; Sharmila Lydia, I. Photocatalytic Degradation of Rhodamine B Using Zinc Oxide Activated Charcoal Polyaniline Nanocomposite and Its Survival Assessment Using Aquatic Animal Model. *ACS Sustainable Chemistry & Engineering* **2018**, *6* (1), 258–267. https://doi.org/10.1021/acssuschemeng.7b02335.

101. El-Khouly, S. M.; Mohamed, G. M.; Fathy, N. A.; Fagal, G. A. Effect of NanosizedCeO_2 or ZnO Loading on Adsorption and Catalytic Properties of Activated Carbon. *Adsorption Science & Technology* **2017**, *35* (9–10), 774–788. https://doi.org/10.1177/0263617417698704.

102. Shanmugapriya, B.; Shanthi, M.; Dhamodharan, P.; Rajeshwaran, K.; Bououdina, M.; Manoharan, C. Enhancement of Photocatalytic Degradation of Methylene Blue Dye Using Ti3+ Doped In2O3 Nanocubes Prepared by Hydrothermal Method. *Optik* **2020**, *202*, 163662.

103. Márquez, G.; Rodríguez, E. M.; Beltrán, F. J.; Álvarez, P. M. Solar Photocatalytic Ozonation of a Mixture of Pharmaceutical Compounds in Water. *Chemosphere* **2014**, *113*, 71–78. https://doi.org/10.1016/j.chemosphere.2014.03.093.

104. Kıdak, R.; Doğan, Ş. Medium-High Frequency Ultrasound and Ozone Based Advanced Oxidation for Amoxicillin Removal in Water. *Ultrasonics Sonochemistry* **2018**, *40*, 131–139. https://doi.org/10.1016/j.ultsonch.2017.01.033.

105. Nikolaou, A.; Meric, S.; Fatta, D. Occurrence Patterns of Pharmaceuticals in Water and Wastewater Environments. *Analytical and Bioanalytical Chemistry* **2007**, *387* (4), 1225–1234. https://doi.org/10.1007/s00216-006-1035-8.

106. Ahmadijokani, F.; Ahmadipouya, S.; Molavi, H.; Rezakazemi, M.; Aminabhavi, T. M.; Arjmand, M. Impact of Scale, Activation Solvents, and Aged Conditions on Gas Adsorption Properties of UiO-66. *Journal of Environmental Management* **2020**, *274*, 111155. https://doi.org/10.1016/j.jenvman.2020.111155.

107. MoradiDehaghi, S.; Rahmanifar, B.; Moradi, A. M.; Azar, P. A. Removal of Permethrin Pesticide from Water by Chitosan–Zinc Oxide Nanoparticles Composite as an Adsorbent. *Journal of Saudi Chemical Society* **2014**, *18* (4), 348–355. https://doi.org/10.1016/j.jscs.2014.01.004.

108. Ahmadijokani, F.; Mohammadkhani, R.; Ahmadipouya, S.; Shokrgozar, A.; Rezakazemi, M.; Molavi, H.; Aminabhavi, T. M.; Arjmand, M. Superior Chemical Stability of UiO-66 Metal-Organic Frameworks (MOFs) for Selective Dye Adsorption. *Chemical Engineering Journal* **2020**, *399*, 125346. https://doi.org/10.1016/j.cej.2020.125346.

109. Fallah, Z.; Zare, E. N.; Ghomi, M.; Ahmadijokani, F.; Amini, M.; Tajbakhsh, M.; Arjmand, M.; Sharma, G.; Ali, H.; Ahmad, A.; Makvandi, P.; Lichtfouse, E.; Sillanpää, M.; Varma, R. S. Toxicity and Remediation of Pharmaceuticals and Pesticides Using Metal Oxides and Carbon Nanomaterials. *Chemosphere* **2021**, *275*, 130055. https://doi.org/10.1016/j.chemosphere.2021.130055.

110. Behnam, R.; Morshed, M.; Tavanai, H.; Ghiaci, M. Destructive Adsorption of Diazinon Pesticide by Activated Carbon Nanofibers Containing Al2O3 and MgO Nanoparticles. *Bull Environ ContamToxicol* **2013**, *91* (4), 475–480. https://doi.org/10.1007/s00128-013-1064-x.

111. Demirbas, A. Heavy Metal Adsorption onto Agro-Based Waste Materials: A Review. *Journal of Hazardous Materials* **2008**, *157* (2–3), 220–229. https://doi.org/10.1016/j.jhazmat.2008.01.024.

112. Hua, M.; Zhang, S.; Pan, B.; Zhang, W.; Lv, L.; Zhang, Q. Heavy Metal Removal from Water/Wastewater by Nanosized Metal Oxides: A Review. *Journal of Hazardous Materials* **2012**, *211–212*, 317–331. https://doi.org/10.1016/j.jhazmat.2011.10.016.

113. Fu, F.; Wang, Q. Removal of Heavy Metal Ions from Wastewaters: A Review. *Journal of Environmental Management* **2011**, *92* (3), 407–418. https://doi.org/10.1016/j.jenvman.2010.11.011.

114. Imamoglu, M.; Tekir, O. Removal of Copper (II) and Lead (II) Ions from Aqueous Solutions by Adsorption on Activated Carbon from a New Precursor Hazelnut Husks. *Desalination* **2008**, *228* (1–3), 108–113. https://doi.org/10.1016/j.desal.2007.08.011.

115. Chingombe, P.; Saha, B.; Wakeman, R. J. Surface Modification and Characterisation of a Coal-Based Activated Carbon. *Carbon* **2005**, *43* (15), 3132–3143. https://doi.org/10.1016/j.carbon.2005.06.021.

116. Liu, Q.-S.; Zheng, T.; Wang, P.; Jiang, J.-P.; Li, N. Adsorption Isotherm, Kinetic and Mechanism Studies of Some Substituted Phenols on Activated Carbon Fibers. *Chemical Engineering Journal* **2010**, *157* (2–3), 348–356. https://doi.org/10.1016/j.cej.2009.11.013.

117. Bhatnagar, A.; Hogland, W.; Marques, M.; Sillanpää, M. An Overview of the Modification Methods of Activated Carbon for Its Water Treatment Applications. *Chemical Engineering Journal* **2013**, *219*, 499–511. https://doi.org/10.1016/j.cej.2012.12.038.

118. Umamaheswari, C.; Lakshmanan, A.; Nagarajan, N. S. Green Synthesis, Characterization and Catalytic Degradation Studies of Gold Nanoparticles against Congo Red and Methyl Orange. *Journal of Photochemistry and Photobiology B: Biology* **2018**, *178*, 33–39. https://doi.org/10.1016/j.jphotobiol.2017.10.017.

119. Bhaskar, S.; Awin, E. W.; Kumar, K. C. H.; Lale, A.; Bernard, S.; Kumar, R. Design of Nanoscaled Heterojunctions in Precursor-Derived t-ZrO2/SiOC(N) Nanocomposites: Transgressing the Boundaries of Catalytic Activity from UV to Visible Light. *Scientific Reports* **2020**, *10* (1), 430. https://doi.org/10.1038/s41598-019-57394-8.

120. Jiang, C.; Moniz, S. J. A.; Wang, A.; Zhang, T.; Tang, J. Photoelectrochemical Devices for Solar Water Splitting—Materials and Challenges. *Chemical Society Reviews* **2017**, *46* (15), 4645–4660. https://doi.org/10.1039/C6CS00306K.

121. Rambhujun, N.; Salman, M. S.; Wang, T.; Pratthana, C.; Sapkota, P.; Costalin, M.; Lai, Q.; Aguey-Zinsou, K.-F. Renewable Hydrogen for the Chemical Industry. *MRS Energy & Sustainability* **2020**, *7* (1), 33. https://doi.org/10.1557/mre.2020.33.

122. Luo, B.; Liu, G.; Wang, L. Recent Advances in 2D Materials for Photocatalysis. *Nanoscale* **2016**, *8* (13), 6904–6920. https://doi.org/10.1039/C6NR00546B.

123. Liu, P.; Liu, Y.; Ye, W.; Ma, J.; Gao, D. Flower-like N-Doped MoS $_2$ for Photocatalytic Degradation of RhB by Visible Light Irradiation. *Nanotechnology* **2016**, *27* (22), 225403. https://doi.org/10.1088/0957-4484/27/22/225403.

124. Li, Z.; Meng, X.; Zhang, Z. Recent Development on MoS2-Based Photocatalysis: A Review. *Journal of Photochemistry and Photobiology C: Photochemistry Reviews* **2018**, *35*, 39–55. https://doi.org/10.1016/j.jphotochemrev.2017.12.002.

125. Ali, A.; Dutta, A. Transition Metal Chalcogenide–Based Photocatalysts for Small-Molecule Activation. In *Photocatalytic Systems by Design*; Elsevier, 2021; pp 297–331. https://doi.org/10.1016/B978-0-12-820532-7.00017-5.

8 Nano–Metal Oxide–Based Materials for Contaminated Soil Treatment

Rayees Ahmad Zargar, Saleem Ahmad Yatoo, S. A. Bhat, S. A. Shameem, Azhar Jameel Bhat, and Manju Arora

CONTENTS

8.1 INTRODUCTION

The importance of soil in plant growth, which is a life-supporting system, cannot be overstated. Nature is more complex and dynamic. It is made up of both organic and inorganic components. Inorganic oxide, silicate, phosphate, carbonate, sulfate rock minerals, microorganisms, and organic waste biomaterials make up the majority of soil. There are four basic components of natural soil, like minerals, organic matter, water, and air. A typical soil contains approximately 45%minerals, 5%organic matter, 25%water, and 25%air, as well as microflora and fauna such as protozoans, mites, nematodes, aerobic/anaerobic bacteria, algae, fungi, insects, and larvae, among others.

DOI: 10.1201/9781003323464-8

It serves as the earth's outer cover layer, or blanket, which aids in the survival and flourishing of life on the planet by providing a variety of agricultural products, fruits, vegetables, medicinal plants for humans, grass for herbivorous animals, and shelter for worms, insects, bees, microorganisms, protozoan, algae, fungi, and bacteria within or on its surface. It is extremely important in the food chain. India is known as an agricultural country because it has 45% agricultural land and agriculture is a major source of its gross domestic product. Crop yield per hectare in agriculture is determined by the type and quality of soil and seeds, as well as climatic and other natural factors. Climate, natural vegetation, and rocks all have an impact on the nature of soil. Similarly, chemical fertilizers and pesticides enter the soil ecosystem by way of runoff from agricultural ecosystems. Contaminated soil is considered as a prominent source that affects human health, ecosystems, and agricultural crops and thus threatens the entire environment. Therefore, the remediation of these contaminated soils has become a major environmental concern [1, 2].

Nanoparticle research can be useful to a wide variety of fields, including automobiles, cosmetics, agriculture, food, textiles, space, defense, engineering, clinical fields, and the atmosphere. Nanotechnology is defined as "the understanding and control of matter in dimensions between 1 and 100 nanometers," according to the US National Nanotechnology Initiative, nanotechnology encompasses nanoscale analysis, design, and innovation, as well as representation, estimation, replication, and control at this scale. This is due to its small particle size, high surface area–to–volume ratio, simple infusion at the site of action, and flexibility for in situ and ex situ application nanotechnology is a well-known phenomenon for removing contaminants [3]. In last few years, the advent of nanotechnology has been the focus of comprehensive research, intersecting with a variety of other fields of science and having an impact on society [4]. Nanotechnology offers an infinite number of benefits to the environment in terms of conservation and remediation, which are divided into three categories: treatment and remediation, observations and detections, and contaminant avoidance. The basic nanotechnologies discussed here are focused onsite remediation and soil treatment for various pollutants. Smaller molecules allow for the placement of smaller sensors that can be used successfully in remote locations. When the demand for food grows in response to the increasing human population, it increases the demand for and consumption of available land to meet those needs. However, contamination of the soil by heavy metals, pesticides, and persistent organic contaminants (POPs) is becoming a limiting factor that must be addressed, and nanoscience appears to be a viable option. Manganese oxide nanoparticles (17B-estradiol MnO_2) have been shown to remove 88% of estrogens from the soil. It is well known that decreasing the injection speed and increasing the concentration of nanoparticles accelerates estrogen degradation [5].

Nano-remediation of soils may provide a long-term solution for replenishing depleted soil resources. Nanotechnology has provided applications that are efficient and cost-effective, as well easy to use, and thus, they propose increasingly successful treatment and remediation strategies that can significantly reduce soil contamination [6]. The combining of microorganisms with nanoparticles has become a rapidly growing field of research in green nanotechnology around the world, with various organic elements being investigated and used in the synthesis of nanoparticles, providing a reliable alternative to nanoparticles created using traditional compound and physical techniques [7–9]. In our economy, nanotechnologies are vectors of foundational applications. New techniques for assessing nanotechnology-based development should be developed in order to better understand it. Recent advancements in this field have brought together a number of innovations into a single system. Nanotechnology has the potential to decontaminate a site in an environmentally friendly and timely manner [10–12]. The goal of sustainable remediation is to reduce the grouping of contaminants to the precise levels determined by the pollutant type and its hazard. Nano-bioremediation is a coordinated combination of nanotechnology and bioremediation to accomplish a more productive, less time-consuming, and environmentally friendly remediation than separate procedures, which is in contrast to conventional bioremediation. Palladium nanoparticles have been shown to decontaminate pentachlorobiphenyl-polluted soil. At 200 atm pressure and all temperature ranges, palladium nanoparticles combined with supercritical liquid CO_2 are capable of separating all polychlorinated biphenyls (PCBs) from the soil. Cheng et al. [13] demonstrated that graphene oxide silver nanoparticles (rGO-Ag) can degrade organic compounds photocatalytically.

This chapter provides an overview of nanotechnology and decontamination of adulterated soils using nanoremediation. It also discusses how to remove pollutants from soil matrices using polymer-based nanomaterials, iron oxide nanomaterials, carbon-based nanomaterials, and other nanomaterials.

8.2   ORIGIN OF HEAVY METALS IN CONTAMINATED SOILS

Since heavy metals are present in soil for millions of years and are mostly found in the soil profile as a result of pedogenic processes such as weathering of parent materials but are considered trace amounts (1000 mg/kg) and only rarely toxic [14]. Soils may accumulate one or more heavy metals in rural and urban areas above the defined background values that are high enough to sufficiently affect human health, flora and fauna, and different ecosystems, as a result of humans' disruption and an acceleration of nature's slowly occurring geochemical cycle of metals [15]. Heavy metals essentially become contaminants in soil environments because of the following reasons (1) their different rates of generation via man-made cycles are faster than natural cycles, (2) they are transferred from mines to random environmental sites with higher potential for direct exposure, (3) usually the concentrations of the heavy metals in unwanted products are comparatively high to those in the receiving environment, and (4) the chemical form (species) in discarded products is different from those in the receiving environment. Anthropogenic heavy metals in the soil are more mobile and thus bioavailable than pedogenic or lithogenic heavy metals [16]. Metal-bearing solids at polluted locations come from many different man-made sources, like metal mine tailings, unsecured protected landfills, and leaded gasoline besides the sources such as use of lead-based paints, land application of fertilizers, animal manures, sewage sludge, compost, pesticides, coal combustion residues, petrochemicals, and atmospheric deposition [17].

8.3   EFFECTS OF TOXIC HEAVY METALS/IONS PRESENT IN POLLUTED SOIL ON HUMAN BEINGS

Heavy metals are very toxic and carcinogenic to humans and belong to the group of trace elements (metals and metalloids) with elemental density >5 g/cm^3: e.g. As, Cd, Cu, Ni, Zn, Pb, Hg and others. Toxic metal ions of natural origin enter the soil from volcanic eruptions, soil erosion, and the weathering of rocks and minerals, as these mostly come from anthropogenic-based activities like burning fuel, processing minerals, landfills, urban runoffs, mining, industrial effluents, agricultural activities, metal finishing and plating, manufacture of printed boards, textile dyes, semiconductor development, etc. Owing to their high stability, migration rate, and solubility in water, they induce environmental changes and health effects. Arsenic is introduced into water and soil by both natural and man-made sources. In human beings, it causes lung and bladder cancer, and excessive consumption may lead to death also. Acute cadmium inhalation for short intervals of time results in flu-like symptoms and damages the lungs, but its chronic exposure causes kidney, bone, and lung disease. Persistent long-term exposure to copper causes irritation in the eyes, nose, and mouth; headache, stomachache; dizziness; vomiting; and diarrhea, and a high intake may cause kidney and liver problems and sometimes death. Lead has harmful effects and damages the kidney, liver, reproductive system, basic cellular processes, brain functions, and so forth. Although Ni and Zn as micronutrients are required by humans, their excessive concentrations are harmful also. Nickel/compound dust, when inhaled by the workers in refineries or nickel-processing plants, causes chronic bronchitis, nasal sinus, lung problems, and even cancer. Excessive zinc can lead to nausea, vomiting, a loss of appetite, stomach pain, headache, and diarrhea. Mercury (Hg) vapor affects the human nervous, digestive and immune systems; damages the lungs and kidneys; and sometimes us fatal if the exposure is for a long duration. The inorganic salts of mercury are corrosive to the skin, eyes, and gastrointestinal tract and can induce kidney toxicity if ingested.

8.4   USE AND APPLICATIONS OF NANOTECHNOLOGY IN SOIL REMEDIATION

Nanotechnology is regarded as unique innovation that is advancing expediently and spreading its application domains in all aspects. Nanotechnology envelops planning, estimating, and manipulating matter at the nuclear, subatomic, macromolecular, and micromolecular scales, with the end goal being that the resulting material has special and totally various properties from the parent mass material [18]. The particles produced throught his are named as designed nanoparticles (ENPs) or basically nanoparticles. ENPs are characterized as particles having something like one outside aspect somewhere in the range of 1 and 100 nm [19]. Nanotechnology finds applications in a broad scope of areas from farming, food creation and bundling, the medical and diagnostic fields, drugs, the nano-based embodiment of pesticides, hereditary material conveyance in plants, drug conveyance in people, and malignant growth therapy to a lot more [19, 20]. Apart from those, its maximum predicted applications are for remediating environmental problems with water and soil, such as treating wastewater, cleaning groundwater, and remediating soil contaminated with pollution [21–25]. The increasing use and awareness of nano based methods, gadgets, or materials for environmental cleanup possibly stem from the urgent want of a technology that is cleaner, very low cost, and effortless while, at the same time, quicker in turning in effects without an extra burden to the cleanup method for pollutant residues and environmental persistence [26]. Therefore, using nanoparticles or some nanomaterial-containing nanoparticles to remediate contaminated soil with chemical compounds like insecticides, heavy metals, and POPs, which can be chiefly nonbiodegradable in the environment, is being broadly considered. Various applications of nanotechnology-based getting used for soil remediation encompass (1) nano-based substances for converting heavy metals to their less poisonous forms, (2) nanomaterials for pesticide and POP degradation, (3) nano-based sensors for pesticide residue detection in soil, and (4) nanomaterial-based phytoremediation or bioremediation of contaminated soil. nanoparticles, which might be most extensively used for soil remediation, include nanoscale zero-valent iron (nZVI), titanium dioxide (TiO_2), zinc oxide (ZnO), multiwalled carbon nanotubes (MWCNTs), fullerenes, bimetallic nanoparticles, and stabilized nanoparticles [25, 27]. Nano-remediation contains programs of reactive nanomaterials used for the contaminated soil for either converting the pollutants to less harmful substances or cleansing them [28]. Nanomaterials, by using their distinctive features of length and structure, possess much large surface to volume ratio and, as a consequence, have higher sorption sites, making them super-absorbents [29]. Their different beneficial structures encompass decreased temperature modification, shorter inter-particle diffusion distance, greater tunable pore sizes, and specific surface chemistries [30]. These characteristic make them fantastic catalysts that could assist chemical reduction and the abatement of harmful pollution. Therefore, nanomaterials have received great attention and are preferably considered for remediating soil contaminated by notorious pollutants like heavy metals, chlorinated organic solvents, organochlorine insecticides, polycyclic aromatic hydrocarbons (PAHs), and PCBs (Figure 8.1).

8.5   DIFFERENT NANOMATERIALS USED FOR SOIL REMEDIATION

The different types of nanomaterials synthesized have been found to be very effective in remediating various pollutants from soil. These types of nanomaterials have been reported to enhance in situ treatment with a high capacity to remove or degrade pollutants. The main classes of nanomaterials used in soil remediation are

1. polymer-based nanoparticles—these are used for extracting Cu^{2+}, Cd^{2+}, Pb^{2+}, and Zn^{2+} from metal-polluted soils.
2. nanoscale metal oxides—these are used for metal adsorption and for destroying organic compounds mostly being halogenated.
3. carbon-based nanomaterials—these have been also used to cleanup contaminated soil ecosystems with volatile organics, such as petroleum hydrocarbons, some chlorinated solvents, PCBs, and pentachlorophenol.

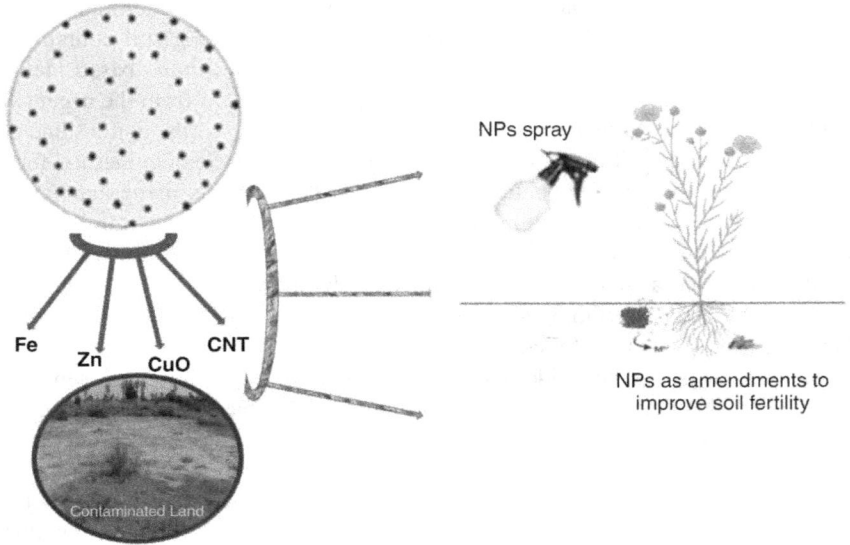

Heavy metals such as Cr(VI), Pb(II), Cd, etc.
can be remediated using different NPs

FIGURE 8.1 Remediation of soils polluted with heavy metals, pesticides, and persistent organic pollutants.

Other nanoparticles, such as manganese oxide and calcium peroxide nanoparticles are used for removing chromium and degrading heavy metal contaminants. These methods are novel in approach; however, a few more studies are recommended for their wider applicability [31].

8.5.1 Iron Oxide

Iron is the most abundantly available transition metal on the earth, and the main advantages of iron are that it is nontoxic in nature, highly absorbent, environmentally friendly, and cost-effective. Based on these characteristics, remediation technologies on iron-based are being widely tested and created rapidly, particularly for environmental remediation [32]. In this regard, because of progress in various fields of the nanoscience, Fe-based nanoparticles have been synthesized to remove As-polluted soils, because they show high bonding for As [33]. Due to greater surface area effects, nZVI materials are now widely used as compared to traditional conventional iron powder or iron filings to increase efficiency or adsorption capability [34].

8.5.2 Carbon-Based Materials

Different soils at different locations are considered to be the main sink for various pollutants rather than water and air [35]. The microbiota has a key role in altering the fate of different pollutants present in the soil environment, and carbon nanotubes (CNTs) may aid in the process [36]. Therefore, it is very much desired to study the impact of CNTs on managing soil ecosystems [37] as it is advocated that CNTs may strongly fluctuate the composition and functioning of the soil ecosystems that will help remediate soils polluted with various pollutants. However, at the same time, the toxicity of CNTs has been reported by various researchers regarding growing concerns about health [38]. Furthermore, low removal efficiency, restricted selectivity and complexity [39], and low absorption capacities are the other concerns with CNTs.

In this context, activated carbon (AC) is the ideal carbon material for removing heavy metals in soil because it is chiefly highly efficient and environmentally friendly and can be derived from waste or by-products or be minimally processed [40]. Because of its excellent properties, AC is recommended to reduce the phytotoxicity of several residues of herbicides and other harmful chemicals in agricultural soils [41]. During soil bioremediation, AC has been also used to clean up soil ecosystems contaminated with volatile organics, like petroleum hydrocarbons, chlorinated solvents, PCBs, and pentachlorophenol.

8.5.3 Polymer-Based Materials

Polymer nanocomposites (PNCs) are composed of materials that use a polymer as a host material on which nanoparticles are interspersed or coated. This substance incorporates the required characteristics of both the polymer host (i.e., excellent mechanical resistance) and the nanoparticles (i.e., high reactivity from their large surface). Different polymers were also used in the manufacture of membranes that integrate metal and metal oxide nanoparticles for environmental remediation applications [42–46]. Polymers are usually used as host materials in polymeric nanocomposites, and certain compounds of the structure, such as nanoparticles, are responsible for contaminant remediation [9]. Functionalized polymeric nanomaterials were, however, also identified as the principal agents responsible for the remediation. Polymer-based nanomaterials can be used for the design of target-specific remediation technologies for the compounds extracted from gaseous mixtures to avoid off-target fouling that could otherwise lead to a decrease in material performance [9].

However, soil conditions can be properly maintained and improved with the help of nanofertilizers [12]. It has been researched that during the application of nanofertilizers only a small amount or percentage of the fertilizer can be applied and used by the crop, which is quite contrary to conventional fertilizers. This technique has helped retain the nutrient uptake in the rhizosphere, increase the uptake of beneficial molecules, and repair damaged soils. Particles smaller than 2 micrometers belong to colloidal fractions, and there is a possibility of recognizing particles within a range of 1 to 100 nanometers as well that can be promoted for remediating soil. Nano-electron acceptors and donors, like fulvic and humic acids, can play a great role in the biodegradation of contaminants in soil. Most of the nanoparticles and nanomaterials are present in deserts as it is estimated that 50% of the minerals in the aerosol category come from the world's deserts which include oxides, hydroxides, and oxyhydroxides of metallic elements, such as Al, Fe, and Mn, from which different types of nanoparticles have been synthesized [47]. Gibbsite and boehmite in soil are aluminum nanostructures generated by geological processes, and iron hydro-oxides are natural nanoparticles in soil that have a role in processing nutrients' absorption without causing any internal damage to the soil [48]. The amalgamation of nanoscience with biohydrometallurgical processes facilitated the degradation of minerals. Moreover, other interesting processes are bioaccumulation, bioflotation, bioprecipitation, and biomineralization and are now considered to be eco-friendly and efficient alternatives for contaminated soil bioremediation [49].

8.6   FUTURE PROSPECTS

Pollution in the environment of living beings, particularly by humans, requires no further explanation. As a result, traditional as well as scientific approaches to solving this serious problem must be used. Biodegradable materials provide a very cumbersome and sophisticated application for the treatment of pollutants as part of a scientific approach. This is also the more efficient technology in terms of consumer acceptance and non-waste production. Nanotechnology is being used in conjunction with physical and chemical changes to materials' surfaces to create better and suitable mixture materials for contaminant remediation. When developed under various favorable conditions that are sometimes very difficult to achieve under normal environmental factors, nanomaterials are relatively target-specific, cost-effective, nontoxic, and recyclable. Some operations are also required to prevent agglomeration, improve monodispersity, and improve stability. The use of toxic nanoparticles is also limited due to the production of toxic nanoparticles as a by-product of remediation sites. As a result, a more prominent nanotechnological bioremediation method is required, one that is simple to implement, efficient, and produces few or no by-products.

8.7   SUMMARY

At present, nanotechnology is no doubt have countless benefits for human beings and thought to be advanced and innovative technology and opens up the opportunity for changing the characteristics of materials in a new way and thus can be more efficient compared with traditional technology. Nano-based remediation opens up efficient potential for decreasing the overall costs (low-cost) and time required for eliminating heavy metals from contaminated soil. Moreover, nanoparticles can be effectively used for on-site remediation of pollutants, thereby eliminating the need for transportation, then treatment, and finally soil disposal after the remediation process ends, and this is due to the strong adsorption property and large surface area. Nanomaterials, because of their small size, increased mobility, and deliverability in soil, can stabilize the heavy metals, thus converting them to less toxic species in soil. Most nanotechnology-based ecosystem studies are conducted at a laboratory scale, and therefore, much effort should be given to field-scale remediation as far as good health is concerned. There is still a need for further research and technology efforts to put these kinds of innovations into industry for full implementation. However, there is still a lack of comprehensive knowledge in science about the synergetic effect of nanoparticles and polymers, iron oxide nanomaterials and carbon-based nanoparticles during a nano-based bioremediation process, and how these are combined to respond the contaminants of a diverse nature. Since, nanoparticles present different advantages over metallic nanoparticles, such as their biodegradability, which produces less environmental impact. Current nanotechnologies may be used to decontaminate soil environments, but more cost-effective manufacturing methods for different polymers need to be conceived.

REFERENCES

1. Cundy, A. B.; Hopkinson, L.; Whitby, R. L. D. Use of Iron-Based Technologies in Contaminated Land and Groundwater Remediation: A Review. *Sci Total Environ.*, **2008**, *400*, 42–51.
2. Li, L.; Fan, M.; Brown, R. C.; Leeuwen, J. H. V.; Wang, J.; Wang, W.; Song, Y.; Zhang, P. Synthesis, Properties, and Environmental Applications of Nanoscale Iron-Based Materials: A Review. *Crit Rev. Environ. Sci. Technol.*, **2006**, *36*, 405–431.
3. Masciangoli, T.; Zhang, W. Environmental Technologies. *Environ. Sci. Technol.*, **2003**, *37*, 102–108.
4. Vaseashta, A.; Vaclavikova, M.; Vaseashta, S.; Gallios, G.; Roy, P.; Pummakarnchana, O. Nanostructures in Environmental Pollution Detection, Monitoring, and Remediation. *Sci. Technol. Adv. Mater.*, **2007**, *8*, 47–59.
5. Han, B.; Zhang, M.; Zhao, D. In-Situ Degradation of Soil-Sorbed 17β-Estradiol Using Carboxymethyl Cellulose Stabilized Manganese Oxide Nanoparticles: Column Studies. *Environ. Pollut.*, **2017**, *223*, 238–246.
6. Han, B.; Zhang, M.; Zhao, D. In-Situ Degradation of Soil-Sorbed 17β-Estradiol Using Carboxymethyl Cellulose Stabilized Manganese Oxide Nanoparticles: Column Studies. *Environ. Pollut.*, **2017**, *223*, 238–246.
7. Perelo, L. W. In Situ and Bioremediation of Organic Pollutants in Aquatic Sediments. *J Hazard Mater.*, **2010**, *177*(1–3), 81–89.
8. Saxena, G.; Bharagava, R. N. Organic and Inorganic Pollutants in Industrial Wastes, Their Ecotoxicological Effects, Health Hazards and Bioremediation Approaches. In: *Environmental Pollutants and their Bioremediation Approaches*, Bharagava, R. N. (Ed.); 1st edn. CRC Press, Taylor & Francis Group, Boca Raton; **2017**; pages 23–56.
9. Saxena, G.; Chandra, R.; Bharagava, R. N. Environmental Pollution, Toxicity Profile and Treatment Approaches for Tannery Wastewater and Its Chemical Pollutants. *Rev Environ ContamToxicol.*, **2016**, *240*, 31–69.
10. Bharagava, R. N.; Saxena, G.; Chowdhary, P. Constructed Wetlands: An Emerging Phytotechnology for Degradation and Detoxification of Industrial Wastewaters. In: *Environmental Pollutants and Their Bioremediation Approaches*; Bharagava, R. N. (Ed.); CRC Press, Taylor & Francis Group, Boca Raton; **2017**; pages 397–426.
11. Chandra, R.; Saxena, G.; Kumar, V. Phytoremediation of Environmental Pollutants: An Eco-Sustainable Green Technology to Environmental Management. In: *Advances in Biodegradation and Bioremediation of Industrial Waste*; Chandra, R. (Ed.); 1st edn. CRC Press, Taylor & Francis Group, Boca Raton; **2015**; pages 1–30.
12. Gautam, S.; Kaithwas, G.; Bharagava, R. N.; Saxena, G. Pollutants in Tannery Wastewater, Pharmacological Effects and Bioremediation Approaches for Human Health Protection and Environmental Safety. In: *Environmental Pollutants and Their Bioremediation Approaches*; Bharagava, R. N. (Ed.); CRC Press, Taylor & Francis Group, Boca Raton; **2017**; pages 369–396.

13. Cheng, J.; Fan, M.; Wang, P.; Su, X. O. The Twice-Oxidized Graphene Oxide/Gold Nanoparticles Composite SERS Substrate for Sensitive Detection of Clenbuterol Residues in Animal-Origin Food Samples. *Food Anal. Methods*, **2020**, *13*(4), 1–9.

14. Kabata-Pendias, A.; Pendias, H. *Trace Metals in Soils and Plants*;2nd edn. CRC Press, Boca Raton;**2001**.

15. D'Amore, J. J.; Al-Abed, S. R.; Scheckel, K. G.; Ryan J. A. Methods for Speciation of Metals in Soils: A Review. *J. Environ. Qual.*, **2005**, *34* (5), 1707–1745.

16. Kuo, S.; Heilman, P. E.; Baker, A. S. Distribution and Forms of Copper, Zinc, Cadmium, Iron, and Manganese in Soils Near a Copper Smelter. *Soil Sci.*, **1983**, 135, 101–109.

17. Zargar, R. A.; Arora, M.; Bhat, R.A. Study of Nanosized Copper Doped ZnO Dilute Magnetic Semiconductor Thick Films for Spintronic Device Applications. *J. Appl. Phys–A*, **2018**, *124*(36).

18. Bhatt, I.; Tripathi, B. N. Interaction of Engineered Nanoparticles with Various Components of the Environment and Possible Strategies for Their Risk Assessment. *Chemosphere*, **2011**, *82*(3), 308–317.

19. Shi, J.; Abid, A. D.; Kennedy, I. M.; Hristova, K. R.; Silk, W. K. To Duckweeds (Landoltiapunctata), Nanoparticulate Copper Oxide Is More Inhibitory Than the Soluble Copper in the Bulk Solution. *Environ. Pollut.*, **2011**, *159*(5), 1277–1282.

20. De Oliveira, J. L.; Campos, E. V. R.; Bakshi, M.; Abhilash, P. C.; Fraceto, L. F. (2014). Application of Nanotechnology for the Encapsulation of Botanical Insecticides for Sustainable Agriculture: Prospects and Promises. *Biotechnol. Adv.*, **2014**, *32*(8), 1550–1561.

21. Gogos, A.; Knauer, K.; Bucheli, T. D. Nanomaterials in Plant Protection and Fertilization: Current State, Foreseen Applications, and Research Priorities. *Journal of Agricultural and Food Chemistry*, **2012**, *60*(39), 9781–9792.

22. Karn, B.; Kuiken, T.; Otto, M. Nanotechnology and in Situ Remediation: A Review of the Benefits and potential Risks. *Environ. Health Perspect.*, **2009**, *117*(12), 1813–1831.

23. Kuiken, T. Cleaning Up Contaminated Waste Sites: Is Nanotechnology the Answer? *Nano Today*, **2010**, *5*(1), 6–8.

24. Shi, J.; Votruba, A. R.; Farokhzad, O. C.; Langer, R. Nanotechnology in Drug Delivery and Tissue Engineering: From Discovery to Applications. *Nano Lett.*, **2010,** *10*(9), 3223–3230.

25. Cai, C.; Zhao, M.; Yu, Z.; Rong, H.; Zhang, C. Utilization of Nanomaterials for In-Situ Remediation of Heavy Metal (loid) Contaminated Sediments: A Review. *Sci. Total Environ.*, **2019**, *662*, 205–217.

26. Yan, W.; Lien, H. L.; Koel, B. E.; Zhang, W. X. Iron Nanoparticles for Environmental Clean-Up: Recent Developments and Future Outlook. *Environmental Science: Processes & Impacts*, **2013**, *15*(1), 63–77.

27. Pan, B.; Xing, B. Applications and Implications of Manufactured Nanoparticles in Soils: A Review. *Eur. J. Soil Sci.*, **2012**, *63*(4), 437–456.

28. Otto, M.; Floyd, M.; Bajpai, S. Nanotechnology for Site Remediation. *Remediation Journal*, **2018**, *19*(1), 99–108.

29. Gong, X.; Huang, D.; Liu, Y.; Peng, Z.; Zeng, G.; Xu, P.; Cheng, M.; Wang, R.; Wan, J. Remediation of contaminated soils by biotechnology with nanomaterials: Bio-behavior, applications, and perspectives. *Crit. Rev. Biotechnol.*, **2018**, *38*(3), 455–468.

30. Tang, W. W.; Zeng, G. M.; Gong, J. L.; Liang, J.; Xu, P.; Zhang, C.; Huang, B. B. Impact of Humic/Fulvic Acid on the Removal of Heavy Metals from Aqueous Solutions Using Nanomaterials: A Review. *Sci. Total Environ.*, **2014**, *468*, 1014–1027.

31. Zargar, R.A.; Arora, M.; Alshahrani, T.; Shkir, M. Screen Printed Novel ZnO/ MWCNTs Nanocomposite Thick Film. *Ceramic Int.*, **2021**, *47*, 6084–6093.

32. Zargar, R.A. ZnCdO Thick Film: A Material for Energy Conversion Devices. *Mater. Res. Express*, **2019**, *6*, 095909.

33. Su, B.; Lin, J.; Owens, G.; Chen, Z. Impact of Green Synthesized Iron Oxide Nanoparticles on the Distribution and Transformation of as Species in Contaminated Soil. *Environ. Pollut.*, **2020**, *258*, 113668.

34. Zargar, R.A.; Kumar, K.; Arora, M.; Shkir, M.; Somaily, H. H.; Algarni, H.; Al Faify, S. Structural, Optical, Photoluminescence, and EPR Behaviour of Novel Zn0 80Cd0·20O Thick Films: An Effect of Different Sintering Temperatures. *J. Luminescence*, **2022**, *245*, 118769.

35. Klaine, S. J.; Alvarez, P. J.; Batley, G. E.; Fernandes, T. F.; Handy, R. D.; Lyon, D. Y.; Lead, J. R. Nanomaterials in the Environment: Behavior, Fate, Bioavailability, and Effects. *Environmental Toxicology and Chemistry: An International Journal*, **2008**, *27*(9), 1825–1851.

36. Madsen, E. L. Microorganisms and Their Roles in Fundamental Biogeochemical Cycles. *Curr. Opin. Biotechnol.*, **2011**, *22*, 456–464.

37. Shrestha, B.; Acosta-Martinez, V.; Cox, S. B.; Green, M. J.; Li, S.; Canas-Carrell, J. E. An Evaluation of the Impact of Multiwalled Carbon Nanotubes on Soil Microbial Community Structure and Functioning. *J. Hazard. Mater.*, **2013**, *261*, 188–197.

38. Muller, J.; Huaux, F.; Lison, D. Respiratory Toxicity of Carbon Nanotubes: How Worried Should We Be? *Carbon.* **2006**, *44*(6), 1048–1056.

39. Moghaddam, H. K.; Pakizeh, M. Experimental Study on Mercury Ions Removal From Aqueous Solution by MnO2/CNTs Nanocomposite Adsorbent. *J. Ind. Eng. Chem.***2015**, *21*, 221–229.

40. Nethaji, S.; Sivasamy, A.; Mandal, A. B. Preparation and Characterization of Corn Cob Activated Carbon Coated with Nano-Sized Magnetite Particles for the Removal of Cr(VI). *Bioresour. Technol.*, **2013**, *134*, 94–100.

41. Strek, H. J.; Weber, J. B.; Shea, P. J.; Mrozek J.; Overcash, E. Reduction of Polychlorinated Biphenyl Toxicity and Uptake of Carbon-14 Activity by Plants through the Use of Activated Carbon, *J. Agr. Food Chem.*, **1981**, *29*, 288–293.

42. Ng, L. Y.; Mohammad, A. W.; Leo, C. P.; Hilal, N. Polymeric Membranes Incorporated with Metal/Metal Oxide Nanoparticles: A Comprehensive Review. *Desalination*, **2013**, *308*, 15–33.

43. Xu, J.; Bhattacharyya, D. Membrane-Based Bimetallic Nanoparticles for Environmental Remediation: Synthesis and Reactive Properties. *Environ. Prog.*, **2005**, *24*, 358–366.

44. Zhao, X.; Lv, L.; Pan, B.; Zhang, W.; Zhang, S.; Zhang, Q. Polymer-Supported Nanocomposites for Environmental Application: A Review. *Chem. Eng. J.*, **2011**, *170*, 381–394.

45. Campbell, M. L.; Guerra, F. D.; Dhulekar, J.; Alexis, F.; Whitehead, D. C. Target-Specific Capture of Environmentally Relevant Gaseous Aldehydes and Carboxylic Acids with Functional Nanoparticles. *Chem. A Eur. J.*, **2015**, *21*, 14834–14842.

46. Chen, H.; Yada, R. Nanotechnologies in Agriculture: New Tools for Sustainable Development. *Trends Food Sci. Technol.*, **2011**, *22*(11), 585–594.

47. Bakshi, S.; He, Z. L.; Harris, W. G. Natural Nanoparticles: Implications for Environment and Human Health. *Critical Reviews in Environmental Science and Technology*, **2015**, *45*(8), 861–904.

48. Shi, Z.; Shao, L.; Jones, T. P.; Lu, S. Microscopy and Mineralogy of Airborne Particles Collected during Severe Dust Storm Episodes in Beijing, China. *J. Geophys. Res. Atmos.*, **2005**, *110*, Article number D01303.

49. Pollmann, K.; Kutschke, S.; Matys, S.; Kostudis, S.; Hopfe, S.; Raff, J. Novel Biotechnological Approaches for the Recovery of Metals from Primary and Secondary Resources. *Minerals*, **2016**, *6*(2), 54.

9 Recent Advancements and Applications of Nanosensors in Various Fields

*Briska Jifrina Premnath, Bichandarkoil Jayaram
Pratima, Ragunath Ravichandiran, Manoj
Kumar Srinivasan, and Namasivayam Nalini*

CONTENTS

DOI: 10.1201/9781003323464-9

9.1   INTRODUCTION

"The cosmos is incomplete without nanotechnology." [1] Nanotechnology is a discipline of science concerned with altering matter on an atomic scale. One of the primary uses of nanotechnology is in the field of nanosensors. [2] A sensor is a gadget that notices a capricious quantity, generally electronically, turning the measurement into definite signals. The most extraordinary, significant necessities for sensors are diversity, sensitivity, information extraction, precision, choosiness, and durability. [3] Signals might be biological, optical, electronic, electrical, physical, or mechanical. [2]

Nanosensors are sensing devices with at least one dimension less than 100 nm that collect data on the nanoscale and change it into data for analysis. These sensors can also be characterized as "a chemical or physical sensor designed with nanoscale components, often microscopic or submicroscopic in size." [1]

Nanosensors are very small elements used to recognize a specific molecule, biological component, or environmental situation. They are exact, portable, and inexpensive and operate at a far lower detection level than their macroscale equivalents. [4]

Nanomaterials are employed in nanotechnology in various industries, including medicine, energy transportation, electronics, information technology, polymers, environmental science, and more. Some primary uses include national security, aircraft, and integrated circuits, among others. Nanosensors are in numerous varieties, have different manufacturing methods, and are used for various purposes. [1]

The essential components of nanosensors and their different types and uses in various fields are briefly overviewed in this chapter.

9.2 BASIC COMPONENTS OF NANOSENSORS

A standard nanosensor device function consists of three main mechanisms.

9.2.1 SAMPLE PREPARATION

The sample could be a simple or complex gas, a liquid, or a solid-state suspension. The sample comprises specific chemicals, functional groupings of molecules, or organisms that the sensors can be aimed at. The analyte is the targeted molecule/organism. It can be molecules (dyes/colors, toxicants, pesticides, hormones, antibiotics, vitamins, etc.), biomolecules (enzymes, DNA/RNA, allergens, etc.), ions (metals, halogens, surfactants, etc.), gas/vapor (O_2, CO_2, quickly evaporated compounds, water vapors, etc.), organisms (bacteria, fungi), moisture, temperature, light, pH, weather, and the like.

9.2.2 APPRECIATION

Some molecules/elements in the sample recognize the analytes. These identified molecules include antibodies, aptamers, chemical legends, enzymes, and others with high affinity, peculiarity, and selectivity to their analytes to measure them to acceptable levels.

9.2.3 SIGNAL TRANSMISSION

Different signal transduction approaches have classified these tiny gadgets as optical, electrochemical, piezoelectric, pyroelectric, electronic, and gravimetric biosensors. They turn noticed actions into quantifiable indications, which progress in generating information.[4]

9.3 CLASSIFICATION OF NANOSENSORS

Nanosensors are categorized based on their energy source, structure, and uses. Table 9.1 depicts the classification of nanosensors. [5]

9.3.1 ENERGY SOURCE

Nanosensors are divided into two types: (1) active nanosensors that necessitate a power source, such as a thermistor, and (2) passive nanosensors that don't need a power source, such as a thermocouple or a piezoelectric sensor.

TABLE 9.1

Classification of Nanosensors [5]

S. No	Classification of Nanosensors	
	Stimuli	**Properties**
1.	Mechanical	Position, acceleration, stress, strain, force, pressure, mass, density, viscosity, moment, torque acoustic wave amplitude, phase, polarization, velocity
2.	Optical	Absorbance, reflectance, fluorescence, luminescence, refractive index, light scattering
3.	Thermal	Temperature, flux, thermal conductivity, specific heat
4.	Electrical	Charge, current, potential, dielectric constant, conductivity
5.	Magnetic	Magnetic field, flux, permeability
6.	Chemical	Components (identities, concentrations, states)
7.	Biological	Biomass (identities, concentrations, states)

9.3.2 STRUCTURE

Structure can be used to distinguish four types of sensors. [6]

9.3.2.1 Optical Nanosensors

Optical sensors can be used to monitor chemical analysis. The optical properties of nanomaterials dictate them. They have a diverse range of approaches in different sectors, including the chemical industry, biotechnology, medicine, environmental sciences, and human safety. The initial optical nanosensor to be revealed was dependent on fluorescein encased in a polyacrylamide nanoparticle and was used to measure pH. The advantage of this fundamental technique is that it minimizes the cell's physical disruption. A downside of free dye is the chemical interference between the dye and the cell induced by protein binding, cell sequestration, and toxicity. An additional possibility is to use tagged nanoparticles with an attached reporter molecule. The primary variance between tagged nanoparticles and free dye is that the former is solid while the latter is fluid. As with the free dye, the tagged nanoparticles move simply, and the reporter molecules come into touch with the components inside the cell. Although outer-tagged particle sensors have been utilized for sensing within the cell, they suffer from the same drawbacks as unbound fluorescent dyes. The signal is generated by receptor molecules exposed to their cellular surroundings. [3]

9.3.2.2 Fiber Optic Nanosensors

Fiberoptic nanosensors can do in vivo analysis of critical biological processes. The intercommunication of the specific molecule (A) and the receptor (R) is meant to generate a physicochemical disturbance that can be translated to an electrical signal or further quantifiable signals. The optical probe then collects this quantifiable signal and transmits it to the database. The difficulties of the color-free approach are overcome by placing the optical fiber arm between the surrounding and sensitive areas. An added benefit of the optical nanosensor is that it achieves a lesser degree of invasion. [3, 7]

9.3.2.3   Electromagnetic Nanosensors

Depending on their recognition mechanisms, electromagnetic nanosensors are divided into two distinct types.

9.3.2.3.1   Current Measurement

The benefit of this procedure is that it is non-marking and does not require dyes. Geng et al. explored the interaction of hydrogen sulfide gas molecules with gold nanoparticles. [8] Each sensor cell contains a chromium electrode and a gold electrode that serves as the source and drain. A distinctive gap width of approximately 40–60 nm has been established between the two electrodes. Au nanoparticles are erratically distributed throughout the gap area. Creating a sulfide shell hinders the transmission of "e" charges between nanoparticles, that is, the so-called bounding phenomenon. The bounding of electrons is determined by measuring the current and voltage across Cr and Au electrodes in the presence of an applied electrical field. [3]

9.3.2.3.2   Magnetism Measurement

Magnetic nanosensors were developed to spot exact biomolecules such as proteins, enzymes, and disease-causing agents (e.g., viruses) with reactivity in the low femtomolar range (0.5–30 fmol). Magnetic nanosensors are nanoparticle-based devices (iron oxide) that interact with their chemical target and develop stable nano assemblies. It decreases the spin–spin relaxation time (T2) of the adjacent H_2O molecules, which can be observed by techniques of nuclear magnetic resonance imaging (NMR/ MRI). [9]

9.3.2.4   Mechanical Nanosensors

Mechanical nanosensors outperform optical and electromagnetic nanosensors to detect nanoscale mechanical qualities. Mechanical nanosensors come in various configurations, including carbon nanotube-based fluidic shear-stress sensors and nanomechanical cantilever sensors. Binh et al. projected the first mechanical nanosensor to measure the shuddering and flexible characteristics linked to a tapered cantilever. [10] They play a critical role in fabricating nanodevice components and nanoscale subassemblies for microelectronic devices. [3]

9.3.3   Application-Based Classification

According to their uses, four categories of sensors are identified. [11]

9.3.3.1   Chemical Nanosensors

This sensor can be used to evaluate a molecule or chemical. Numerous optical chemical nanosensors were employed to determine the potential of hydrogen and different ion concentrations.

9.3.3.2   Deployable Nanosensors

These are employed by the military and for various sorts of nationwide safety, such as Sniffer STAR. It is described as a lightweight, portable chemical detection device that syndicates a nanomaterial for sample collection with a micro-electromechanical finder for concentration measurement.

9.3.3.3 Electrometers

They are composed of a mechanical resonator, a detection electrode, and a gate electrode. The detection and gate electrodes pair the charges to the mechanical element.

9.3.3.4 Biosensors

A biosensor is a gadget that alters a signal from a biotic process into one that can be studied and evaluated. It consists of a living sensing element or bioreceptor (enzymes, antibodies, nucleic acids, and so on) and a physical transducer (e.g., optical, mass, or electrochemical). The transducer section gathers data on physicochemical changes (i.e., electron transfer, heat transfer, pH change, mass change, uptake, or release of specific ions or gases) caused by the contact of the bio-element and the analyte under test. The transducer then transforms this mechanical transformation into electrical impulses. The electrical signal is augmented, analyzed, and eventually presented under the analyte concentration in the sample. Biosensors combine the computational capabilities of microchips with the susceptivity and specific features of biotic systems in an interdisciplinary design. Figure 9.1 illustrates the fundamental components of a biosensor.

According to the type of transducer utilized in their manufacturing, biosensors are categorized into the following types. Figure 9.2 explains the diverse categories of transducers utilized in biosensors.

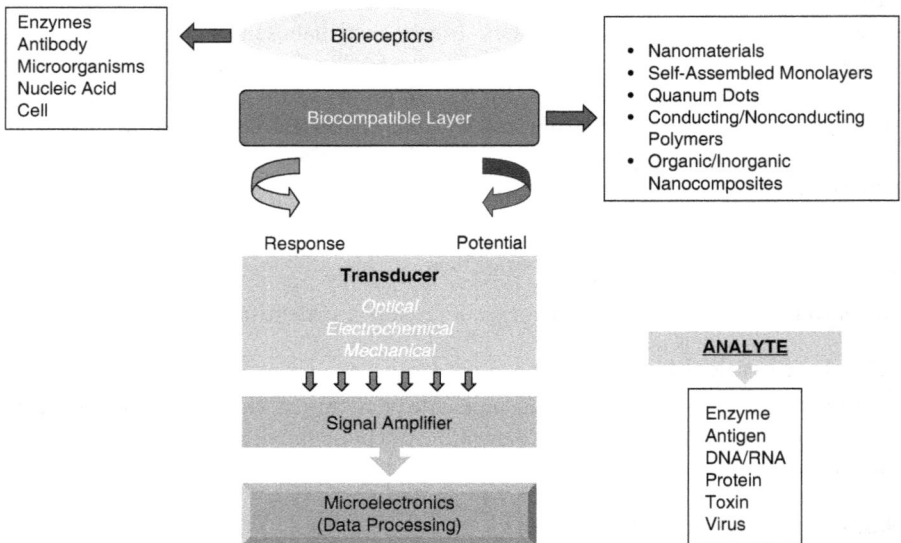

FIGURE 9.1 Fundamental principle of a biosensor.

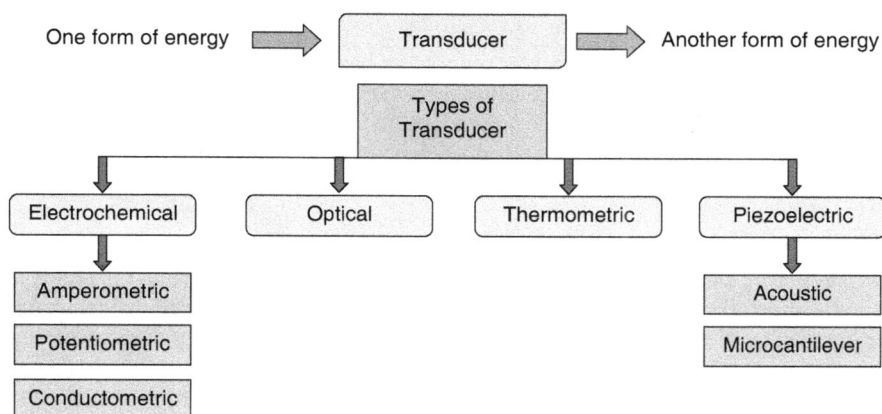

FIGURE 9.2 Categories of transducers utilized in biosensors.

9.3.3.4.1 Electrochemical Biosensor

The electrochemical biosensor's fundamental idea is that the generation or exhaustion of ions or electrons during a chemical reaction between the immobilized biomolecule and the target analyte influences the solution's quantifiable electrical properties (e.g., electric current or potential). According to its working principle, electrochemical biosensors can convert chemical information into a detectable amperometric signal using amperometric, conductometric, or potentiometric transducers.

9.3.3.4.1.1 Amperometric Amperometric biosensors are self-contained, integrated devices that measure the generated electronic current and offer quantitative analytical information when an electroactive biological ingredient is oxidized or reduced. The amplitude of the electric current generated at the electrode's surface due to the electrocatalytic reaction of the electroactive species is directly proportional to the analyte concentration in the sample.

9.3.3.4.1.2 Potentiometric A potentiometric biosensor converts the biological reaction to an electric signal using ion-selective electrodes. These biosensors combine a biorecognition element with a transducer that detects changes in the concentration of an ionic species and then records the analytical signal that is logarithmically associated with the analyte concentration.

9.3.3.4.1.3 Conductometric Owing to the analyte's presence, a conductometric biosensor can distinguish changes in the electrical conductance of a sample solution. The concentration of ionic species varies due to the reaction between the biomolecule and the analyte, which results in a change in the solution's electrical conductivity or current flow.

9.3.3.4.2   Optical Biosensors

The optical biosensors' fundamental premise is based on optical measures such as fluorescence, absorbance, and chemiluminescence. Photons are used to determine the analyte concentration through optical transducers. These biosensors work by converting the target analyte to a product that is either oxidized or reduced at the working electrode's exterior.

9.3.3.4.3   Thermometric Biosensor

Thermometric or calorimetric biosensors are based on the fundamental factor of endothermic or exothermic biotic responses. Thermistors are used to determine the temperature difference between the substrate and product by the result of heat received or released in the reaction media. Thermal biosensors are capable of detecting even minute changes in temperature.

9.3.3.4.4   Piezoelectric Biosensors

Since piezoelectric biosensors function based on acoustics (sound vibrations), they are also called acoustic biosensors. The fundamentals of these biosensors are piezoelectric crystals and their vibrations with positive and negative charges exhibiting distinctive frequencies. Electronic equipment can monitor changes in the resonance frequencies caused by certain molecules adhering to the crystal surface. [12]

9.4   APPLICATION OF NANOSENSORS

9.4.1   Nanowires and Nanotubes

Compared to bulk planar devices, nanomaterials with a single dimension such as nanowires and nanotubes can be employed as nanosensors. They can act as both transducers and cables for signal transmission. In addition, due to their compact size, they enable the multiplexing of individual sensor units within a single device.

9.4.1.1   Semiconductor Nanowires Are Used as Detection Elements in Sensors

Semiconductor nanowires are employed as detection elements in sensors that detect chemical vapors. When molecules attach to semiconducting nanowires such as zinc oxide, the conductance of the wire changes. The amount and direction of conductance change are determined by the molecule bound to the nanowire.

9.4.1.2   Semiconducting Carbon Nanotubes

Functionalize carbon nanotubes by connecting them to metal molecules, such as gold, to notice chemical vapors. Chemical molecules then form bonds with the metal, altering the conductivity of the carbon nanotube.

9.4.1.3   Carbon Nanotubes and Nanowires for Bacteria and Virus Detection

These materials can also detect bacteria and viruses. Functionalize the carbon nanotubes by coating them with an antibody; when the matching bacteria or virus binds to an antibody, the conductance of the nanotube changes. This technology has several intriguing applications, one of which is bacteria detection in hospitals. By identifying contaminated germs, healthcare personnel may lower the number of patients who acquire complications such as staph infections.

9.4.1.4 Nanocantilevers

These devices are being used to produce single-molecule sensors. Nanocantilevers oscillate at a resonance frequency that changes in response to the weight of a molecule that lands on it. [2]

9.4.2 ENVIRONMENTAL DISCIPLINE

Nanosensors have a wide range of environmental applications. Environmental agencies are mainly concerned with the ability to detect dangerous compounds and microbes in the air and water. Nanosensors will revolutionize how air and water quality are analyzed due to their small size, thickness, and high precision of measurements. Detecting mercury, zinc, or other dangerous compounds in any source (such as air or water) using dandelion-like Au/polyaniline (PANI) nanoparticles in aggregation with surface-enhanced Raman spectroscopy (SERS) nanosensors. [13] A novel technique for air sampling is to use nanosensors to assess the quality of the air, particularly for contaminants. Nanosensors have previously been employed to monitor solar irradiance, aerosol–cloud interactions, climate forcing, and other biogeochemical cycles in East Asia and the Pacific. This type of gear was advantageous for monitoring air pollution levels in Beijing during the Summer Olympic Games. [14]

9.4.3 AGROECOSYSTEMS

The last few decades have seen a lot of problems, such as a growing population, changes in the climate, and more people competing for limited resources. All of these have put a lot of people at risk and made it more critical for the world to have food security. Existing agricultural methods use a lot of resources, advanced machinery, and many agrochemicals to meet people's needs for food. These methods have caused a lot of damage to soil, air, and water resources, which has led to more pollution in agricultural settings and has harmed human and animal health. Because pesticide residues remain in the surroundings for a long duration, they pollute the soil, making people concerned about its function, biodiversity, and food safety. Furthermore, there are countless stories about pesticides making their way into the food chain and accumulating in the bodies of people who eat them, causing major health problems. Pesticides are cytotoxic and can cause cancer in people. In addition, they can cause a wide range of neurological and bone marrow problems and infertility, respiratory, and immunological problems. It is vital to monitor pesticide residues in the environment. Monitoring the residual pesticides helps identify the dosage within or over the legal limits. [15] Lethal heavy metals like Cd, Hg, Cu, Zn, Ni, Pb, and Cr are another big problem for agroecosystems. These metals are thought to cause long-term and substantial damage to living systems by interfering with biological functions at the cellular level, such as photosynthesis, mineral absorption,

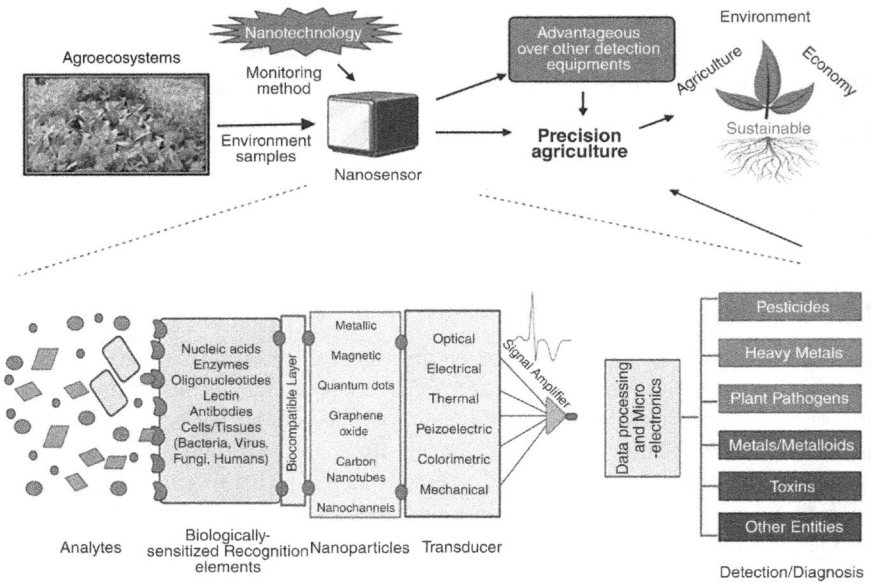

FIGURE 9.3   The component of nanosensors to monitor agroecosystems.

electron transport chain interruptions, and the induction of lipidic acid. While these elements are found naturally in the earth's crust, uncontrolled human activities have significantly impacted these elements' geochemical cycling and biological balance. As a result, the number of these metals in different parts of plants has increased. It's essential to find ways to detect heavy metals at low concentrations in environmental samples because they can negatively affect diverse ecosystems. As a result, there is a huge need for basic, quick, and gainful ways to check for agricultural toxins. [16] Figure 9.3 illustrates the components of nanosensors in agroecosystems.

9.4.4 AGRICULTURE

Nanotechnology has the potential to pave the path for environmentally friendly agriculture, and the condensed working illustration of nanofertilizers and nanopesticides is described in Figure 9.4.

9.4.4.1 Nanofertilizers

Chemical fertilizers are required for current agricultural systems to function. Yet, synthetic chemical product efficacy has declined for decades, resulting in water pollution, soil contamination, and greenhouse gas emissions. [17]

Nanofertilizers are mineral nutrients produced primarily through nanoparticle encapsulation and are classed as macronutrients or micronutrients. [18] Macronutrients such as C, N, K, P, Ca, S, and Mg have been enclosed in various nanomaterials to enhance crop fertilizer absorption and reduce fertilizer outflow.

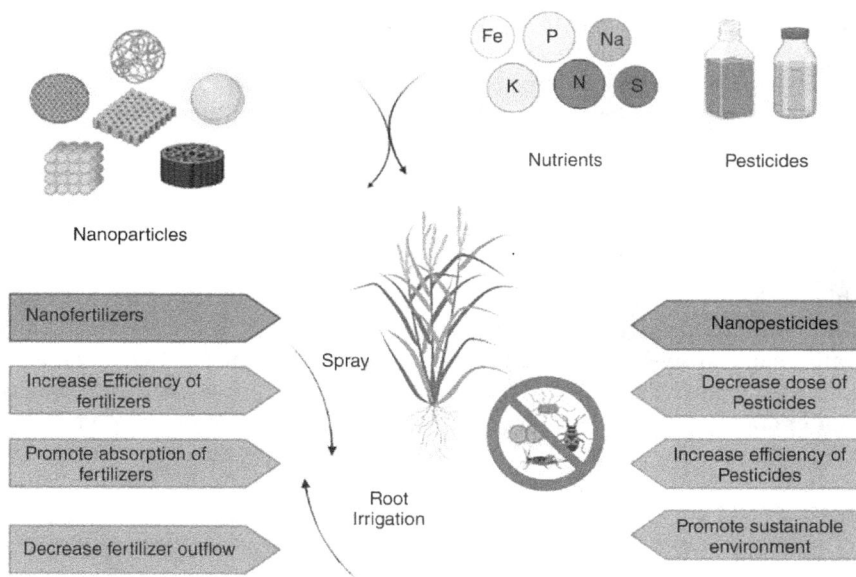

FIGURE 9.4   Condensed working illustration of nanofertilizers and nanopesticides.

[19, 20] The large definite surface area, high constancy, and outstanding biocompatibility of nanoparticles contribute to the improved release efficiency of NPs fertilizer composites. [21] For instance, urea-hydroxyapatite (HA) nanoparticles have demonstrated significant capacity for extending release duration and dropping nitrogen fertilizer consumption. Urea achieves the benefits of nanoparticles through interaction with the amine and carbonyl groups of HA nanoparticles. Field trials have demonstrated that nanohybrids of urea and HA boost agronomic nitrogen use effectiveness by roughly 30% compared to pure urea. Additionally, multiple kinds of research have shown that the high specific surface area and density of nanoparticles confer a significant degree of reactivity on nanohybrids. [20] Nanofertilizers hold substantial promise for boosting fertilizer absorption and agricultural production. [19] Recent research indicates that binding N, P, and K into chitosan nanoparticles improves N, P, and K acquisition by 17.04%, 16.31%, and 67.50%, respectively, in cultivated coffee plants. Magnesium oxide (MgO) nanoparticles sprayed on cotton considerably boosted seed cotton output by 42.2% compared to unsprayed controls. [22]

9.4.4.2   Nanopesticides

The nanoformulation or encapsulation of insecticides, herbicides, fungicides, and bactericides in nanomaterials has immense capability to reduce pesticide doses, boost crop productivity, and support maintainable growth. Polymeric nanoparticles (such as chitosan and solid lipids), inorganic nonmetallic nanoparticles (such as silica nanoparticles and nanoclays), and metallic nanoparticles (such as Cu nanoparticles and ZnO nanoparticles) all serve as nanocarriers for nanopesticides. [20] Numerous studies have demonstrated that nanoparticle pesticides are more effective in destroying pests and are less possible to produce adverse human reactions. For example, spinosad- and permethrin-loaded chitosan nanoparticles applied to *Drosophila melanogaster* demonstrated enhanced bioavailability even at lower doses than free spinosad and permethrin, respectively; nanocomposites induced fewer adverse reactions on humans and the ecological circumstance. [23] When nano-insecticide particles are encapsulated in nanoparticles, they develop slighter and further centralized, giving them firmness and a gradual ability to release. These features enhance the pesticide action while lowering its toxicity to humans. Additionally, the biological toxicity of highly concentrated nanoparticles enables the direct inhibition of pests, germs, and viruses. For example, numerous nanoinsecticides make use of the toxicity of metallic nanoparticles. Compared to a bulk aluminum oxide (Al_2O_3) treatment, aluminum oxide (Al_2O_3) nanoparticles demonstrated a more excellent capability for eradicating *Sitophilus oryzae* on stored rice. [24] Pheromones have been documented as a viable and effective means of insect population control. The combined forms of nanocarriers and pheromones increase the benefits of sex pheromones. [25] For example, when applied to guava orchards, nanogels containing methyl eugenol boosted snare catches equated to the control group holding only methyl eugenol. [18]

9.4.5 TOXICITY OF HEAVY METAL IONS IN WATER

There has been a long-running problem with the toxicity of heavy metal ions like Hg_{21}, Pb_{21}, Cu_2, and Zn_2. These ions are toxic. Hg is released into the environment as a by-product of fossil fuel combustion, mining, volcanic eruptions, and the burning of garbage and waste. Hg and Pb affect the ecology of wildlife and the health of people, as well. They turn inorganic mercury ions into neurotoxic compounds that build up in plants and the food chain. Mercury (II) and copper (II) ions can be detected with bare Au-gated and Au-gated AlGaN/GaN high-electron-mobility transistors (HEMTs). Sensors with thioglycolic acid were detected in less than 5 seconds. The addition of thioglycolic acid made the surface more sensitive to mercury detection by 2.53 compared to the surface that had no thioglycolic acid on it. Attention is necessary to find mercury (II) ions. The detection limit was 1027 M, but the selectivity for finding them over Na or Mg was more than 100. The sensors can be cleaned with deionized water. [26]

9.4.6 FOOD INDUSTRY

Nanosensors are bioanalytical devices designed utilizing a combination of nanomaterials and biological receptors. Nanosensors are critical in the food business and have garnered considerable interest owing to their rapid recognition capability, honesty, and low cost. Due to their great sensitivity and specificity, nanosensors have the quality to be combined with a wide variety of analytes. As a result of the conjugation of nanomaterials, such as carbon nanotubes, nanoparticles (metallic, nonmetallic, metal oxide), semiconductor nanoparticles, nanorods, nanowires, nanobiofilms, nanofibers, and quantum dots, these objects have a high surface-to-volume ratio

TABLE 9.2

Nanosensors in the Identification of Toxins

Nanosensor-Based Applications	Nanomaterial Used	Analyte Detected	Method of Detection
Detection of toxins	Magnetic nanoparticles	Mycotoxin	Immunoassay and enzyme-linked immunosorbent assay
	Quartz nanopipettes	Zearalenone and HT-2	Ion nanogating and enzyme-linked immunosorbent assay
	Ionic liquids (gold and graphene oxide), cerium dioxide and zinc oxide nanoparticles	Ochratoxin-A	Cyclic voltammetry and impedance
	Gold nanoparticles	Botulinum neurotoxin type B and brevetoxins	Enzyme-linked immunosorbent assay, cyclic voltammetry, and immune-chromatographic assay
	Single-walled and multiwalled carbon nanotubes	Palytoxin and Microcystin-LR	Electro-chemiluminescence and immunoassay
	Gold, iron oxide, and superparamagnetic nanoparticles	Aflatoxins B1 and aflatoxin M1	Immunoassay and enzyme-linked immunosorbent assay

and outstanding optical and electrical properties. Now, nanosensors are utilized to detect foodborne infections, contaminants, toxins, chemicals, and pesticides extant in various foods. They were also utilized to screen the freshness of food and the integrity of food packaging. Nanobiosensing tools, including cyclic voltammetry, surface plasmon resonance, differential pulse voltammetry, interdigitated array microelectrode-based impedance examination, amperometry, flow injection analysis, and bioluminescence, are used to detect a variety of disease-causing agents, toxins, and contaminants that takes place in foods hastily and precisely. Tables 9.2 through 9.5 summarize the possible applications of nanosensors in several food business sectors. [27]

9.4.7 FLUORESCENCE-BASED SENSORS

One of the first ways to look at the movement of biomolecules of interest inside a living system was to use fluorescence-based sensors. [28] Since the late twentieth century, Roger Tsien et al. have made calcium-sensitive fluorescent dyes. Along with later-made genetically encoded proteins, these dyes are the building blocks of

TABLE 9.3

Nanosensors in the Detection of Microbes

Nanosensor-Based Applications	Nanomaterial Used	Analyte Detected	Method of Detection
Detection of microbes	Single-walled carbon nanotubes	*Salmonella infantis* and *E. coli*	Field-effect transistor and fluorescence microscopy
	Core–shell nanoparticles (zinc sulfite–coated cadmium selenide)	*E. coli* and *S. typhimurium*	Fluorescence microscopy
	Polypyrrole nanowires	*Bacillus globigii*	Linear sweep voltammetry
	Tris-hexahydrate-doped silica nanoparticles	*E. coli, S. typhimurium,* and *B. cereus*	Spectro-fluorometry and flow cytometry
	Gold nanoparticles	*E. coli, Staphylococcus aureus, Vibrio parahaemolyticus, Salmonella enterica,* and *Salmonella typhi*	Cyclic voltammetry, surface plasmon resonance, and differential pulse voltammetry
	Bismuth nanofilm, iron oxide nanoparticles, and peptide nanotubes	*E. coli, S. typhimurium,* and *L. monocytogenes*	Interdigitated array micro electrode-based impedance analysis, cyclic voltammetry, amperometry, flow injection analysis, and bioluminescence

quantitative sensing and are used in a wide range of biological studies. Both molecular markers and proteins that light up in the dark have flaws. Molecular probes that reach the space within the cell through acetoxymethyl esters can leave some types of cells in minutes. Another thing to remember about genetically encoded indicators is that they may change how biochemical processes work for the analytes or bioactivities. Exogenous fluorescent nanosensors can be added to the existing range of pointers, making them a good choice for studying the dynamics of biomolecules and their living processes. [29, 30]

9.4.8 DNA-BASED NANOSENSORS

Another way to classify fluorescent nanosensors is by the parts of them that they use to find things. For example, DNA has been used as the backbone or "recognition moiety" in sensor design for a long time because it is very modular and can be easily changed. So, it makes it possible to build complex structures with a low chemical variation. [31] Nanosensors that use DNA as a vital part of the sensor structure can detect intracellular Cl, messenger RNA (mRNA), enzyme catalytic activity, and

TABLE 9.4

Nanosensors in the Recognition of Pesticides and Chemicals

Nanosensor-Based Applications	Nanomaterial Used	Analyte Detected	Methods of Detection
	Poly (ethylene glycol dimethacrylate-N methacryloyl-1-histidine methylester)	Chloramphenicol	Surface plasmon resonance and ultraviolet—visible spectroscopy
Detection of pesticides and chemicals	Multi-walled carbon nanotubes, iron oxide nanoparticles, and graphene	Sudan I	Cyclic voltammetry and high-performance liquid chromatography
	Single-walled carbon nanotubes, multiwalled carbon nanotubes conjugated with silica, platinum and zinc oxide nanoparticles, and ionic liquids of multiwalled carbon nanotubes	Cadmium ions, sunset yellow, Bisphenol A, and tartrazine	Field-effect transistor and cyclic voltammetry
	Zinc sulfide–cadmium selenide; liposome; gold, cadmium, and selenide zirconium dioxide nanoparticles	Parathion, paraoxon, and carbamate pesticides	Square wave voltammetry. photoluminescence, colorimetry, fluorescence-based ultraviolet–visible spectroscopy
	Cobalt nitroprusside	Sulfite	Cyclic voltammetry
	Silver and gold nanoparticles	Melamine	Fluorescence- and colorimetric-based ultraviolet–visible spectroscopy

cancer biomarkers. Krishnan et al. made the Clensor nanodevice to notice Cl in the Drosophila melanogaster cells. [30, 32]

Clensor's DNA sequence, with a Cl-sensitive fluorophore and a reference fluorophore, precisely targets and tracks lumenal lysosomal Cl levels through a ratiometric signal output. The snail's anatomy inspired another interesting DNA nanosensor for detecting messenger RNA in single cells. The nano-SNail-inspirEd Locator (nano-SNEL) has a molecular beacon for mRNA detection inside a protective DNA nanoshell. The nanoshell model of the sensors kept them from being damaged by enzymes, allowing them to keep track of live-cell RNA transcription for a more prolonged period. Tan's group has done two studies recently that show how nanomachines made of three-dimensional DNA can be used to find cancer biomarkers [33] and messenger RNAs in living cells. [34]

TABLE 9.5

Nanosensors in Detection of Unstable Key Food Ingredients

Nanosensor-Based Applications	Nanomaterial Used	Analyte Detected	Methods of Detection
Detection of unstable key food ingredients	Diphenylalanine peptide nanotubes, multiwalled carbon nanotubes, gold and nickel oxide nanoparticles	Ascorbic acid, acetaminophen, glucose, and tryptophan	Amperometry and cyclic voltammetry
	Platinum-cobalt, single-walled, double-walled, and multiwalled carbon nanotubes	Folate and vitamin B9	Cyclic voltammetry
	Silver, zirconium dioxide, iron, nickel-platinum, chitosan, gold, tin-dioxide nanoparticles, and Prussian blue–gold, and cuprous oxide conjugated single-walled carbon nanotubes	Hydrogen peroxide, glucose, fructose, sucrose, glutamic acid, and succinic acid	Ultraviolet–visible spectroscopy, cyclic voltammetry, and amperometry
	Silver-tin dioxide nanoparticles	Ethanol	Adsorption
	Gold nanoparticles	Caffeic acid, gallic acid catechol, and chlorogenic acid	Amperometry and cyclic voltammetry

9.4.9 PEPTIDE NANOSENSORS

Peptides are another type of biomolecule that can be used as a nanosensor's "recognition moiety." The development of peptide-based nanosensors to detect disease-related protease activity has increased a lot. One way to use peptide-based sensors is to put fluorophore-labeled proteins on nanoparticle substrates that will be split by the target protease in the body and then analyzed after the urinal has been cleared. As a result of this sensor design, many problems that arise when imaging through body tissues can be solved by taking the signal reporters out of the body. The ability of this method to work has been shown by sensing thrombin in mice by using iron oxide nanoparticles that have been coated with thrombin-sensitive peptides that have been linked to fluorescent reporters. A recent study grabbed this even further by putting together a library of fluorescent nanosensors that can sense the activity of proteases to classify prostate cancer based on how aggressive it is. [35] The nanosensors were very good at categorizing malignancies and outstripped traditional serum cancer biomarkers, which could be helpful in the future. [36]

9.4.10 Plasmonics-Based Nanosensors

Even though there have been a lot of improvements in fluorescence-based sensing, the mechanism often has a short observation time because of the sensor's vulnerability to photodamage. Nanosensors that use plasmonic nanomaterials, such as noble metal NPs, can help to lessen this apprehension. [37] Localized surface plasmon resonance (LSPR) or plasmon from noble metal nanoparticles, united fluctuations of free electrons in nanoparticles when they're illuminated, are the only plasmonic materials that this review discusses. Plasmon nanoparticles have bigger optical cross-sections at their resonance wavelengths than fluorescent materials, but each specific dye molecule has a limited photon emission rate of about 5 kHz, limiting signal intensity. [38] Furthermore, plasmonic nanoparticles have very stable optical outputs because the signal depends on light scattering, making them very stable. Thus, the nanoparticles never fade or blink, allowing for long-duration, intermittent-free sensing with no limits. However, organic fluorophores can only last a certain count of excitation/emission cycles before they fade. [39] The optical response of plasmonic nanomaterials also changes depending on their size, shape, and surroundings, [40] which are essential parts of different plasmonic-based sensing systems for biological applications. [41]

9.4.11 Biosensors

Using it is one of the most common things people do because it can help them find early signs of cancer and other diseases. It can also be used to look for a certain type of DNA. Biosensors are usually thought of as part of the group of chemical sensors because the way they get information, or the "sensor platforms," are the same. [42] Field-effect transistors (FETs) are one of the best biosensor technologies that have been made. They have a lot of advantages, like being very sensitive, being able to make a lot of them, and being cheap to make. [43] There are a lot of FET-based biosensors out there; the most common ones are ion-sensitive FETs (ISFETs), silicon nanowires, organic FETs, graphene FETs, and compound-semiconductor FETs. [44] In this case, the textile-based wearable nanobiosensors can spot neurological signals and look for abnormalities to diagnose specific neurological and cardiovascular disorders. [45]

9.4.11.1 Biosensors-on-Chip

Microfluidic biosensors (biosensors-on-chip or lap-on-chip) are important for making point-of-care diagnostics that are both reliable and cost-effective. [46] When microfluidic and biosensor technologies are used together, they make it possible to combine chemical and biological components into a single platform. That allows for new biosensing factors, such as compactness, disability, real-time detection, unique precisions, and instantaneous investigation of distinct analytes in one device. [47] A way to look for nucleic acids found in the blood of people who have cancer. [48]

9.4.12 Detection of Microorganisms

9.4.12.1 Bacterial Identification

It is imperative in medical identification to quickly and accurately figure out which bacteria are dangerous. Some standard discernment methods have problems, like being too sensitive or taking long to learn. Many powerful techniques have already been tested, like ferrofluid magnetic nanospheres and ceramic nanospheres. An in-situ pathogen count bioassay that can detect one bacterium takes only 20 minutes. Nanoparticles can be used easily when a molecule is being recognized because they have a lot of fluorescence. Still, it is a flaw because quantum dots give qualitative information but not quantitative information. However, the nanoparticle-based colorimetric test could make the test more sensitive by more than a few times. When *Salmonella enteric* bacteria stick to the silicon nitride cantilever surface, there is a slight change in the surface level, which helps identify a minimal amount of the bacteria. It can be seen with an electron microscope that less than 25 organisms can be absorbed to be able to tell them apart. A nanotechnology-based technique called SEPTIC is used to monitor electrical changes and bacteria. The nano equipment has two antennas that act as electrodes. Iron-reducing bacteria (IRB) is a highly potent bacteria found everywhere. They can work in the opposite direction of bacteria that eat iron. Iron-oxidizing bacteria change ferrous ions into ferric ions. In contrast, IRB changes this reaction and turns ferric ions into ferrous ions. [49]

9.4.12.2 Detection of Viruses

Choosing viruses efficiently with an adequate response is crucial. Plaque and immunological tests, transmission electron microscopy (TEM) and polymerase chain reaction (PCR) based testing of viruses are some methods that can be used to do a more thorough search. These techniques need a lot of control, which isn't apt for infectious agents or for finding them quickly. It has been possible to get a real-time electrical diagnosis by using nanowire FETs and single-virus particles to look at the electricity. [49] Fluorescently tagged influenza was used to make both electrical and optical measurements at the same time to show clearly which individual viruses caused changes in conductance at the nanowire level. It has been a few years since magnetic nanoparticles were used as an antimicrobial agent and a new type of material to make things. *Listeria monocytogenes* is the leading cause of listeriosis, and it plays an important role in public health. One of the unique things about this bacterium is that it can attach itself to the host cell membrane and make molecular changes in the body. [50]

9.4.13 Magnetic Resonance Imaging–Based Nanosensors

Fluorescence and plasmonic nanosensors use photons to show how the analyte or activity changes. However, magnetic resonance imaging (MRI)–based nanosensors use the magnetism of superparamagnetic particles, which are magnetic only when there is an outside field, to change the relaxivity of water molecules near them. [51] Many people will want to use MRI-responsive sensors because the instruments will become more compact, portable, and cheap to run. There are a lot of benefits to using

MRI to find things. MRI isn't affected by the depth of the tissue. [52] MRI isn't invasive and doesn't lose its signal gradually, which is a problem with fluorescence-based sensors. [53] Unassisted MRI has a very low sensitivity, which means that it needs to be used with contrast agents and other tools to make it more sensitive. Nanoscale probes have a lot of advantages for magnetic-based sensing, like better penetrability and maintenance, which means more nanomaterial is delivered to malignant tissue than ordinary tissue. [54] The nanomaterial's core and exterior are programmable, allowing for changes in the sensors' physical and chemical properties (such as magnetic anisotropy energy barrier and hydrophilicity/hydrophobicity) to better their performance at interacting with certain types of organisms. [55] Recent breakthroughs in creating MRI-based nanosensors for biological applications have helped us better understand how the brain and bodywork and have made medical imaging safer and more precise. MRI is a safe and reliable way to look at the brain, but it has drawbacks. For example, imaging neurotransmitters with MRI is hard because they change quickly and have low concentrations in the brain. Luo et al. have made a nanosensor that is very good at detecting acetylcholine in the brain. Magnetic Ca^{2+}-responsive nanoparticles can be used to look at real-time changes in Ca^{2+} dynamics in the brain. It will help scientists better to know extracellular Ca^{2+}signaling and distinguish between normal and abnormal Ca^{2+} responses. [56, 57] The sensors were made with magnetic nanoparticles and proteins found in natural Ca^{2+} responsive machinery. Nanosensors that can detect tiny molecules and ions have also made progress. Nanosensors that can detect macromolecules have also been made. [58, 59] Zabow et al. came up with a way to study biological processes in the subsurface using nanosensor musters that work at NMR radio frequencies. Arrayed sensors also make it possible to measure ion gradients in a specific location and at a specific time. The real-time screening of drug release kinetics has been getting a lot of attention since the development of chemotherapies. [52, 60] There can be both under- and overdoses if medicines aren't adequately measured in the tissues where they're needed. MRI is an excellent imaging tool for drug screening because it allows for the spatial location of NPs no matter how deep into the tissue they are. Flexible near-infrared (NIR) triggered nanoparticle drug delivery and monitoring system by loading anticancer medicines into hollow-structure nanocomposites. Even though MRI-based nanosensors have a lot of advantages, their ability to be delivered noninvasively to the site of interest is very limited. [61]

9.5 CONCLUSION

Today, nanosensors are utilized in most fields of science due to their nanostructure and benefits. Many research papers and studies on nanosensor applications are released each year. However, only a few nanosensors for detecting heavy metals, herbicides, plant pathogens, and other chemicals have been commercialized. Since these scholarly outputs are not appropriately changed/transmitted to commercial or other controlling platforms. Specific systematic and nonsystematic problems impede the commercialization of these nanosensors. These elements include scale-up and real-world application (technical), authentication and compliance (supervisory), administration significances and pronouncements (political), calibration (legal), cost, demand, and intellectual property safety (economic), as well as safety and security (environmental health and safety). Therefore, it is critical to encourage passionate researchers and research and development institutes to progress nanosensors for monitoring, product justification, conceptual property protection, and social acceptance and execution. Addressing these elements will aid in developing and deploying nanosensor products.

REFERENCES

1. A. H. Deshpande, J. M. Weldode, and J. S. Pise. Applications of Nanosensors in various Fields: A Review. *International Journal of Management, Technology and Engineering*, November 2018.https://www.ijamtes.org/gallery/174-nov.pdf.
2. Asim Kumar. Nanosensors: Applications and Challenges. *International Journal of Science and Research (IJSR), Maulana Azad College of Engineering and Technology, Neora*, 8(7), July 2019.
3. R. Abdel-Karim, Z. Y. Reda, and A. Abdel-Fattah. Review—Nanostructured Materials-Based Nanosensors. *Journal of the Electrochemical Society*, 8(XI), 167037554, 2020.
4. Pankaj Sharma, Vimal Pandey, Mayur Mukut Murlidhar Sharma, Anupam Patra, Baljinder Singh, Sahil Mehta, and Azamal Husen. A Review on Biosensors and Nanosensors Application in Agroecosystems. *Nanoscale Research Letters*, 16, 136, 2021.
5. V. K. Khanna. Nanosensors. In *Physical Chemical and Biological*, CRC Press, 1st ed., Chap. 2, 2011.
6. S. Agrawal, and R. Prajapati. Nanosensors and Their Pharmaceutical Applications: A Review. *International Journal of Pharmaceutical Sciences and Nanotechnology*, 4, 1528, 2012.
7. S. M. Mousavi, S. A. Hashemi, and M. Zarei. Nanosensors for Chemical and Biological and Medical Applications. *Med Chem (Los Angeles)*, 8, 205, 2018.
8. J. Geng, M. D. Thomas, D. S. Shephard, and B. F. Johnson. Suppressed Electron Hopping in a Au Nanoparticle/H 2 S System: Development Towards a H 2 S Nanosensor. *Chemical Communications*, (14), 1895–1897, 2005.
9. J. M. Perez, L. Josephson, and R. Weisleder. Use of Magnetic Nanoparticles as Nanosensors to Probe for Molecular Interactions. *ChemBioChem*, 5, 261, 2004.
10. V. T. Binh, N. Garcia, and A. L. Levanuyk. A Mechanical Nanosensor in the Gigahertz Range: Where Mechanics Meets Electronics. *Surface Science*, 301(1–3), L224–L228, 1994.
11. R. K. Saini, L. P. Bagri, and A. K. Bajpai. *New Pesticides and Soil Sensors*, ed. A. Grumezescu, Elsevier, Chap. 14, 2017.
12. Vinita Hooda, Anjum Gahlaut, Ashish Gothwal, and Vikas Hood. *Bilirubin Enzyme Biosensor: Potentiality and Recent Advances Towards Clinical Bioanalysis*, Springer Science + Business Media B.V., 2017.
13. Fredrik Schedin, A. K. Geim, S. V. Morozov, E. W. Hill, P. Blake, M. I. Katsnelson, and K. S. Novoselov. Detection of Individual Gas Molecules Adsorbed on Graphene. *Nature Materials*, 6(9), 652, 2007.
14. Padmavathy Tallury, Astha Malhotra, Logan M. Byrne, and Swadeshmukul Santra. Nanobioimaging and Sensing of Infectious Diseases. *Advanced Drug Delivery Reviews*, 62(4–5), 424–437, 2010.
15. K. M. Giannoulis, D. L. Giokas, G. Z. Tsogas, and A. G. Vlessidis. Ligand-Free Gold Nanoparticles as Colorimetric Probes for the Non-Destructive Determination of Total Dithiocarbamate Pesticides After Solid Phase Extraction. *Talanta*, 119, 276–283, 2014.

16. R. Bala, S. Dhingra, M. Kumar, K. Bansal, S. Mittal, R. K. Sharma, and N. Wangoo. Detection of Organophosphorus Pesticide-Malathion in Environmental Samples Using Peptide and Aptamer Based Nanoprobes. *Chemical Engineering Journal*, 311, 111–116, 2017.
17. Y. Shang, M. K. Hasan, G. J. Ahammed, M. Li, H. Yin, and J. Zhou. Applications of Nanotechnology in Plant Growth and Crop Protection: A Review. *Molecules*, 24, 2558, 2019.
18. H. Guo, J. C. White, Z. Wang, and B. Xing. Nano-Enabled Fertilizers to Control the Release and Use Efficiency of Nutrients. *Current Opinion in Environmental Science & Health*, 6, 77–83, 2018.
19. H. Singh, A. Sharma, S. K. Bhardwaj, S.K. Arya, N. Bhardwaj, and M. Khatri. Recent Advances in the Applications of Nanoagrochemicals for Sustainable Agricultural Development. *Environmental Science: Processes & Impacts*, 23, 213–239, 2021.
20. C. Liu, H. Zhou, and J. Zhou. The Applications of Nanotechnology in Crop Production. *Molecules*, 26, 7070, 2021.
21. F. Zulfiqar, M. Navarro, M. Ashraf, N. A. Akram, and S. Munne-Bosch. Nanofertilizer Use for Sustainable Agriculture: Advantages and Limitations. *Plant Science*, 289, 110270, 2019.
22. D. Kanjana. Foliar Application of Magnesium Oxide Nanoparticles on Nutrient Element Concentrations, Growth, Physiological, and Yield Parameters of Cotton. *Journal of Plant Nutrition*, 43, 3035–3049, 2020.
23. A. Sharma, K. Sood, J. Kaur, and M. Khatri. Agrochemical Loaded Biocompatible Chitosan Nanoparticles for Insect Pest Management. *Biocatalysis and Agricultural Biotechnology*, 18, 101079, 2019.
24. S. Das, A. Yadav, and N. Debnath. Entomotoxic Efficacy of Aluminium Oxide, Titanium Dioxide and Zinc Oxide Nanoparticles Against Sitophilus Oryzae (L.): A Comparative Analysis. *Journal of Stored Products Research*, 83, 92–96, 2019.
25. H. K. Dweck, S. A. Ebrahim, M. Thoma, A. A. Mohamed, I. W. Keesey, F. Trona, S. Lavista-Llanos, A. Svatoš S. Sachse, M. Knaden, and B. S. Hansson. Pheromones Mediating Copulation and Attraction in Drosophila. *Proceedings of the National Academy of Sciences,* 112(21), E2829–E2835, 2015.
26. Jiancheng Yang, Patrick Carey, IV, Fan Ren, Brian C. Lobo, Michael Gebhard, Marino E. Leon, Jenshan Lin, and S.J. Pearton. Nanosensor Networks for Health-Care Applications. *Nanosensors for Smart Cities*, 405–417, 2020.
27. Mehwish Shafiq, Sumaira Anjum, Christophe Hano, Iram Anjum, and Bilal Haider Abbasi. Review An Overview of the Applications of Nanomaterials and Nanodevices in the Food Industry. *Foods*, 9(2), 148, 2020.
28. F. Ma, Y. Li, B. Tang, and C. Y. Zhang. Fluorescent Biosensors Based on Single-Molecule Counting. *Accounts of Chemical Research*, 49, 1722–1730, 2016.
29. R. Y. Tsien. Constructing and Exploiting the Fluorescent Protein Paintbox (Nobel Lecture). *Angewandte Chemie International Edition,* 48(31), 5612–5126, 2009.
30. Guoxin Rong, Erin E. Tuttle, Ashlyn Neal Reill, and Heather A. Clark. Recent Developments in Nanosensors for Imaging Applications in Biological Systems. *Annual Review of Analytical Chemistry (Palo Alto Calif)*, 12(1), 109–128, June 12, 2019.
31. Kasturi Chakraborty, Aneesh T. Veetil, Samie R. Jaffrey, and Yamuna Krishnan. Nucleic Acid–Based Nanodevices in Biological Imaging. *Annual Review of Biochemistry*, 85, 349–373, 2016.
32. B. Krishnan, S. E. Thomas, R. Yan, H. Yamada, I. B. Zhulin, and B. D. McKee. Sisters Unbound is Required for Meiotic Centromeric Cohesion in Drosophila Melanogaster. *Genetics*, 198(3), 947–965, 2014.
33. Ruizi Peng, Xiaofang Zheng, Yifan Lyu, Liujun Xu, Xiaobing Zhang, Guoliang Ke, Qiaoling Liu, Changjun You, Shuangyan Huan, and Weihong Tan. Engineering a 3D DNA-Logic Gate Nanomachine for Bispecific Recognition and Computing on Target Cell Surfaces. *Journal of the American Chemical Society*, 140, 9793–9796, 2018.

34. Lei He, Danqing Lu, Hao Liang, Sitao Xie, Xiaobing Zhang, Qiaoling Liu, Quan Yuan, and Weihong Tan. mRNA-Initiated, Three-Dimensional DNA Amplifier Able to Function Inside Living Cells. *Journal of the American Chemical Society*, 140, 258–263, 2018.

35. Y. Z. Lin, and P. L. Chang. Colorimetric Determination of DNA Methylation Based on the Strength of the Hydrophobic Interactions between DNA and Gold Nanoparticles. *ACS Applied Materials & Interfaces*, 5(22), 12045–12051, 2013.

36. Jaideep S. Dudani, Maria Ibrahim, Jesse Kirkpatrick, Andrew D. Warren, and Sangeeta N. Bhatia. Classification of Prostate Cancer Using a Protease Activity Nanosensor Library. *Proceedings of the National Academy of Sciences*, 115, 8954–8959, 2018.

37. Philip D. Howes, Rona Chandrawati, and Molly M. Stevens. Colloidal Nanoparticles as Advanced Biological Sensors. *Science*, 346, 1247390, 2014.

38. Tae-Hee Lee, Lisa J. Lapidus, Wei Zhao, Kevin J. Travers, Daniel Herschlag, and Steven Chu. Measuring the Folding Transition Time of Single RNA Molecules. *Biophysical Journal*, 92, 3275–3283, 2007.

39. A. V. Zvyagin, V. K. A. Sreenivasan, E. M. Goldys, V. Y. Panchenko, and S. M. Deyev. Photoluminescent Hybrid Inorganic-Protein Nanostructures for Imaging and Sensing in Vivo and in Vitro. In A. Boker and P. van Rijn (eds.), *Bio Synthetic Hybrid Materials and Bionanoparticles: A Biological Chemical Approach Towards Material Science*, RSC, pp. 245–284, 2015.

40. L. H. Guo, J. A. Jackman, H. H. Yang, P. Chen, N. J. Cho, and D. H. Kim. Strategies for Enhancing the Sensitivity of Plasmonic Nanosensors. *Nano Today*, 10, 213–239, 2015.

41. A. B. Taylor, and P. Zijlstra. Single-Molecule Plasmon Sensing: Current Status and Future Prospects. *ACS Sensors*, 2, 1103–22, 2017.

42. D. Gonçalo, C. João, V. Bruno, G. Leticia, A. Carina, A. Maria, R. João, and V. Pedro. Noble Metal Nanoparticles for Biosensing Applications. *Sensors*, 12, 1657, 2012.

43. J. R. Stetter, W. R. Penrose, and S. Yao, Sensors, Chemical Sensors, Electrochemical Sensors, and ECS. *Journal of the Electrochemical Society*, 150, S11, 2003.

44. A. K. Pulikkathodi, I. Sarangadharan, Y.-H. Chen, G.-Y. Lee, J.-I. Chyi, G.-B. Lee, and Y.-L. Wang. A Comprehensive Model for Whole Cell Sensing and Transmembrane Potential Measurement Using FET Biosensors. *ECS Journal of Solid State Science and Technology*, 7, Q3001, 2018.

45. P. Rai, S. O. P. Shyamkumar, M. Ramasamy, R. E. Harbaugh, and V. K. Varadan. Nano-Bio-Textile Sensors with Mobile Wireless Platform for Wearable Health Monitoring of Neurological and Cardiovascular Disorders. *Journal of the Electrochemical Society*, 161, B3116, 2014.

46. M. H. Shamsi and S. Chen. Biosensors-on-chip: A topical review. *Journal of Micromechanics and Microengineering*, 27, 083001, 2017.

47. G. Luka, Ali Ahmadi, Homayoun Najjaran, Evangelyn Alocilja, Maria DeRosa, Kirsten Wolthers, Ahmed Malki, Hassan Aziz, Asmaa Althani, and Mina Hoorfar. Microfluidics Integrated Biosensors: A Leading Technology Towards Lab-on-a-Chip and Sensing Applications. *Sensors*, 15, 30011, 2015.

48. J. Das, I. Ivanov, L. Montermini, J. Rak, E. H. Sargentand, and S. O. Kelley, Ivaylo Ivanov, Laura Montermini, Janusz Rak, Edward H. Sargent, and Shana O. Kelley. An Electrochemical Clamp Assay for Direct, Rapid Analysis of Circulating Nucleic Acids in Serum. *Nature Chemistry*, 7, 569, 2015.

49. S. M. Mousavi, S. A. Hashemi, M. Zarei, A. M. Amani, and A. Babapoor. Nanosensors for Chemical and Biological and Medical Applications. *Med Chem (Los Angeles)*, 8, 205–217, 2018.

50. Alireza Ebrahiminezhad, Sara Rasoul-Amini, Soodabeh Davaran, Jaleh Barar, and Younes Ghasemi. Impacts of Iron Oxide Nanoparticles on the Invasion Power of Listeria Monocytogenes. *Current Nanoscience*, 10(3), 382–388(7), 2014, Bentham Science Publishers.

51. J. B. Haun, Tae-Jong Yoon, Hakho Lee, and Ralph Weissleder. Magnetic Nanoparticle Biosensors. *Wiley Interdisciplinary Reviews: Nanomedicine and Nanobiotechnology*, 2, 291–304, 2010.
52. G. Zabow, S. J. Dodd, and A. P. Koretsky. Shape-Changing Magnetic Assemblies as High-Sensitivity NMR-Readable Nanoprobes. *Nature*, 520, 73–77, 2015.
53. G. Liu, J. Gao, H. Ai, and X. Chen. Applications and Potential Toxicity of Magnetic Iron Oxide Nanoparticles. *Small*, 9, 1533–1545, 2013.
54. Y. Nakamura, A. Mochida, P. L. Choyke, and H. Kobayashi. Nanodrug Delivery: Is the Enhanced Permeability and Retention Effect Sufficient for Curing Cancer? *Bioconjugate Chemistry*, 27, 2225–2238, 2016.
55. N. Lee, D. Yoo, D. Ling, M. H. Cho, T. Hyeon, and J. Cheon. Iron Oxide-Based Nanoparticles for Multimodal Imaging and Magnetoresponsive Therapy. *Chemical Reviews*, 115, 10637–10689, 2015.
56. Satoshi Okada, Benjamin B. Bartelle, Nan Li, Vincent Breton-Provencher, Jiyoung J. Lee, Elisenda Rodriguez, James Melican, Mriganka Sur, and Alan Jasanoff. Calcium-Dependent Molecular fMRI Using a Magnetic Nanosensor. *Nature Nanotechnology*, 13, 473–477, 2018.
57. Y. Luo, E. H. Kim, C. A. Flask, and H. A. Clark. Nanosensors for the Chemical Imaging of Acetylcholine Using Magnetic Resonance Imaging. *ACS Nano*, 12(6), 5761–5773, 2018.
58. Yue Yuan, Shuchao Ge, Hongbin Sun, Xuejiao Dong, Hongxin Zhao, Linna An, Jia Zhang, Junfeng Wang, Bing Hu, and Gaolin Liang. Intracellular Self-Assembly and Disassembly of 19F Nanoparticles Confer Respective "Off" and "On" 19F NMR/MRI Signals for Legumain Activity Detection in Zebrafish. *ACS Nano*, 9, 5117–5124, 2015.
59. JuanGallo, Nazila Kamaly, Ioannis Lavdas, Elizabeth Stevens, Quang-De Nguyen, Marzena Wylezinska-Arridge, Eric O. Aboagye, and Nicholas J. Long. CXCR4-Targeted and MMP-Responsive Iron Oxide Nanoparticles for Enhanced Magnetic Resonance Imaging. *Angewandte Chemie International Edition*, 53, 9550–9554, 2014.
60. S. S. Liow, Q. Dou, D. Kai, Z. Li, S. Sugiarto, C.Y.Y.Yu, R.T.K. Kwok, X. Chen, Y.L. Wu, S.T. Ong, and A. Kizhakeyil. Long-Term Real-Time in Vivo Drug Release Monitoring with AIE Thermogelling Polymer. *Small*, 13, 1603404, 2017.
61. J. Liu, J. Bu, W. Bu, S. Zhang, L. Pan, W. Fan, F. Chen, L. Zhou, W. Peng, K. Zhao, and J. Du. Real-Time in Vivo Quantitative Monitoring of Drug Release by Dual-Mode Magnetic Resonance and Upconverted Luminescence Imaging. *Angewandte Chemie International Edition*, 53, 4551–4555, 2014.

10 Review on Attributes and Device-Designing Technologies on Sensors (Chemical, Biochemical, Nano- and Bionanosensors)

History, Importance, Key Challenges, and Future Prospects along with Selective Case Studies

Vinars Dawane, Jayandra Kumar Himnashu, Pankaj Kumar, Sunil Kumar Patidar, Satish Piplode, Sonu Sen, Saleem Ahmad Yatoo, Man Mohan Prakash, and Bhawana Pathak

CONTENTS

DOI: 10.1201/9781003323464-10

10.1 INTRODUCTION

In the last few decades, sensors and sensing technologies have taken the centerstage of the modern era of automation along with various applications that have made this field a more fascinating and burning topic [1]. The abilities of sensors in multiple applications directly belong to the unique physical-chemical attributes and specific device designs [2]. The design of unique sensor devices allows the detection of events or changes in the surroundings and the exchange of that information with other electronic modules or other forms of signals that are linked with a sensing device to get the required output in the desired shape and size necessities under an applied working methodology [3]. Thus, working with sensing methodology is very important in the design of sensing devices because it is directly linked with their intended purposes, applications, materials, and manufacturing processes. Important factors like cost, accuracy, and range of the sensor should be considered for device design attributes. The various signaling phenomenon/conversation phenomena of input–output such as electrochemical, amperometric, thermoelectric, magnetic, piezoelectric, and optical [4], which are fluorescence-based; potentiometric; surface plasmon resonance (SPR)–based, acoustic-based, and quartz crystal–based, among others, and has been directly associated with the physical and chemical sensor device designing technology. This working principle, along with sensors' major components, such as the sample input assembly with powerful algorithms, transducers, and electronic systems, has been a key target for any sensing device technology [5, 6]. These directly influence the sensor shape, size, cost, working, and applicability of the sensor along with device design [7]. The use of microchips, tiny materials, bioassays, nanomaterials, and nanobiomaterials made a remarkable change and progress in micro-sensing, *in vivo* sensors, and sensing device design technologies in the past few years and focused on more compactness, cheapness, sensitivities, and multifunctionalities [8].

Based on the signaling/conversation phenomena of input and output, the physics of sensors has been an important driver due to technological advancement and development. In the physical operation of electrochemical sensors, it produces an electrical signal that is proportional to the gas concentration when it reacts with the gas of interest [9]. Various electrochemical sensors might seem to be identical, yet they are consisting of different materials that include key features like hydrophobic barrier porosity, electrolyte composition and sensing electrodes [10]. However, various electrochemical sensors are made to be reactive to the target gas by using additional electrical power. The overall properties of the sensors are determined by all the sensors' components.

Conductometric, potentiometric and amperometric are the types of electrochemical biosensors [11]. Modifications in the conductivity or resistivity of the electrolytic solution's ionic particles in contact with the electrode are recorded by conductometric biosensors. Such modifications come from the synthesis or intake of ionic particles through various reactions and can thus identify the presence of a biological process [12]. Amperometric biosensors work by applying a voltage and monitoring the current proportional to electroactive particle oxidation and reduction on the electrodes. Changes in the production and consumption rate of electroactive particles cause current variations, which represent a biological phenomenon [13]. Potentiometric biosensors are intended to measure the electrical potential difference between a common reference electrode (when the value of current flowing through the system is zero) and a working electrode. The analyte concentration is then used to correlate electrical potential values [14].

Based on thermoelectric phenomena, thermocouples are sensors that are made to detect variations in temperature [15]. When two different types of conductor or semiconductor materials at two different contact temperatures, say, T1 and T2, are put together in a closed circuit; thus, due to the influence of the thermoelectric effect, electromotive forces are observed in the circuit. The thermoelectric effect generates two electromotive forces: a temperature-difference electromotive force and a contact electromotive force. The electromotive force that is generated between the ends of the same conductor or semiconductor due to the difference in temperature is known as a temperature-difference electromotive force. Contact electromotive force is that electromotive force that is generated at the contact of two conductors or semiconductors. It is produced due to dissimilar free electron concentrations, which is a material property of two different materials.

A piezoelectric sensor is a device that converts force, temperature, acceleration, pressure or strain into an electrical charge using the piezoelectric effect [16]. Initially, between the two metal plates, a piezoelectric crystal, like quartz or tourmaline, is placed. The crystal is in perfect balance in this instant and does not conduct any electrical charge. Now, with the help of the metal plates, mechanical pressure is applied to the crystal; thus, the crystal is not in a balanced-charge condition, resulting in the generation of an electrical charge across the faces of a piezoelectric crystal.

Optical radiations are converted into electrical signals by an optical sensor. It calculates the actual amount of optical radiation and converts it into a form that can be processed by an electronic device. When something changes in the optical radiation, the optical sensor works as a photoelectric trigger, causing a decrease or increase in the electrical output. Retro-reflective sensors, such as beam sensors, point sensors, intrinsic sensors, extrinsic sensors, distributed sensors and diffuse reflective sensors, are some of the optical sensor types [17].

The concept of a surface plasmon resonance sensor using Kretschmann's configuration based on the attenuated total reflection within a detection range [18]. A connecting prism, which can be substituted by an optical fiber core, a metallic sheet of width "d", dielectric function "$\varepsilon 1$", and an outer absorbing material of dielectric function "$\varepsilon 2$" makes up this arrangement. The cladding around the core of an optical fiber is removed and replaced with a thin metal film, like gold, which is then wrapped with the sensing medium.

The physical properties of sensors are classified in static accuracy and dynamic accuracy [19]. The sensor's static accuracy describes how well the sensor output adequately describes the measured amount once it has stabilized. Repeatability, resolution, full-scale drift, zero drift, linearity, sensitivity and range are all important static features of sensors. The temporal responsiveness of various sensors is represented by their dynamic properties. Delay time, rise time, settling time, steady-state error, percentage error, and peak time are all frequent dynamic responses of sensors. Information about the static and dynamic accuracy is crucial for the effective use and application of any sensor.

These physical shapes, unique designs and multifunctionalities can be seen in hospitals, industries, manufacturing units, construction sites, business parks, information technology (IT) sectors, shopping centers, offices, schools, houses, children's toys and vehicles, among others, and are used in efforts to make lives easier, simpler and more comfortable [3]. The selective and necessary device designs made it possible for the sensors to become useful from home appliances to tuning the picture and sound in home theaters, making the surrounding environment suitable by adjusting the temperature of air conditioners, adjusting the color and brightness of lights, turning gadgets on by detecting presence and rotation of fans to the spinning of washing machines, cooling of refrigerators, setting fire alarm systems and detecting thumb impressions, among others [1].

A schematic flow chart for the major components of a typical sensor on which the sensor device design depends is shown in Figure 10.1. A sample is first considered to identify which analysis will be performed; it might be environmental contaminants, a human sample or a food sample. The next phase is the reaction component detection, which determines whether the sample attribute is physical, chemical or biological. After the sample is detected, the transducer section begins to work, and the actual sensor circuit is installed according to the technique. Finally, the collected data are processed, and the result is displayed in the desired format at the output assembly.

Thus, the present chapter targets the benefits, importance and scope of sensing device design technology. The chapter focuses on a brief history and update in this domain by using selective case studies on the physics and devices, such as chemical sensors, biosensors, nanosensors and nanobiosensors. The chapter also explains key

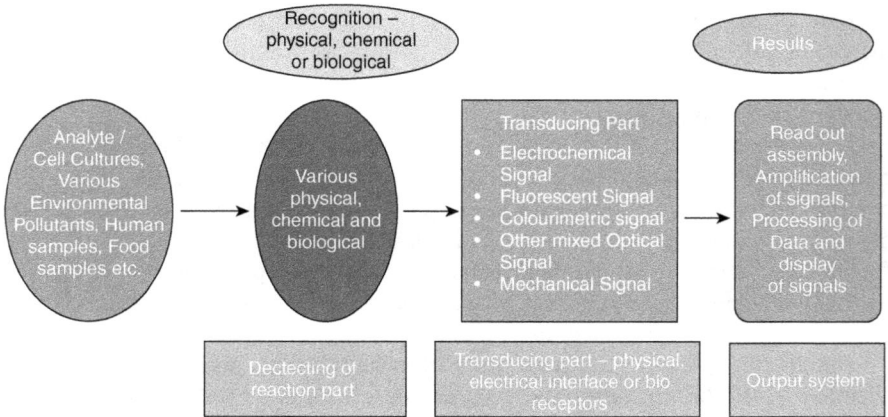

FIGURE 10.1 Major components of a typical sensor on which the sensor device design depends.

challenges and the future prospects of sensor device design technologies to understand the potential of this topic in detail. The authors hope that this chapter succeeds in developing for readers a core understanding of the importance of physics, design and the associated attributes in sensors/sensing technology and their advancements.

10.2   A BRIEF HISTORY AND DEVELOPMENT OF SENSORS AND DEVICES

Advances in materials science and engineering demonstrated key aspects in developing sensor technology. Sensor technology has become commonplace and has harvested a lot of attention in the recent decade. Sensors have been used in various sectors, such as healthcare, agriculture, forest, automotive and marine monitoring. Numerous recent developments in the sensing sector have been prompted by nanotechnological multidisciplinary advancements, offering several novel options for extremely constructed devices with great features. Sensors show an important character in meeting public demands and their necessary designs allow them to be applicable in various fields, like threat exposure [20], contamination and ecosystem conservation [21], energy generation [22] and storage [23] and medical cures [24].

In brief, electrical resistance was discovered in 1860 and used by Wilhelm von Siemens for temperature sensitivity to construct a copper resistor based on a temperature sensor. The great resonance strength of single-crystal quartz and its piezoelectric capabilities have enabled the development of an unusually broad variety of advanced-performance, low-cost sensors that have become an important part in ordinary life and defense systems [25].

The objective of a sensor device has been associated to identify variations in the surroundings and transmit the data to its display units; hence, proper electronics (assembly) have been a vital necessity for sensing instruments. Initially, sensing instruments were applied to assessing physicochemical and biological factors were large and cumbersome. Several environmental activities, such as heat, motion, light, temperature and others, might be the particular inputs. The information received from the sensors may be wrong sometimes because the user had to interpret the output manually [26].

Sensors have grown ubiquitous and indispensable in today's industrial environment. Their applications span from complex engineering procedures to everyday customer items. In several ways, the industrial sector has pioneered the use of sophisticated sensing instruments to monitor and manage manufacturing operations [27].

The terms *sensor* and *transducer* are sometimes used interchangeably. A transducer is defined in the ANSI standard MC6.1 as "a useful output in reaction to a particular measurand" [28]. The result of the process can be defined as output, and the measurement of any physical conditions and property can be defined as "ameasurand". The ANSI standard of 1975 indicates that *transducer* was preferable over *sensor*. However, since the scientific community did not typically accept the ANSI standards, the word *sensor* is now used the most often used term [28].

Many papers by National Materials Advisory Board from 1980s onward recognized sensor techniques as a vital zone to promote advances in resource handling. Researchers have addressed various topics such as bioprocessing [29], heat management [30], combined handling schemes [31], metals treating [32], nondestructive assessment [33] and purifying [34] in these publications [35]. These studies arose to combine criteria and build a comprehensive research and development methodology capable of meeting critical sensor material demands [27].

Middlehoek and Noorlag carefully tried to categorize sensors by representing the input and output energy just as the transduction principle and ignored some "internal" or multiple transduction influences that can occur [36].

In early times, the sensing devices, especially the chemical sensors created for evaluating the surrounding chemical, physical or biological parameters, were large in shape and massive in size; also they were also often imprecise and often provided incorrect results. The manual operations made the applications and precision of the results more complicated in the history of sensing technology [26]. But regular updates and technological shifts have made significant contributions to overcoming these limitations for chemical sensor technologies. The attributes such as area, length, angular/ linear accelerations, mass flow and pressure have been considered as important to design mechanical modes [37]. Thermal sensor devices focus on heat, energy of electromagnetic radiations, flow and matter [38]. Electrical sensor device design focuses on the inductance, polarization and electric field, among others [39]. Magnetic sensor device design focuses more on intensity, moment, penetrability and other aspects [40]. Radiant sensors focus on polarity, transmission and thick film refractive indices [41], and chemical sensor design focuses on configuration and reaction rate pH oxidation, among others [37]. Thus, sensor device technology shifted more toward the specific evaluation and sensitive analysis along with precision [42–44].

More recently, the introduction of large-scale silicon processing ushered in sensor techniques, permitting the misuse of silicon to produce novel techniques for

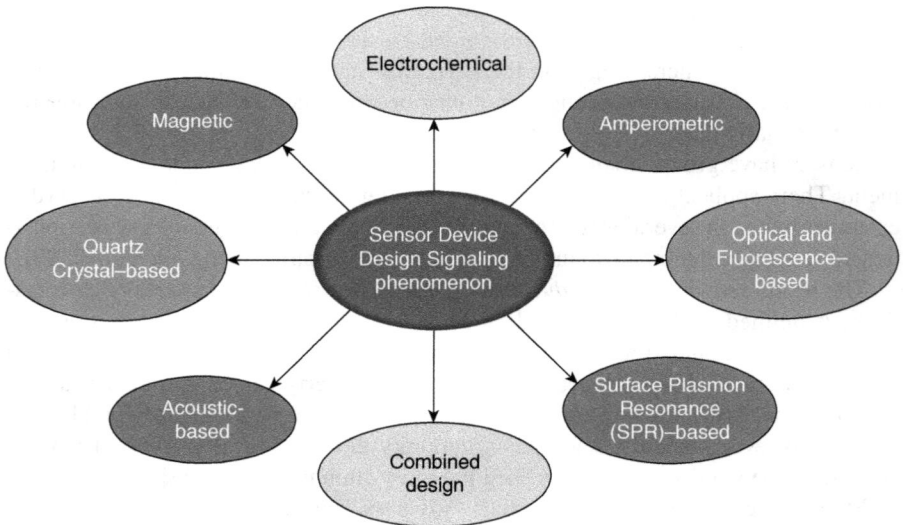

FIGURE 10.2 Signaling phenomenon/conversation phenomena of input–output.

transducing physical action into electrical productivity that may be easily handled by a personal computer. Continuous advancements in materials technology have allowed for improved regulation of their characteristics, opening the door for novel sensors with innovative properties, such as higher reliability, cheaper price and enhanced dependability [45].

Nanotechnological interdisciplinary progressions have prompted numerous fresh improvements in the sensing arena, introducing many new resolutions for extremely engineered expedients with outstanding features [26, 46]. Thus, the sensor device design technologies shifted from normal transducers to advanced nanomaterial transducers that can be combined into the instruments [47]. The general attention in nanomaterials has been focused on their looked-for properties; specifically, the capacity to modify the size and assembly, and henceforth the assets of, nanomaterials proposed exceptional visions for scheming innovative sensing systems and enhanced the presentation of the sensor [48]. Significant developments have been made with nanowires, quantum dots, nanomembranes, and carbon nanotubes, as well as the biomimetic nature of various bio-nanomaterials, and they are the face of cutting-edge sensor device design technology in this era, which is continuously expanding and achieving new horizons [49].

10.3 PHYSICAL AND CHEMICAL DEVICE DESIGNS AND CASE STUDIES ON CHEMICAL SENSORS AND NANOSENSORS

Plenty of reports, work and case studies have assigned different classes to chemical and nanosensors, along with their device design technologies. Typically, an active chemical sensor or nanosensor uses an external signal known as an excitation signal for its working and responses, while a passive sensor does not require any excitation signal. In the wide range of sensing devices, a unique type of sensor has to be credited with the contribution of chemical sensors. Research on chemical sensors has been an emerging discipline formed in recent years by their mutual combination and penetration into chemistry, biology, optics [4], microelectronics technology, thin film technology, semiconductor technology, thermal, electricity, mechanics, acoustics and other disciplines. A chemical sensor is an assembly of an electronic device that shows sensitivity to different chemical constituents and converts their concentration into electrical signals (other signals depend on the phenomena and mechanism applied to the design of the sensor) for detection; thus, the associated device design also depends on this assembly. When a nanomaterial is associated with an overall working sensor for any detection, the latter can be called a nanosensor and nanosensor device design. The device design technology for these chemical sensors and nanosensors directly lies in the various phenomena used for the detection of the signals. Optical signals, thermal signals, electrical signals and mass-based signals have been among the most popular phenomena to design chemical signals and/or nanochemical sensors. The literature is filled with various examples and case studies that have shown these selective phenomena for designing a sensor device. An electrochemical sensor design works as an assembly of electronic devices that transform electrochemical data into a signal that can be analyzed. These electrochemical sensors can be further divided into three broader categories based on their working sub-phenomena, such as potentiometric (works by measuring voltage), where the electromotive force produced by ions dissolved in the electrolyte solution stand-in on the ion electrode is extracted as the output of the sensor so that the ions can be detected. Another one can be an amperometric device (works by measuring current), where the boundary between the electrode and the electrolyte solution is kept at a constant potential, the measured object is directly oxidized or reduced and the current curving through the external route has booked as the output of the sensor so that the detection to realize the application of chemical substances. The third can be conductometric (works by measuring conductivity), which extrudes the variation in the conductivity of the electrolyte solution following oxidation or a reduction of the analyzer as the sensor's output, allowing the material to be recognized. Especially for chemical sensors, optical, thermochemical and mass chemical sensor phenomena have been considered as preferential working mechanisms for designing devices in the various reported case studies [50].

The major applications in chemical and/or nano-chemical measurement, commonly used in environmental pollution monitoring, production process analysis, and can be used for meteorological observation, industrial automation, real-time monitoring, mineral resource detection, telemetry, medical remote diagnosis, agricultural, forest conservation and fish detection and more [51].

In this manner, Rahaman and co-workers in 2018 constructed an efficient metal oxide-based nanomaterial as a chemical sensor for the detection of 2-nitrophenol. Nitrophenols and their derivatives are highly toxic not only for humans but also for animals and vegetation. Therefore, they are classified as hazardous chemicals by the U.S. Environment Protection Agency. The hydrothermally prepared nanosheets of Ag_2O/CuO were utilized for preparing better working electrodes along with a glassy carbon electrode (GCE) and a commercial binder. The designed chemical sensor electrochemically works at pH 7.0 in a phosphate buffer medium. This fabricated chemical sensor is a novel approach and is preferred for detecting 2-nitrophenol along with a sensitivity of 28.6392 mAmM^{-1} cm^{-2}. The calculated lower detection limit (3.31 ± 0.17 pM) for 2-nitrophenol at a signal-to-noise ratio of 3 makes an efficient chemical sensor for detecting environmental hazards. They proposed a possible mechanism for the reduction of 2-nitrophenol to 2-aminophenol (Scheme 1) during the chemical sensing process.

The detection of antibiotics in simulated water using a highly sensitive and selective fluorescent-based chemical sensor was achieved by Zhu and co-workers in 2018 [52]. They prepare Zn(II)-based metal organic framework (MOF) chemical sensor for detecting various sulfonamide-based antibiotics. In the current scenario, the extensive use of antibiotics for the well-being of humans and animals, as well as pharmaceutical wastes of the drugs, might be responsible for the spread of environmentally polluting antibiotic drugs. Sulfonamide-based antibiotics are poorly metabolized and adsorbed in human, as well as animal, bodies. Therefore, a quick and effective tool to detect such antibiotics is of utmost importance. They synthesized a fluorescent-based Zn-II chemical sensor ({[Zn$_3$(μ_3-OH)(HL) L(H$_2$O)$_3$]·H$_2$O} n) and provided experimental, as well as quantum chemical, insights to illustrate the detailed mechanism of detection. The fluorescent-quenching efficiency at various concentrations of antibiotics was calculated using a prepared suspension of fluorescent chemical sensors. This MOF-based chemical sensor is almost unaffected by heavy metal ions and works well in the range of pH = 3.0 to 9.0 in simulated antibiotic wastewater.

In another remarkable work on germanium-doped ZnO nonmaterial was done by Rahman to develop an effective chemical sensor for detecting 4-aminophenol. The 4-aminophenol is commonly used intermediate for synthesis of paracetamol drug and in preparing fungicides, insecticides and various types of dyes. 4-aminophenol is a hazardous chemical that is contaminating the environment (Figure 10.3). The fabrication of Ge-doped ZnO(Ge/ZnO) nanoparticles was utilized to prepare a chemical sensor along with a GCE and Nafion binder. The Ge/ZnO nanoparticles were prepared by a wet-chemical process. The electrochemical approach was employed to detect the 4-aminophenol in room temperature conditions. This Ge/ZnO-based chemical sensor is highly selective for the detection of 4-aminophenol, along with a sensitivity of 0.5063 μA cm^{-2} μM^{-1}, and it shows a lower detection limit (0.5925 ± 0.02 nM). The sensing process for 4-aminophenol is based on the electrochemical oxidation reaction on the surface of semiconductor-based nanomaterials. In the first step, the dissolved oxygen chemisorbes on the surface of Ge-doped ZnO/GCE/ Nafion nanoparticles gets converted into O_2 and O ionic forms. The 4-aminophenol reacts with the absorbed oxygen and converts it into the p-quinoneimine (PQI) with release of electrons. In the next step, the p-quinone (PQ) and ammonia (NH_3) are formed in an aqueous medium [53] (Figure 10.4).

In recent work, Espro and his co-workers developed a biomass-based chemical sensor derived from hydrochar of orange peels. The solid hydrochar was prepared by a hydrothermal carbonization technique. The solid–liquid mass of the orange peel waste and the deionized water was heated at a temperature of 180–300°C and separated out to develop different types of hydrochars. The developed various types of hydrochar are based on temperature variations. They utilize high temperature (300°C)–derived hydrochar for preparing a conductometric-based chemical sensor for NO_2 gas. The detection limit of this hydrochar-based chemical sensor for the detection of NO_2 gas, calculated at 100°C, is 50 ppb. The hydrochar-modified screen-printed carbon electrode (SPCE) shows lower detection limits of 0.18 μM for dopamine. The hydrochar-based chemical sensor shows high performance, and its low cost and biodegradable waste material–based feature make it a unique chemical sensor for detecting environmental pollutants [54].

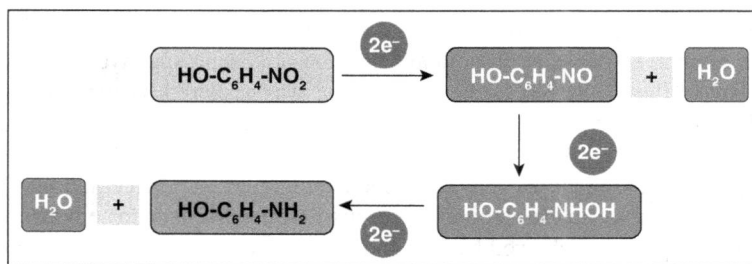

FIGURE 10.3 Chemical sensing mechanism for the detection of 2-nitrophenol.

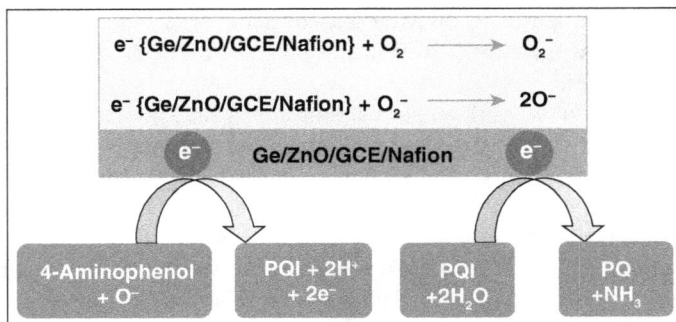

FIGURE 10.4 Schematic presentation of reaction mechanism for detecting 4-aminophenol in the presence of a Ge/ZnO chemical sensor.

In influential work, Hu and co-workers designed a novel chemical sensor for detecting NH_3 gas at room temperature using hollow nickel oxide (NiO) and polyaniline as sensing materials. The NiO (p-type semiconductor) can detect NH_3 gas at high temperatures with great selectivity and sensitivity. However, polyaniline is a conducting material, which can detect NH_3 at room temperature with poor selectivity. The perfect combination of these NiO and polyaniline makes a highly efficient chemical sensor for detecting environment-polluting NH_3 gas at room temperature. The hydrothermally synthesized solid Ni-Cu glyceratenanosphere was used to prepare a hollow NiO-CuO sphere using a calcination method at 400°C. The hollow NiO-polyaniline composite chemical sensor was prepared by in situ polymerization of a hollow NiO-CuO sphere in an acidic medium. The polymerization of aniline at the hollow NiO surface enhanced the surface area and adsorption properties for sensing NH_3. The hollow NiO-polyaniline-based chemical sensor shows a response of 43% at 10 ppm of NH_3. The electron-donating NH_3 reacts with the protonated form of polyaniline, and it gets converted into the NH_{4+} form. These protonation and deprotonation mechanisms are responsible for sensing NH_3 gas. The efficiency of the hollow NiO-polyaniline chemical sensor is enhanced due to the formation of a p-p-type heterojunction and a large surface area [55].

In a very significant work, Singh and co-workers developed a chemical sensor for detecting NH_3 in an extremely humid (RH = 95%) environment at room temperature. They synthesized a MoS_2/MoO_3 heterostructure chemical sensor by the hydrothermal method. The MoS_2/MoO_3 nanocomposite shows approximately 55% and 15% sensing response at 50 ppm and 10 ppm of NH_3, respectively. The sensing mechanism of the MoS_2/MoO_3 chemical sensor was explored by the density functional theory method. The adsorption energy calculation and Bader charge analysis confirm the selectivity of a composite toward NH_3 gas. The feasibility of adsorption from physical to chemical was also examined using theoretical calculations [56].

The fabrication of an Au-decorated porous graphene sensor for environmental-polluting NO_2 gas was achieved by Fan and co-workers. The porous surface of graphene nanomaterial abruptly responded toward a lower concentration (50×10^{-9}) of NO_2; this response was further enhanced up to a ppb level by the decoration of the Au nanoparticle. The fabricated chemical sensor was prepared on polyethylene terephthalate substrate; therefore, it shows mechanical stability and flexibility with a high bending property. The simple fabrication process, high sensitivity and flexible behavior of the prepared Au-decorated porous graphene sensor make it a unique chemical sensor for the detection of NO_2 gas [57].

In noteworthy work, Jung and co-workers designed a chemical sensor decorated with Ag_2S nanoparticles on the surface of a porous graphene material for detecting acetone. The ultrasonic irradiation method was used for preparing the Ag_2S nanoparticles. The prepared nanoparticles were coated onto the surface of porous graphene. The fabricated chemical sensor was tested to evaluate its sensing property for various organic volatile compounds. This sensor shows high response (~660%) for acetone compared to other studied gases. The remarkable sensibility of the chemical sensor to acetone was further justified by a theoretical approach. The higher binding energy and greater electron transfer value between the fabricated chemical sensor and the acetone compared to other gases confirmed the sensing efficacy of Ag_2S decorated porous graphene sensor for acetone at room temperature [58].

A cerium oxide (CeO_2)–based highly efficient chemical sensor for detecting acetylacetone was developed by Umar and co-workers. The nanoparticles of CeO_2 were utilized with a GCE for detecting hazardous acetylacetone. The developed chemical sensor shows high sensitivity (262 mA·mM^{-1} cm^{-1}) to acetylacetone. The CeO_2 nanoparticles were prepared by a hydrothermal process of cerium chloride and hexamethylenetetramine solution. The electrocatalytic sensing property of the fabricated sensor was observed by cyclic voltammograms. The proposed mechanism for the sensing behavior of CeO_2-based chemical sensors suggests that the n-type nature of the CeO_2 semiconductor provides electrons for the chemosorption of oxygen. The produced anionic oxygen, when it comes in contact with acetylacetone, gets readily oxidized into carbon dioxide and a water molecule [59] (see Figure 10.5).

The chemical sensor for detecting toxic m-tolyl hydrazine hydrochloride (m-THyd) was fabricated by Alamry and co-workers. They prepared cadmium oxide

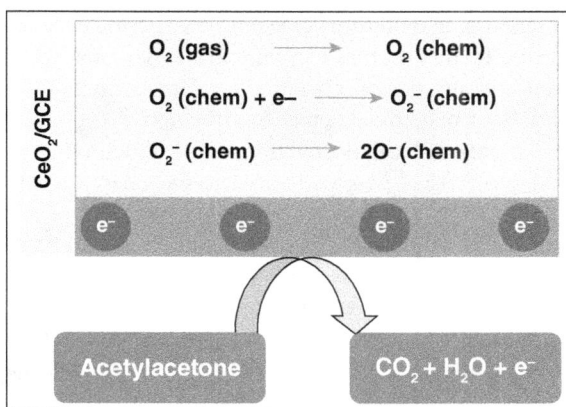

FIGURE 10.5 Chemical sensor mechanism for detecting acetylacetone.

(CdO) nanoparticles using a wet-chemical method and used with multiwalled carbon nanotubes and GCE for fabricating an efficient chemical sensor for detecting m-THyd. The good sensitivity (25.7911 $\mu A \mu M^{-1}$ cm^{-2}) and the lower detection limits (4.0 ± 0.2 pM) for m-THyd suggest a higher capability of the sensor for detecting hazardous and environmental-polluting chemicals [60].

Thus, previously selected case studies highlighted the important aspects of materials, fabrication of physical attributes and unique sensor device design to create remarkable chemical and nanochemical sensors for diverse applications.

10.4 PHYSICAL AND CHEMICAL DEVICE DESIGNS AND CASE STUDIES IN BIOSENSORS AND NANOBIOSENSORS

In last few years or even the last decade, remarkable progress has been made in biosensing and nanobiosensing technology that reflected the special influence on sensor device designs and their associated attributes. This technological progress merged important changes in physical sensor device designs for detecting various biological agents, such as plant/animal metabolites, microbes/pathogens, antibiotics/bioactive chemicals and nucleic acids/serum proteins, among others, and needs the implementation of special bioreceptors and bioanalyte-binding places to detect them.

A sensitive biosensor device thus, in turn, modulates the physiochemical responses associated with the bio-binding and further process the signal toward the transducer. Later, the transducer captures and translates the physiochemical signal into another form like colored signals or fluorescent signals or electrical signal and so on. As far as the nanobiosensor device or associated nanostructures are concerned in this scenario, they simply act as an intermediate layer between the biological agents and the physicochemical detector apparatuses or biological representatives. Thus, in turn, the physical attributes of the sensor change accordingly, such as much of the time, the nanomaterials have been conjugated with the working transducer, to shape the signal detection and build a biosensor device [61].

The biosensor's and/or nanobiosensor's physical attributes and device designs directly depend on the factors such as bio-sample size/volume, the sensitivity of the bioagent, the presence or absence of other bioagents, the nature or biomimetic nature of the nanomaterial, electrochemical detections (the mass, temperature and viscosity being monitored), the present biomolecule transduction methodologies, the intensity of developed signal or electrical potential (current) and so on, thus the overall working principal [62].

The overall working principal is thus extremely important in finalizing the device design. The major working principals include transduction methodologies, such as label-based or non-label-based, the detection assembly (optical detection, fluorescent detection, electrochemical detection, mass-based detection, radioactive detection, electrical detection, combination or more than one detections), materials for biosensing, nanomaterials and biomimetic nature of nanomaterials for nanobiosensing (carbon or non-carbon-containing nanoparticles, nanowires, nanorods, nanotubes, quantum dots, nanomembranes, etc.) [62].

Serrano and co-workers developed an efficient biosensor for the detection of saxitoxin ($C_{10}H_7N_7O_4$). Saxitoxin is a neurotoxin that is very harmful for human health. It was found in toxic phytoplankton species. The biosensor was prepared by the immobilization of an APT (M-30f) oligonucleotide sequence on an Au electrode. The prepared system was further incubated for 1 hr in 1mM solution of 6-mercapto-1-hexanol (MCH). The modified Au/APT/MCH electrode has been used for detecting saxitoxin. The fabricated biosensor has been found to be highly selective for the detection of saxitoxin, and lower detection limits (0.3 mg/L), make an efficient biosensor for detecting toxins. The overall performance of toxin detection has been based on electrochemical parameters. The authors have performed experiments on real water samples for the detection of saxitoxin [63].

The detection of copper ions in sweat and serum using highly sensitive and selective biosensors based on Aunano particle–decorated paper was achieved by Arduini and co-workers. Excessive high and low levels of copper in the human body cause several diseases. The detection of copper ions using Au nanoparticle–decorated paper is facilitated to overcome the requirement of a sophisticated laboratory. They prepared user-friendly paper-based electroanalytical sensors for copper ion detection. The Au nanoparticle–decorated filter paper was found to be capable of detecting copper ions down to 3 ppb in biological fluids (sweat and serum). The paper was based on fabricated an Au-nanoparticle biosensor that later was found to be user-friendly, eco-friendly, cost-effective and capable of playing an important role in the growth of sustainable development approaches [64].

Early and rapid diagnosis in very essential in the case of prostate cancer because it is a major cause of death in men age of 55 years and above [65]. A rapid, sensitive and integrated lab-on-a-chip-based cancer biomarker sensor prototype (MiSens) was described by Uludag and co-workers [66]. The prototype was fully automated and capable of biomarker testing of cancer. It consists of real-time amperometric (works on measuring of current) analysis during the movement of enzyme and microfluidic system. To detect the level of prostate-specific antigens that are commonly used for diagnosing prostate cancer, this prototype has been used. The authors collected

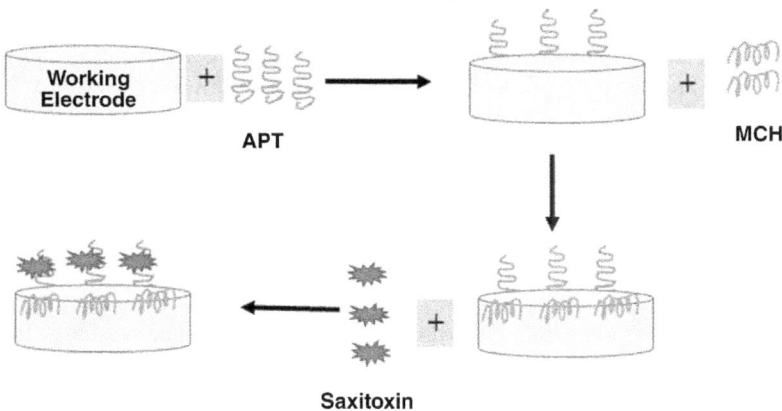

FIGURE 10.6   Preparation steps of biosensor device for detecting saxitoxin.

samples from Kartal Lütfi Kırdar Education and Research Hospital (İstanbul, Turkey) and analyzed them using the MiSens prototype. They also compared the MiSens results with hospital results and found that their prototype is very useful for fast and rapid testing in the case of early diagnosis.

The deficiency of folic acid in the body is the major root of anemia. Xiao and co-workers [67] develop a voltametric (works on measuring of voltage) method for the determination of folic acid by using single-walled nanotube (SWNT) ionic liquid paste electrodes. By using 1-octyl-3-methylimidazolium hexafluorophosphate (OMIMPF$_6$), SWNT carbon electrodes were prepared. Using the voltametric approach, folic acid shows an anodic peak with irreversible oxidation in a pH 5.5 phosphate buffer solution. The authors have applied this method on real samples for the determination of folic acid.

Lee and co-workers synthesized bio-Ag nanoparticles for detecting 4-nitrophenol in a tomato sample. The green tea, grapefruit peel and mangosteen peel extracts were used to prepare bio-Ag nanoparticles. The bio-nanoparticles were prepared using a modified electrochemical method. The efficiency of the prepared biosensor depends on so many factors, such as its morphology, distribution and crystalline nature. The fabricated green tea–Ag nanoparticles show high efficiency compared to the grapefruit–Ag nanoparticles and the mangosteen peel–Ag nanoparticles. The prepared green tea–Ag nanoparticles show high sensitivity (1.2 μA μM^{-1} cm^{-2}) and lower detection limits (0.43 μM) toward 4-nitrophenol [68].

Zheng et al. [69] developed a "turn-off" fluorescent biosensor for mercury detection. In this method, direct, on-site monitoring of mercury is achieved by using a chromophore environment of fluorescent protein m Cherry L199C. In this method, *Escherichia coli* cells were used as the biological compound. As a result, cell-alginated hydrogel-based paper easily detected the mercury from the environment within 5 minutes.

Sunanatha and Vasudevan [70] worked on the bacterial biosensor for perfluorooctanoic acid (PFOA) and perfluorooctane sulfonate (PFOS) in water. They used defluorinase and gfp gene as the regulating and reporter genes, respectively. They used liquid chromatography with mass spectroscopy for detection. Various pollutants, including chlorinated compounds, polycarbonated hydrocarbon and pesticides, among others, were easily detected using this method in both the presence and absence of PFOA and PFOS. They reported a 10ng/L to 1000 mg/L linear range for the biosensor.

Guo et al. [71] developed a novel Au-tetrahedral aptamer nanostructure for acetamiprid detection. This is an electro-chemo-luminescence apta-sensor for pollutant detection in the environment. Luminol and hydrogen peroxide were co-reactant in this method. Due to the luminescence property of these two co-reactants, acetamiprid was easily detected in the environment. They reported 0.0576 pM detection limit for acetamiprid under optimum conditions. An alkaline environment has been suitable for pollutant sensing.

In recent study, Xiong et al. [72] worked on an aptamer-based biosensor for aflaoxin B1 and ochratoxinA detection. In this method, dual DNA tweezers were developed for detecting substances. This biosensor worked on the "turn-off" mechanism. Aptamer of mycotoxin is locked by the dual DNA tweezers. They reported 3.5 × 10^{-2} ppb and 0.1 ppb detection limits for aflaoxin B1 and ochratoxin A, respectively.

In a remarkable work, He et al. [73] reported CdSe@CdS quantum dot–based biosensor for Hg(II) detection. The prepared device was found to be a type of electrochemiluminescence sensor. MoS2 and polycationic poly (diallyldimethylammonium chloride) were used for the construction of biosensor. They reported a linear range from 1×10^{-2} M to 1×10^{-6} M and 1×10^{-13} lowest detection limit for Hg(II).

In significant work, Li et al. [74] developed a gold-doped carbon dot (CDAu)–based biosensor for Pb(II) detection. This biosensor has synthesized using microwave conditions. They reported 0.0005–0.46 µmol/L linear detection range and 0.25 nmol/L lowest detection limit for Pb(II) ions. They also compared these results with undoped carbon dots. The CDAu exhibited superior activity over the undoped carbon dot.

In a recent study, Zhao et al. [75] investigated an aptasensor for acetamiprid detection. This aptasensor was made up of single-stand DNA, G-quadruplexes and grapheme oxide. The detection mechanism was based on the change in the confirmation of DNA through the process. They reported a 0–500nM and 5.73 nM linear range and low detection limit, respectively.

Similarly in another study, Nashuka et al. [76] developed a paper-based biosensor for Hg detection. This was an equipment-free method for detecting Hg. In this method, iodine played an important role for the detection of Hg. The digital images of the Hg and iodine were used for detection purposes. They reported 50–350 mg/L and 20 mg/L linear range and low detection limit, respectively.

Thus, the previously selected case studies highlighted the important aspects of materials, fabrication of physical attributes and unique sensor device design to create remarkable biosensors and nanobiosensors for diverse applications.

10.5 KEY CHALLENGES

Sensing technologies face a broad array of key challenges. Modern progress in sensing technologies has caused comprehensive forms of applications; therefore, in recent times, the sensing field has extended, and device, shapes and applicability are being explored for controlling nearly of the noteworthy limitations in manufacturing a responsible and cost-effective sensor device [77]. Thus, the key challenges for advanced sensing device technology are directly allied with its price issue, the compactness of device design, the materials used, its multifunctionality, the response to sensitivity and selectivity, its reply time, its recognition mechanism, how long it lasts and its harmfulness issues [78, 79] (Figure 9.6).

Major challenging factors associated with sensing device design technologies are the components and the structure of the sensor mechanism because the key benefits of any sensor are directly associated with them [80]. The materials and the structure of a receptor or a bioreceptor depend on what type of samples are being analyzed, the mechanism of recognition involved, the type of transducer used for signal conversion and the detector for response capturing and are very influential in any design [81]. Additionally, flexibility and sensitivity are complementary abilities for sensors, and both have shown great potential in determining sensor device issues [82, 79].

The directions are therefore continuously considering further sensitive and selective detection approaches and determining principles as well as new investigative procedures to advance the current sensing expedients and instruments. From a scientific point of view, the main tasks of sensor devices are connected to reduce cost, size, and energy consumption [26, 79]. Moreover, supplement arypains in the design and expansion of nanoscale sensing constituents have to be completed to accomplish better device presentation. Another task, occasionally ignored and misjudged, is the consistent amalgamation of innovative materials and constructions into sensor design. Their incorporation must be fit the manufacture of marketable, scalable and industrial devices; otherwise, their usage will be restricted to laboratory scales [26].

The nature of materials used to generate a sensor is a very important key challenge to making an effective sensor device. The excellent conductive properties of metallic and/or metallic nanomaterials can allow improvements in flexibility and fine detection along with compactness. Searching for and making multifunctional nanomaterials that can be incorporated into biological sensing can surely enhance the results [83]. Thus, the search for unique and applicable materials is the most desirable key challenge for making a sensor device (Figure 10.7)

Since most of the recent sensing applications are either medical or nonclinical, therefore, it would be a good idea to evaluate the toxicity of the materials used for many physical and/or chemical sensors or, more precisely, the size-dependent toxicity for nanomaterial-related sensors [84–87]. As it is already clear that the toxicity of the sensor directly depends on the material used; thus, their applications have been among the relatively critical key challenges. Therefore, various questions have always been under investigation, such as which selective materials should be used; why these materials should used; what fabrication means will be needed to obtain them; what kind of characteristics do they have that can allow them to symbolize the intent; what level of engineering will be required to control and assess all the

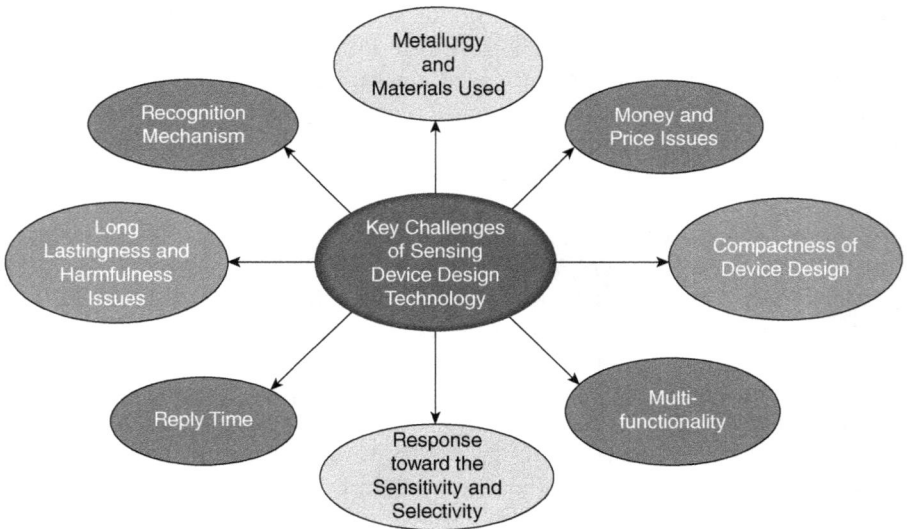

FIGURE 10.7 Key challenges.

influencing parameters involved in the sensing technology; how the overall sensor device will be able to evaluate, design and manipulate the processes involved; how long will the materials or products last; and will they maintain the same qualities or desired catalytic activities when shifted to another medium or are isolated from their original mediums, as well as a calculation of their applicability and fate with the admiration of harmfulness [81].

Like a central nervous system in the body, receptors and transducers are the two weighty functions that are usually involved in sensing mechanisms; thus, the recognition mechanism is also a very important key challenge in sensor device design technology. Other factors like adsorption–desorption kinetics, physicochemical properties, surface property, Gibbs free energy and thermodynamic and kinetic stability [88, 89] are the most important domains involved in sensing mechanisms for almost every type of device design.

Cost and pricing issues are critical for any new technology, and thus, they are great key challenges for sensing device design technology too. Since biosensors and nanobiosensing are relatively recent technologies in device design domains and very high level of engineering is involved in forming compact, as well as sensitive, sensors, cost always becomes a critical connected challenging step [90]. But the progression of material sciences and engineering, as well as the convenience of new, cheap materials, is creating great expectations for the future possibility regarding cost-saving opportunities [81, 79].

Despite the effortlessness and extensive sensing applications, majority of the sensing mechanisms have not yet been fully understood because of the complexity involved in the various parameters that can affect sensors' sensitivity [89]. The term sensitivity many times used with a description of the limit of detection (LOD) and limit of quantification (LOQ) of any sensor device. The LOD and LOQ are the smallest or lowest amount of sample concentration that can be consistently identified by the sensor and quantified by the sensor, respectively. Thus, the LOD and LOQ are important key challenges in figuring out the merit of a sensor and can be understood as the same interlinked parameters of device design technology of any sensor. For this reason, a daunting task often comes to the fore when dealing with their ability to detect trace amounts of analyte-diverse samples in foods, soil, water bodies, biofluids and air [90]. As the detection limit and response or recovery time greatly fluctuating, they can affect sensor efficiency, so sensitivity is always a significant key challenge for any sensor.

Moreover, for an exact sample, the data accomplished from a sensor must have adequate precision so that they can be repeated within a certain range. A sensor device design must avoid systematic errors to ensure sure the accuracy of results and their expected value. Therefore, suitable standard models and standard protocols for device design must be used to get good calibration.

Thus, the key challenges of sensing devices in terms of sensitivity, selectivity, resolution, accuracy and precision are continuously being amended. At the same time, they are possible in terms of manipulation, and their applications are also quickly expanding. Providing data in real time so that the status of key parameters can be tracked is among the greatest challenges in this era as well as exchanging into with other procedures while learning the overall working of the system [26].

10.6 FUTURE PROSPECTS

Accurate, fast and precise sensing is the need of the hour especially in biomedical and environmental domains. In this time of various pandemics like COVID-19 [91], different sanitary emergencies [92], diverse environmental fluctuations and reliable assessments [93], it is unmistakable that sensors that can offer perfect, speedy data have a vibrant part to play in reducing the distribution of deadly and/or contagious microorganisms and their contaminations, predicting quick environmental responses and analyzing diverse samples like food articles and human body fluids, thereby saving numerous lives, costs and times [26]. Thus, the actual necessity and future dynamics in this domain belong toward augmenting key sensor performance attributes such as sensitivity, selectivity, constancy and usability [94]. Therefore, the future prospects have been focusing on evolving the design of sensing mechanisms and developing novel active sensing constituents. Nanotechnology and material science have been playing a significant role in this respect and are uninterruptedly following revolutions for improved sensor devices [26].

Many chemical sensors that are noninvasive and wearable have already been discovered for various types of monitoring. At present, the main demand for chemical sensors as biomarkers is that these should be flexible, real time and biocompatible and should be prepared with comfortable substrates. Multidisciplinary approaches from chemistry, biology, material sciences, computer sciences and others are noteworthy for establishing efficient sensors to keep track of health conditions. With the advent of noninvasive wearable chemical sensors that can allow personalized and early diagnosis of diseases and their prevention at home, thereby leading to enhanced life quality [95]. Improving the sensitivity and stability of sensors is still a big problem being faced by research groups all around the world working in this direction. One of the crucial existing challenges is constructing a sensor that can detect more than one biomarker with high sensitivity and at a low concentration of analytes [79]. To obtain the optimum biomedical application, a sensor device that integrates the chemical, physical and electronic parts of the sensor is a need of society. The analytes and the chemical part of the sensor should be constructed in a way that they remain flexible to allow the sensor to overcome mechanical stress due to various body movements [96]. The theranostic approach can also be thought of as a future perspective in health sector by incorporating a drug delivery system into various wearable chemical sensors. The stability and lifetime issues of sensors must also be taken into consideration because most of enzyme-based chemical sensors cannot be utilized long term as they tend to degrade overtime [97]. As biomarkers are present in trace levels in our body, the sensitivity of the chemical sensors should also be enhanced, which can be done by incorporating novel and highly potential nanomaterials [98].

Emphasis has to be given to developing wearable chemical sensors with continuous, stable and prolonged energy supplies [99]. To achieve this, novel material and technologies need to be discovered for enhanced energy harvesting, storage and supply for different wearable sensors [100]. Analytical instrumentation such as flow-injection analyzers or chromatographs can be coupled to existing sensors which will lead to improved selectivity as compared to conventional chromatography techniques

[101]. Additionally printed wearable chemical sensors need to be explored further which would enable the mass production of flexible and uniform chemical sensors [102]. This way, powerful chemical sensors can be developed, promising a future with a wide array of wearable and portable sensing devices with applications in the health sector and environmental monitoring. Efforts have to be made to construct low-cost and disposable sensors, which will be preferred over reusable ones in many cases [103–105].

The development of efficient chemical biosensors is a collaborative work involving scientists from diverse and different disciplines; for this reason, an interdisciplinary approach will be the future of sensor device design technology [8]. For example, scientists carrying on research in the energy field should focus on discovering biocompatible power sources with high energy density and long lifetime with highly efficient energy-harvesting mechanisms. Engineers working on developing wireless communication should wearable chemical sensors that will work uninterruptedly with high bit rates. But, with the generation of huge personal data, data security and privacy will be at risk, so cryptologists should work in this direction to develop next-generation algorithms that will secure the personal data of users utilizing wearable chemical sensors. The construction of the device should be such that the chemical-sensing region should be visible to the biofluids, and the auxiliary electronics must be kept away from any acquaintance to moisture [104].

In the upcoming decade, nano-based diagnostics devices will be extensively useful owing to their potential of undertaking thousands of measurements in a quick span of time and in an inexpensive manner with special emphasis on miniaturizing biochip technology to the nanoscale level. Thus, a remarkable biomedical application of sensor devices in biomedical fields will emerge a great future application. In a physiological application, the systemic circulation blood flow is the reflection of most of the organ functioning; hence, efforts are focusing on making and designing blood molecular fingerprints, which can be achieved by employing molecule-level electronics and nanoscale sensors [106]. Plausible capacity estimates of the devices were used to assess their presentation for representative chemicals out into the blood by tissues in response to a localized wound, allowing them to readily differentiate single-cell-sized chemical foundation from the contextual chemical attentiveness *in vivo*, yielding high-resolution sensing. The present methods utilized for blood analysis are not efficient enough to differentiate from the background upon dilution in the blood volume [106]. The future trend focuses on designing diagnostic devices from the fundamental building blocks, and thus, nanobiosensors are promising candidates for this domain. But it discourages the employment of fluorescent labeling because going to the nanoscale reduces signal intensity. But efforts are being made to incorporate fluorescent labeling methods efficiently. Non–polymerase chain reaction technology is also used for diagnostic purposes. Nanotechnology can also contribute to single-cell genetic diagnoses. In the upcoming decades, these nanotechnology-based biosensors will be explored for the development of personalized medicines. Its application will further be exploited in the domain of cancer detection and diagnosis because with the conventional diagnostic methods of cancer, it becomes too late to control metastasis. Nanobiosensors offer an amazing capability for the early detection of cancer, which allows for an easy and successful treatment of cancers.

Additionally, these technologies are quite feasible. A nanoapproach could be utilized as a prophylactic way for individuals showing no obvious manifestation of cancer, and remote monitoring, which should be biodegradable and safe, could be done for cancer surveillance. These remote monitoring options could lead to early-stage detection of diseases, and accordingly, therapy could be given to the patient [106].

Thus, the near-future prospects and cutting-edge directions will be shifting toward nanotechnology and tiny-level engineering (micro-meso-nano level engineering) in sensor device design to push the boundaries of different diagnoses and molecular analyses and enable point-of-care opinions and the integration of therapeutics that can help in the development of personalized care and advanced biomedical applications [107].

10.7 CONCLUSION

Thus, sensing technology has a remarkable history and has made progress and systematic developments. The literature and case studies have been replete with physical and chemical device design in various technologies such as chemical sensors, nanosensors, biosensors and nanobiosensors. Today, sensing devices and designing technologies are expanding into new horizons every second, and they are truly impressive because of the advancements in interdisciplinary sciences such as materials sciences, nanotechnology and nanobiotechnology. The sensitivity, selectivity, multifunctionality, safety, cost and compactness are among the critical key challenges associated with the advanced sensing device design technology. The future of this great technology belongs to more advanced materials and tiny constituents, specifically nanomaterials.

ACKNOWLEDGMENT

Authors are thankful to the director of ANCHROM HPTLC Labs, Mumbai and Central University of Gujarat for providing the necessary facilities.

CONFLICTS OF INTEREST

All authors have no conflicts of interest to declare.

REFERENCES

1. Channi, H.K. and Kumar, R., 2022. The role of smart sensors in smart city. In *Smart Sensor Networks* (pp. 27–48). Springer, Cham.
2. Helton, K.L., Ratner, B.D. and Wisniewski, N.A., 2011. Biomechanics of the sensor-tissue interface—effects of motion, pressure, and design on sensor performance and the foreign body response—part I: Theoretical framework. *Journal of Diabetes Science and Technology*, 5(3), pp. 632–646.
3. Zarshenas, P., 2021. *Sensors & Nano Sensors: From A to Z*. Zarnevesht, Iran (9786222881269). doi: 10.5281/zenodo.6113636.
4. Prasad, S., Bruce, L.M. and Chanussot, J., 2011. Optical remote sensing. In *Advances in Signal Processing and Exploitation Techniques*. Springer, Berlin, Germany.

5. Watro, R., Kong, D., Cuti, S.F., Gardiner, C., Lynn, C. and Kruus, P., 2004, October. Tiny PK: Securing sensor networks with public key technology. *Proceedings of the 2nd ACM Workshop on Security of Ad Hoc and Sensor Networks*, pp. 59–64.
6. Susanto, F., Budi, S., de Souza, P., Engelke, U. and He, J., 2016. Design of environmental sensor networks using evolutionary algorithms. *IEEE Geoscience and Remote Sensing Letters*, 13(4), pp. 575–579.
7. Schwiebert, L., Gupta, S.K. and Weinmann, J., 2001, July. Research challenges in wireless networks of biomedical sensors. *Proceedings of the 7th Annual International Conference on Mobile Computing and Networking*, pp. 151–165.
8. Hafezi, N.L. and Hafezi, F., 2022. Developing affordable, portable and simplistic diagnostic sensors to improve access to care. *Sensors*, 22(3), p. 1181.
9. Kumar, A., Kim, H. and Hancke, G.P., 2012. Environmental monitoring systems: A review. *IEEE Sensors Journal*, 13(4), pp. 1329–1339.
10. Williams, D.E., 2020. Electrochemical sensors for environmental gas analysis. *Current Opinion in Electrochemistry*, 22, 145–153.
11. Lakard, B., 2020. Electrochemical biosensors based on conducting polymers: A review. *Applied Sciences*, 10(18), p. 6614.
12. Jaffrezic-Renault, N. and Dzyadevych, S.V., 2008. Conductometric microbiosensors for environmental monitoring. *Sensors*, 8(4), pp. 2569–2588.
13. Dzyadevych, S.V., Arkhypova, V.N., Soldatkin, A.P., El'Skaya, A.V., Martelet, C. And Jaffrezic-Renault, N., 2008. Amperometric enzyme biosensors: Past, present and future. *IRBM*, 29(2–3), pp. 171–180.
14. Huang, X., Zhu, Y. and Kianfar, E., 2021. Nano biosensors: Properties, applications and electrochemical techniques. *Journal of Materials Research and Technology*, 12, pp. 1649–1672.
15. Li, Z., Zhang, J., Wu, B., Huang, J., Nie, Z., Sun, Y., An, F. and Wu, N., 2013. Examining temporal and spatial variations of internal temperature in large-format laminated battery with embedded thermocouples. *Journal of Power Sources*, 241, pp. 536–553.
16. Gautschi, G., 2006. *Piezoelectric Sensorics: Force Strain Pressure Acceleration and Acoustic Emission Sensors Materials and Amplifiers*. Springer Science & Business Media, Berlin.
17. Deshmukh, K., Goel, N. and Patel, B. C., 2020. Optical sensors: Overview, characteristics and applications. *Advances in Modern Sensors*, 3.
18. Chien, F.C. and Chen, S.J., 2004. A sensitivity comparison of optical biosensors based on four different surface plasmon resonance modes. *Biosensors and Bioelectronics*, 20(3), pp. 633–642.
19. Greiner, G. and Maier, I., 2002. Anthrylmethylamines and anthrylmethylazamacrocycles as fluorescent pH sensors—a systematic study of their static and dynamic properties. *Journal of the Chemical Society, Perkin Transactions*, 2(5), pp. 1005–1011.
20. Rasheed, T., Bilal, M., Nabeel, F., Iqbal, H.M., Li, C. and Zhou, Y., 2018. Fluorescent sensor based models for the detection of environmentally-related toxic heavy metals. *Science of the Total Environment*, 615, pp. 476–485.
21. Shak, K.P.Y., Pang, Y.L. and Mah, S.K., 2018. Nanocellulose: Recent advances and its prospects in environmental remediation. *Beilstein Journal of Nanotechnology*, 9(1), pp. 2479–2498.
22. Hou, J., Inganäs, O., Friend, R.H. and Gao, F., 2018. Organic solar cells based on non-fullerene acceptors. *Nature Materials*, 17(2), pp. 119–128.
23. Kawai, T., Nakao, S., Nishide, H. and Oyaizu, K., 2018. Poly (diphenanthrenequinone-substituted norbornene) for long life and efficient lithium battery cathodes. *Bulletin of the Chemical Society of Japan*, 91(5), pp. 721–727.
24. Kumar, J. and Liz-Marzán, L.M., 2019. Recent advances in chiral plasmonics—towards biomedical applications. *Bulletin of the Chemical Society of Japan*, 92(1), pp. 30–37.

25. Shokry Hassan, H., Elkady, M.F. and Serour, N.M., 2021. Intelligent nanosensors (INS) for environmental applications. *Handbook of Nanomaterials for Sensing Applications, Micro and Nano Technologies*, pp. 321–344. https://doi.org/10.1016/B978-0-12-820783-3.00017-8.

26. Comini, E., 2021. Achievements and Challenges in Sensor Devices. *Frontiers in Sensors*, p. 7.

27. National Academies of Sciences, Engineering, and Medicine (NASEM), 1995. *Expanding the Vision of Sensor Materials*. The National Academies Press, Washington, DC. https://doi.org/10.17226/4782.

28. Instrument Society of America, 1975.http://worldcat.org/identities/lccn-n79089958.

29. NRC (National Research Council), 1986a. *Bioprocessing for the Energy-Efficient Production of Chemicals*. National Academy Press, Washington, DC.

30. NRC (National Research Council), 1989a. *On-Line Control of Metals Processing*. National Academy Press, Washington, DC.

31. NRC (National Research Council), 1992. *Opportunities in Attaining Fully-Integrated Processing Systems*. National Academy Press, Washington, DC.

32. NRC (National Research Council), 1989b. *Intelligent Process Control Systems for Materials Heat Treatment*. National Academy Press, Washington, DC.

33. NRC (National Research Council), 1986c. *Automated Non-destructive Characterization and Evaluation in Metal and Ceramic Powder Production*. National Academy Press, Washington, DC.

34. NRC (National Research Council), 1986b. *New Horizons in Electrochemical Science and Technology*. National Academy Press, Washington, DC.

35. NRC (National Research Council), 1987. *Control of Welding Processes*. National Academy Press, Washington, DC.

36. Middlehoek S. and Noorlag, D.J.W., 1982. Three-dimensional representation of input and output transducers. *Sensors and Actuators*, 2(1), pp. 29–41.

37. Eaton, W.P. and Smith, J.H., 1997. Micromachined pressure sensors: Review and recent developments. *Smart Materials and Structures*, 6(5), p. 530; Stetter, J.R. and Penrose, W.R., 2002. Understanding chemical sensors and chemical sensor arrays (electronic noses): Past, present, and future. *Sensors Update*, 10(1), pp. 189–229.

38. Corsi, C., 2010. History highlights and future trends of infrared sensors. *Journal of Modern Optics*, 57(18), pp. 1663–1686.

39. Hammock, M.L., Chortos, A., Tee, B.C.K., Tok, J.B.H. and Bao, Z., 2013.25th anniversary article: The evolution of electronic skin (e-skin): A brief history, design considerations, and recent progress. *Advanced Materials*, 25(42), pp. 5997–6038.

40. Mansour, A.M., 2020. Magnetic sensors and geometrical magnetoresistance: A review. *Journal of Metals, Materials and Minerals*, 30(4), pp. 1–18.

41. White, N.M. and Turner, J.D., 1997. Thick-film sensors: Past, present and future. *Measurement Science and Technology*, 8(1), p. 1.

42. Kim, J., Campbell, A.S., de Ávila, B.E.F. and Wang, J., 2019. Wearable biosensors for healthcare monitoring. *Nature Biotechnology*, 37(4), pp. 389–406.

43. Sishodia, R.P., Ray, R.L. and Singh, S.K., 2020. Applications of remote sensing in precision agriculture: A review. *Remote Sensing*, 12(19), p. 3136.

44. Wang, F., Gui, Y., Liu, W., Li, C. and Yang, Y., 2022. Precise molecular profiling of circulating exosomes using a metal–organic framework-based sensing interface and an enzyme-based electrochemical logic platform. *Analytical Chemistry*, 94(2).

45. Stetter, J.R., Hesketh, P.J. and Hunter, G.W., 2006, Spring. Sensors: Engineering structures and materials from micro to nano. *The Electrochemical Society Interface*, 15(1).

46. Garg, S., Kumar, P., Greene, G.W., Mishra, V., Avisar, D., Sharma, R.S. and Dumée, L.F., 2022. Nano-enabled sensing of per-/poly-fluoroalkyl substances (PFAS) from aqueous systems–a review. *Journal of Environmental Management*, 308, p. 114655.

47. Mizsei, J., 2022. Gas sensors and semiconductor nanotechnology. *Nanomaterials*, 12(8), p. 1322.

48. Pandey, P., 2022. Role of nanotechnology in electronics: A review of recent developments and patents. *Recent Patents on Nanotechnology*, 16(1), pp. 45–66.

49. Lee, D., Lee, T., Hong, J.H., Jung, H.G., Lee, S.W., Lee, G. and Yoon, D.S., 2022. Current state-of-art nanotechnology applications for developing SARS-CoV-2-detecting biosensors: A review. *Measurement Science and Technology*, 5.

50. Yonzon, C.R., Stuart, D.A., Zhang, X., McFarland, A.D., Haynes, C.L. and Van Duyne, R.P., 2005. Towards advanced chemical and biological nanosensors—an overview. *Talanta*, 67(3), pp. 438–448.

51. Huang, X.J. and Choi, Y.K., 2007. Chemical sensors based on nanostructured materials. *Sensors and Actuators B: Chemical*, 122(2), pp. 659–671.

52. Zhu, X.D., Zhang, K., Wang, Y., Long, W.W., Sa, R.J., Liu, T.F. and Lü, J., 2018. Fluorescent metal–organic framework (MOF) as a highly sensitive and quickly responsive chemical sensor for the detection of antibiotics in simulated wastewater. *Inorganic Chemistry*, 57(3), pp. 1060–1065.

53. Rahman, M.M., 2020. Selective and sensitive 4-aminophenol chemical sensor development based on low-dimensional Ge-doped ZnO nanocomposites by electrochemical method. *Microchemical Journal*, 157, 104945.

54. Espro, C., Satira, A., Mauriello, F., Anajafi, Z., Moulaee, K., Iannazzo, D. and Neri, G., 2021. Orange peels-derived hydrochar for chemical sensing applications. *Sensors and Actuators B: Chemical*, 341, p. 130016.

55. Hu, Q., Wang, Z., Chang, J., Wan, P., Huang, J. and Feng, L., 2021. Design and preparation of hollow NiO sphere-polyaniline composite for NH3 gas sensing at room temperature. *Sensors and Actuators B: Chemical*, 344, p. 130179.

56. Singh, S., Deb, J., Sarkar, U. and Sharma, S., 2021. MoS_2/MoO_3 nanocomposite for selective NH_3 detection in a humid environment. *ACS Sustainable Chemistry & Engineering*, 9(21), pp. 7328–7340.

57. Fan, Y.Y., Tu, H.L., Pang, Y., Wei, F., Zhao, H.B., Yang, Y. and Ren, T.L., 2020. Au-decorated porous structure graphene with enhanced sensing performance for low-concentration NO_2 detection. *Rare Metals*, 39(6), pp. 651–658.

58. Jang, A. R., Lim, J. E., Jang, S., Kang, M. H., Lee, G., Chang, H., Kim, E., Park, J.K. and Lee, J. O., 2021. Ag2S nanoparticles decorated graphene as a selective chemical sensor for acetone working at room temperature. *Applied Surface Science*, 562, p. 150201.

59. Umar, A., Almas, T., Ibrahim, A.A., Kumar, R., AlAssiri, M.S., Baskoutas, S. and Akhtar, M.S., 2020. An efficient chemical sensor based on CeO2 nanoparticles for the detection of acetylacetone chemical. *Journal of Electroanalytical Chemistry*, 864, p. 114089.

60. Rahman, M.M., Alam, M.M. and Alamry, K.A., 2019. Sensitive and selective m-tolyl hydrazine chemical sensor development based on CdO nanomaterial decorated multi-walled carbon nanotubes. *Journal of Industrial and Engineering Chemistry*, 77, 309–316.

61. Carrara, S. ed., 2010. *Nano-Bio-Sensing*. Springer Science & Business Media, Berlin.

62. Mun'delanji, C.V., Kerman, K., Hsing, I.M. and Tamiya, E. eds., 2015. *Nanobiosensors and Nanobioanalyses*. Springer, New York.

63. Serrano, P.C., Nunes, G.E., Avila, L.B., Reis, C.P., Gomes, A., Reis, F.T., Sartorelli, M.L., Melegari, S.P., Matias, W.G. and Bechtold, I.H. (2021). Electrochemical impedance biosensor for detection of saxitoxin in aqueous solution. *Analytical and Bioanalytical Chemistry*, 413(25), pp. 6393–6399.

64. Bagheri, N., Mazzaracchio, V., Cinti, S., Colozza, N., Di Natale, C., Netti, P.A., Saraji, M., Roggero, S., Moscone, D. and Arduini, F. (2021). Electroanalytical sensor based on gold-nanoparticle-decorated paper for sensitive detection of copper ions in sweat and serum. *Analytical Chemistry*, 93(12), pp. 5225–5233.

65. Ferlay, J., Autier, P., Boniol, M., Heanue, M., Colombet, M. and Boyle, P. 2007. Estimates of the cancer incidence and mortality in Europe in 2006. *Annals of Oncology*, 18(3), pp. 581–592.

66. Uludag, Y., Narter, F., Sağlam, E., Köktürk, G., Gök, M. Y., Akgün, M., Barut, S. and-Budak, S. 2016. An integrated lab-on-a-chip-based electrochemical biosensor for rapid and sensitive detection of cancer biomarkers. *Analytical and Bioanalytical Chemistry*, 408(27), pp. 7775–7783.

67. Xiao, F., Ruan, C., Liu, L., Yan, R., Zhao, F. and Zeng, B. 2008. Single-walled carbon nanotube-ionic liquid paste electrode for the sensitive voltammetric determination of folic acid. *Sensors and Actuators B: Chemical*, 134(2), pp. 895–901.

68. Trang, N.L.N., Nga, D.T.N., Hoang, V.T., Ngo, X.D., Nhung, P.T. and Le, A.T. (2022). Bio-AgNPs-based electrochemical nanosensors for the sensitive determination of 4-nitrophenol in tomato samples: The roles of natural plant extracts in physicochemical parameters and sensing performance. *RSC Advances*, 12(10), pp. 6007–6017.

69. Zheng, Y., Wei, L., Duan, L., Yang, F., Huang, G., Xiao, T., Wei, M., Liang, Y., Yang, H., Li, Z. and Wang, D., 2021. Rapid field testing of mercury pollution by designed fluorescent biosensor and its cells-alginate hydrogel-based paper assay. *Journal of Environmental Sciences*, 106, pp. 161–170.

70. Sunantha, G. and Vasudevan, N., 2021. A method for detecting perfluorooctanoic acid and perfluorooctane sulfonate in water samples using genetically engineered bacterial biosensor. *Science of the Total Environment*, 759, p. 143544.

71. Guo, Y., Yang, F., Yao, Y., Li, J., Cheng, S., Dong, H., Zhang, H., Xiang, Y. and Sun, X., 2021. Novel Au-tetrahedral aptamer nanostructure for the electrochemiluminescence detection of acetamiprid. *Journal of Hazardous Materials*, 401, p. 123794.

72. Xiong, Z., Wang, Q., Xie, Y., Li, N., Yun, W. and Yang, L., 2021. Simultaneous detection of aflatoxin B1 and ochratoxinA in food samples by dual DNA tweezers nanomachine. *Food Chemistry*, 338, p. 128122.

73. He, Z.J., Kang, T.F., Lu, L.P. and Cheng, S.Y., 2020. An electrochemiluminescence sensor based on CdSe@ CdS-functionalized MoS 2 and a GOD-labeled DNA probe for the sensitive detection of Hg (ii). *Analytical Methods*, 12(4), pp. 491–498.

74. Li, D., Yuan, X., Li, C., Luo, Y. and Jiang, Z., 2020. A novel fluorescence aptamer biosensor for trace Pb (II) based on gold-doped carbon dots and DNA zyme synergetic catalytic amplification. *Journal of Luminescence*, 221, p. 117056.

75. Zhao, Y., Zhang, H., Wang, Y., Zhao, Y., Li, Y., Han, L. and Lu, L., 2021. A low-background fluorescent aptasensor for acetamiprid detection based on DNA three-way junction-formed G-quadruplexes and graphene oxide. *Analytical and Bioanalytical Chemistry*, 413(8), pp. 2071–2079.

76. Nashukha, H.L., Sitanurak, J., Sulistyarti, H., Nacapricha, D. and Uraisin, K., 2021. Simple and equipment-free paper-based device for determination of mercury in contaminated soil. *Molecules*, 26(7), p. 2004.

77. Spencer Jr, B.F., Ruiz-Sandoval, M.E. and Kurata, N., 2004. Smart sensing technology: opportunities and challenges. *Structural Control and Health Monitoring*, 11(4), pp. 349–368.

78. Lou, Z., Wang, L. and Shen, G., 2018. Recent advances in smart wearable sensing systems. *Advanced Materials Technologies*, 3(12), p. 1800444.

79. Dawane, V., Piplode, S., Manam, V. K., Ranjan, P. and Chandra, A., 2022. Nano biosensors containing non-carbon-based nanomaterials to access environmental pollution level. In F.M.P. Tonelli, R.A. Bhat, and G.H. Dar (eds.), *Nanotechnology for Environmental Pollution Decontamination Tools, Methods, and Approaches for Detection and Remediation*. AAP, CRC, Boca Raton.

80. Mousavi, S.M., Hashemi, S.A., Zarei, M., Amani, A.M. and Babapoor, A., 2018. Nanosensors for chemical and biological and medical applications. *Med Chem (Los Angeles)*, 8(8), pp. 2161–0444.

81. Bandgar, D.K., Navale, S.T., Nalage, S.R., Mane, R.S., Stadler, F.J., Aswal, D.K., Gupta, S.K. and Patil, V.B., 2015. Simple and low-temperature polyaniline-based flexible ammonia sensor: A step towards laboratory synthesis to economical device design. *Journal of Materials Chemistry C*, 3(36), pp. 9461–9468.
82. Patolsky, F. and Lieber, C.M., 2005. Nanowire nanosensors. *Materials Today*, 8(4), pp. 20–28.
83. Wang, F., Gui, Y., Liu, W., Li, C. and Yang, Y., 2022. Precise molecular profiling of circulating exosomes using a metal–organic framework-based sensing interface and an enzyme-based electrochemical logic platform. *Analytical Chemistry*, 94(2).
84. Cesarini, D., Calvaresi, D., Marinoni, M., Buonocunto, P. and Buttazzo, G., 2015, April. Simplifying tele-rehabilitation devices for their practical use in non-clinical environments. In *International Conference on Bioinformatics and Biomedical Engineering* (pp. 479–490). Springer, Cham.
85. Iqbal, Z., Ilyas, R., Shahzad, W. and Inayat, I., 2018, April. A comparative study of machine learning techniques used in non-clinical systems for continuous healthcare of independent livings. *2018 IEEE Symposium on Computer Applications & Industrial Electronics (ISCAIE)*, pp. 406–411.
86. Chandra, R., Zhou, H., Balasingham, I. and Narayanan, R.M., 2015. On the opportunities and challenges in microwave medical sensing and imaging. *IEEE Tansactions on Biomedical Engineering*, 62(7), pp. 1667–1682.
87. Nair, P.R. and Alam, M.A., 2006. Performance limits of nanobiosensors. *Applied Physics Letters*, 88(23), p. 233120.
88. Korotcenkov, G., 2005. Gas response control through structural and chemical modification of metal oxide films: State of the art and approaches. *Sensors and Actuators B: Chemical*, 107(1), pp. 209–232.
89. Nimal, A.T., Mittal, U., Singh, M., Khaneja, M., Kannan, G.K., Kapoor, J.C., Dubey, V., Gutch, P.K., Lal, G., Vyas, K.D. and Gupta, D.C., 2009. Development of hand-held SAW vapor sensors for explosives and CW agents. *Sensors and Actuators B: Chemical*, 135(2), pp. 399–410; Lee, D., Lee, T., Hong, J.H., Jung, H.G., Lee, S.W., Lee, G. and Yoon, D.S., 2022. Current state-of-art nanotechnology applications for developing SARS-CoV-2-detecting biosensors: A review. *Measurement Science and Technology*, 5.
90. Singh, E., Kumar, A., Mishra, R. and Kumar, S., 2022. Solid waste management during COVID-19 pandemic: Recovery techniques and responses. *Chemosphere*, 288, p. 132451.
91. Shenashen, M.A., Emran, M.Y., El Sabagh, A., Selim, M.M., Elmarakbi, A. and El-Safty, S.A., 2022. Progress in sensory devices of pesticides, pathogens, coronavirus, and chemical additives and hazards in food assessment: Food safety concerns. *Progress in Materials Science*, 124, p. 100866.
92. Kim, M.S., Kim, M.S., Lee, G.J., Sunwoo, S.H., Chang, S., Song, Y.M. and Kim, D.H., 2022. Bio-inspired artificial vision and neuromorphic image processing devices. *Advanced Materials Technologies*, 7(2), p. 2100144.
93. Promphet, N., Ummartyotin, S., Ngeontae, W., Puthongkham, P. and Rodthongkum, N., 2021. Non-invasive wearable chemical sensors in real-life applications. *Analytica Chimica Acta*, 1179, p. 338643.
94. dos Santos, C.C., Lucena, G.N., Pinto, G.C., Júnior, M.J. and Marques, R.F., 2021. Advances and current challenges in non-invasive wearable sensors and wearable biosensors—a mini-review. *Medical Devices & Sensors*, 4(1), p.e10130.
95. Bandodkar, A.J. and Wang, J., 2014. Non-invasive wearable electrochemical sensors: A review. *Trends in Biotechnology*, 32(7), pp. 363–371.
96. Michelusi, N., Stamatiou, K. and Zorzi, M., 2013. Transmission policies for energy harvesting sensors with time-correlated energy supply. *IEEE Transactions on Communications*, 61(7), pp. 2988–3001.

97. Jung, S., Hong, S., Kim, J., Lee, S., Hyeon, T., Lee, M. and Kim, D.H., 2015. Wearable fall detector using integrated sensors and energy devices. *Scientific Reports*, 5(1), pp. 1–9.

98. Tahir, H.E., Xiaobo, Z., Xiaowei, H., Jiyong, S. and Mariod, A.A., 2016. Discrimination of honeys using colorimetric sensor arrays, sensory analysis and gas chromatography techniques. *Food Chemistry*, 206, pp. 37–43.

99. Imani, S., Mercier, P.P., Bandodkar, A.J., Kim, J. and Wang, J., 2016, May. Wearable chemical sensors: Opportunities and challenges. *2016 IEEE International Symposium on Circuits and Systems (ISCAS)*, pp. 1122–1125.

100. Coyle, S., Curto, V.F., Benito-Lopez, F., Florea, L. and Diamond, D., 2014. Wearable bio and chemical sensors. In *Wearable Sensors* (pp. 65–83). Academic Press, Cambridge.

101. Tessarolo, M., Gualandi, I. and Fraboni, B., 2018. Recent progress in wearable fully textile chemical sensors. *Advanced Materials Technologies*, 3(10), p. 1700310.

102. Sempionatto, J.R., Jeerapan, I., Krishnan, S. and Wang, J., 2019. Wearable chemical sensors: Emerging systems for on-body analytical chemistry. *Analytical Chemistry*, 92(1), pp. 378–396.

103. Rajasundari, K. and Ilamurugu, K., 2011. Nanotechnology and its applications in medical diagnosis. *Journal of Basic and Applied Chemistry*, 1(2), pp. 26–32.

104. Bagheri, N., Mazzaracchio, V., Cinti, S., Colozza, N., Di Natale, C., Netti, P. A., Saraji, M., Roggero, S., Moscone, D. and Arduini, F. (2021). Electroanalytical sensor based on gold-nanoparticle-decorated paper for sensitive detection of copper ions in sweat and serum. *Analytical Chemistry*, 93(12), pp. 5225–5233.

105. Comini, E., 2021. Achievements and challenges in sensor devices. *Frontiers in Sensors*. https://doi.org/10.3389/fsens.2020.607063.

106. Rahman, M.M., Alam, M.M., Hussain, M.M., Asiri, A.M. and Zayed, M.E.M., 2018. Hydrothermally prepared Ag_2O/CuO nanomaterial for an efficient chemical sensor development for environmental remediation. *Environmental Nanotechnology, Monitoring & Management*, 10, pp. 1–9.

107. Trang, N.L.N., Nga, D.T.N., Hoang, V.T., Ngo, X.D., Nhung, P.T. and Le, A.T. (2022). Bio-AgNPs-based electrochemical nanosensors for the sensitive determination of 4-nitrophenol in tomato samples: The roles of natural plant extracts in physicochemical parameters and sensing performance. *RSC Advances*, 12(10), pp. 6007–6017.

11 Metal Oxide Decorated CNT Nanosensors
A Brief Overview

Saleem Khan, Vaishali Misra, Ajay
Singh, and Vishal Singh

CONTENTS

11.1 INTRODUCTION

The evolution of living organisms and their existence is inextricably linked to their sensory awareness of the environment. Sharp vision, touch, taste, hearing, and smell have all played a role in the survival of species over time. The olfactory system is one of the earliest sensory systems developed in evolution, allowing creatures with odorant receptors to locate food and detect threats [1]. The naturally developed sensory system in humans has a poor limit-of-detection range, which can be fatal to life in cases of poisonous and explosive gas present in the vicinity. Exposure to poisonous fumes in significant amounts has been shown to have severe implications for the pulmonary system. Various gases are produced in the coal and petroleum industries that cause explosions. The detection of poisonous and explosive gases, from human safety point of view, is of utmost importance [2]. Over the past few decades, new innovative semiconductor fabrication techniques have led to the development of sensors to detect the analyte ranging from ppm (parts per million) to ppb (parts per billion) range. Researchers are adopting and improving device fabrication methods to design the next generation of nanosensors [3].

Nanosensors are being used to monitor physical and chemical anomalies in harsh environments and confined to very low areas. Nanosensors can be classified on the basis of energy source, structure, and application depicted in Figure 11.1. In group I,

DOI: 10.1201/9781003323464-11

classification is the active (which requires energy to operate, such as thermistor) and passive (which doesn't require an energy source, such as thermocouple and piezo-electric) nanosensors. In group II, classification is based on the structural properties, which include optical, electromagnetic, and mechanical nanosensors. The transduction mechanism in optical nanosensors to detect chemical or biological analyte is done by using an optical signal. These sensors are solely dependent on the optical properties of the nanostructured material. Sasaki et al. developed the first pH micro-probing device using a trapping technique for fluorescent material. Polyacrylamide nano particles (NPs) were used to determine the pH of the water [4]. Electromagnetic nanosensors have two monitoring methods, that is, an electrical current measurement and a magnetic measurement. When compared to non-magnetic-based technologies, sensing systems based on a magnetic method provide advantages in terms of analytical figures of merit, such as speedy analysis time, high signal-to-noise ratio, low detection limits, and improved sensitivity. Geng et al. studied the behavior of hydrogen sulfide molecules with Au NPs. A chromium and gold electrode was fabricated on SiO_2 and used as a source and drain. A 40–60-nm channel was created between the electrodes, and Au NPs were spread in the gap. The sensing device fabricated occupied only a 12-μm^2 area [5]. Magnetic nanosensors are preferably used in the detection of femtomolar-range proteins, enzymes, and viruses [6]. Mechanical nanosensors are based on the principle of mechanics to quantify the physical and chemical characteristics to detect biomolecules. These nanosensors have superiority over optical and electromagnetic sensors in the detection of analytes at the nanoscale because of their mechanical detection phenomenon. In the early stage of nanosensor development, Wachter et al. fabricated a microcantilever sensor for detecting mercury vapors with a sensitivity of 1.25 Hz/pg [7]. In microelectronic device development, mechanical nanosensors play a critical role because of their unique property of nanoscale subassemblies.

11.2 NANOSENSORS FOR GAS SENSING

Nanosensors are being developed every day to meet application-specific demands. Gas sensors with high sensitivity, selectivity, and cost-effectiveness with low power usage are being upgraded to focus on today's challenging sensing specifications. A controlled volume of gases is detected by nanosensors [8, 9]. The transducers and receptors are two major components of nanosensors. Receptors are sensing materials that interact with analyte molecules. These molecules cause changes in some parameters, such as resistivity, refractive index, surface morphology, light, heat, work function, and more, in receptor material. Gas sensors can be grouped as resistive and non-resistive. The characteristics of gas sensors are discussed next.

- Response and Sensitivity: The gas sensor response is defined as the ratio of resistance reduction in air to stationary resistance. Normalize conductance follow. The rate of recovery time (τ) is needed to be 90% of the full response.

Conductance = R_a/R_g

The stationary resistance depends on the partial pressure of the target gas. The relationship is given as

$$R_g = cP_g^{\alpha},$$

where α (power index) and c are the constants (power law). The gas response also follows the power law given as

$$R_a / R_g = cP_g^{\alpha} \quad \text{(Inflammable gases)}$$
$$R_g / R_a = cP_g^{\alpha} \quad \text{(Oxidizing gases)},$$

where α varies from ½ to 1.

The correlation between the gas response and P_g gives the sensitivity. When α is unity sensitivity depends on P_g. Also, sensitivity is the power law proportionality constant. The physicochemical values of the material, target gas, and oxygen define the sensitivity.

- Operating Temperature: The operating temperature influences the response and response parameters. As the temperature rises, the rate of response and recovery time also increases. Inflammable gas response increases exponentially with high temperatures because of the surface reaction between gas and adsorbed oxygen and vice versa for oxidizing gases [10, 11].

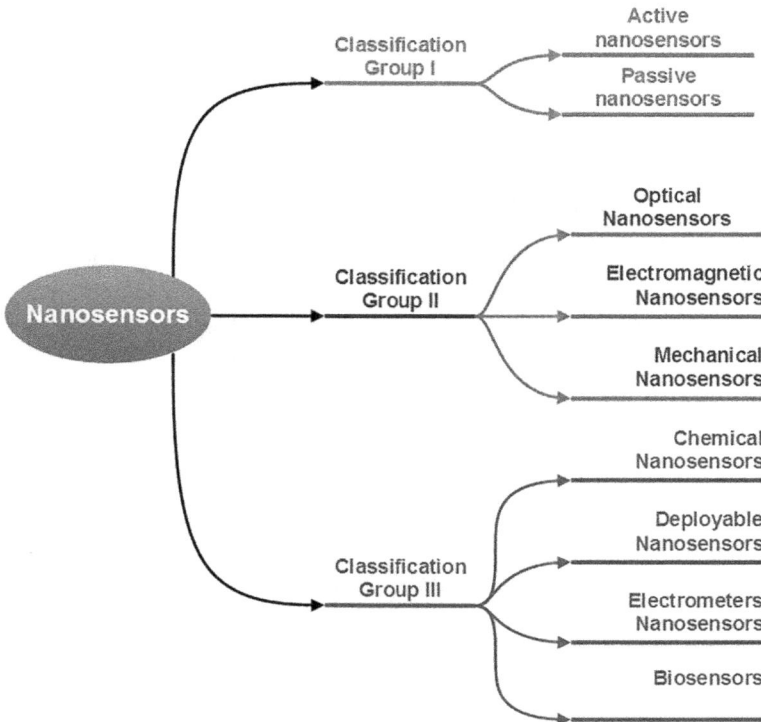

FIGURE 11.1   Classification groups of nanosensors.

11.3 HIGH-PERFORMANCE STRUCTURED NANOMATERIALS

Since the early 1980s, intense research is being carried out to develop nanostructured materials for sensing applications. Nanomaterials are still the subject of research, which includes conducting polymers, metal oxides (MOs), two-dimensional (2D) materials, and carbon nanotubes (CNTs) [12–15]. These materials contain several flaws, in contrast to intriguing characteristics, that must be addressed first. The difficult and time-consuming production of conducting polymers is a hindrance in sensor fabrication. When compared to metal-oxide gas sensors, the lifetime of conducting polymer gas sensors is relatively limited (9–18 months) due to oxidation. Also, the fabrication costs of 2D materials are very high [16].

MO nanomaterials are significant candidates for gas sensor fabrication. MO thin film, nano wires (NWs), NPs, nano flowers (NFs), nanospheres, NTs, and nano rods (NRs) have ultrahigh surface area, high sensitivity, phenomenal stability, and low-cost synthesis make them efficient sensing materials [17–20]. Both dry-chemical and wet-chemical processes, which include solid-state processes [21], sol-gel [22], [23], hydrothermal, solvothermal [24–26], electrospinning [27], and thin film deposition (sputtering, plused laser deposition [PLD], atomic layer deposition [ALD], etc.), are employed to synthesize MO nanostructures. [28–31]. Chemoresistive and catalytic conductivity sensors have been continuously evolving since the 20th century. The charge transfers between analyte molecules and the MO surface complex like H^+, OH^-, O^-, and O^{2-} is the fundamental gas sensing process. MOs have the ability to sense toxic and flammable gases. Table 11.1 summarizes MO-based gas sensors developed over the years. The limitations of MO nanomaterials as sensing materials

TABLE 11.1
MO-Based Gas Sensors and Their Limit of Detection

Nanostructure	Sensing Material	Analytes	Limit of Detection	Ref.
NPs	α-Fe_2O_3	H_2S	0.05 ppm	[33]
NWs	Ga_2O_3	O_2 and CO	O_2 = 5ppm and CO = 500 ppm	[34]
NWs	ZnO	NO_2	10 ppm	[35]
NWs	ZnO	Ethanol	50 ppm	[36]
NRs	ZnO	H_2	5ppm	[37]
NFs	In_2O_3	Ethanol	10 ppm	[38]
Nanosheets	CuO	Ethanol	3 ppm	[39]
Nanoflower	SnO_2	Methanol	100 ppm	[40]
Membranes	Pd-doped Co_3O_4	NO	50 ppm	[41]
Mesoporous	Pt-doped WO_3	CO	100 ppm	[42]
Thin film	Rh-doped WO_3	CH_4	5 ppm	[43]
NWs	CuO/SnO_2	HCHO	50 ppm	[44]
NWs	SnO_2/NiO	H_2	500 ppm	[45]
NWs	α-Fe_2O_3/ZnO	H2S	5 ppm	[46]
Nanosheets	WO_3/SnO_2	NH_3	100 ppm	[47]

are deficient selectivity and resistance drift [32]. A combination of two or more nano-materials to form composites provides superior properties as compared to individual nanomaterials.

Carbon and its derivatives, such as graphite, graphene oxide, diamond, fullerene, and CNTs, have superior and interesting properties compared to other nanomaterials. Among the carbonaceous nanofillers, CNTs have fascinating structural and electronic properties which can be explored for functionalized implementation. There are two types of CNTs: single wall (SW) and multiwall (MW). Single wall carbon nanotube (SWCNTs) are formed by wrapping graphene sheets, and multiwall carbon nanotube (MWCNTs) are formed by wrapping numerous sheets. In MWCNTs, the outermost atoms of the sheets participate in the sensing process. In 1991, MWCNTs were discovered by Iijima [48], and SWCNTs were reported in 1993 by his research group [49]. MWCNTs have an inner diameter of 0.4 nm up to a few nanometers, and their outer diameters vary from 2 nm up to 30 nm. The half-fullerene dome-shaped molecules close both ends of the tube. SWCNTs' diameter ranges from 0.4 nm to 3 nm, and they have a length of up to micrometers. New hybrid MOs and CNTs are being developed by doping CNTs with MO or MO-decorated CNTs, which gives new properties, causing an upgradation in the functionalities [50, 51].

11.4 CNT-BASED GAS SENSORS

The outer surface of CNTs (SWCNTs and MWCNTs) participates in the gas sensing process due to electrochemical and adsorption phenomena. CNTs have very high adsorption capabilities, and exceptional sensitivity makes them a highly usable sensor material [52]. Types of gas sensors fabrication possible using CNTs include the following:

- Sorption gas sensor: These are the predominantly fabricated gas sensors. Adsorption is the principle of operation in these sensors. The adsorbed molecules either transfer or extract electrons from CNTs, which causes a change in the electrical properties of the CNTs. This type of sensor utilizes pristine as well as functionalized SWCNTs and MWCNTs [53, 54].
- Ionization gas sensors: These sensors are designed to detect gas molecules with low adsorption energies. The accelerated ions collide with gas molecules, which determines the gas ionization parameters. The absence of gas molecules and sensitive material interaction molecules with low energies are detected. The disadvantages associated with these sensors are high dimensions and an operating voltage ranging from 102–103 V [55].
- Capacitance gas sensors: In these types of sensors, CNTs are used as sensitive elements acting as the first plate on the capacitor containing disoriented CNTs and silicon as the second plate. The polarization of the adsorbed gas molecule occurs when high external voltage is applied to the capacitor. Poor performance under humid conditions is a drawback of this sensor [56, 57].
- Resonance-frequency-shift gas sensors: Disk resonators with NTs grown on the outside surfaces of the discs could be used as sensitive elements in

TABLE 11.2

MO Functionalized CNTs and Their Limit of Detection

Sensing Material	Functional Material	Analytes	Limit of Detection	Ref.
SWCNTs	–	NO_2	44 ppb	[60]
CNFET	–	NO_2	100 ppm	[61]
CNTs film	–	NO_2 and NH_3	1 ppm = NO_2 and 7 ppm = NH_3	[62]
MWCNTs	SnO_2	LPG and ethanol	100 ppm = LPG and 10 ppm = ethanol	[63]
MWCNTs	SnO_2	NH_3	60 ppm	[64]
MWCNTs	WO_3	NO_2, CO, and NH_3	500 ppb = NO_2, 10 ppm = CO, 10 ppm = NH_3	[65]
MWCNTs	ZnO	NO_2	10 ppm	[66]
MWCNTs	Fe_2O_3	NO_2	20 ppm	[67]
MWCNT	ZnO	CO	25 ppm	[68]
MWCNTs	V_2O_5	CH_4	40 ppm	[69]
MWCNTs	SnO_2–Pt	CH_4	100 ppm	[70]
CNT	WO_3 nanobricks	NH_3	30 ppm	[71]
SWCNTs	Fe_2O_3	H_2S	100 ppm	[67]
CNTs	CuO–SnO_2	H_2S	0.1 ppm	[72]
MWCNTs	α-Fe_2O_3	C_3H_6O	50 ppm	[73]
CNTs	$ZnSnO_3$	C_2H_5OH	100 ppm	[74]

these sensors. The dielectric permeability of the disc containing the NTs changes when the CNTs on the resonator are exposed to gases, causing the resonance frequency to shift. These sensors have good sensitivity and selectivity due to the differences in frequency shifts generated by different gases [58, 59]. Table 11.2 summarizes MO functionalized CNTs for gas sensors.

11.5 CNT/MO COMPOSITE SYNTHESIS METHODS

CNTs and MO composites are synthesized using two approaches: ex situ and in situ. In the ex situ method, CNTs and MO nanomaterials are synthesized separately with defined structures and dimensions. The composite is synthesized using a functional group or linker that combines the materials by a covalent bond, a hydrogen bond, an π–π assembly, weak Van der Waal forces, or electrostatic forces and hydrophobic interlinking. Wang et al. synthesized MgO/MWCNTs composite using hydrophilic interaction. The pyrolysis method was used to synthesize MWCNTs. The composite was prepared by ultrasound waves to create a suspension of MgO and MWCNTs in ethanol, the synthesized composite is shown in Figure 11.2a [75]. Zhou et al. synthesized Pt/CNTs using the π–π assembly shown in Figure 11.2b [76].

The in situ synthesis of MO and CNTs is carried out simultaneously in the same synthesis system. In this process, CNTs provide support to the MO material to settle

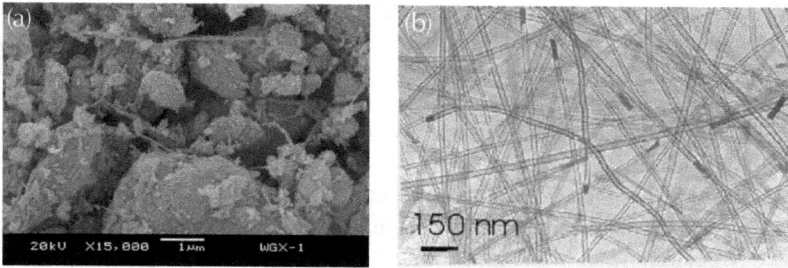

FIGURE 11.2 (a) Scanning electron microscopy (SEM) of MgO/MWCTs composite, Reprinted with permission [75], copyright (2005) Solid State Ionics, Elsevier; (b) SEM image of CNTs decorated with Pt NPs, Reprinted with permission [76], copyright (2007) Chemical Physics Letters, Elsevier.

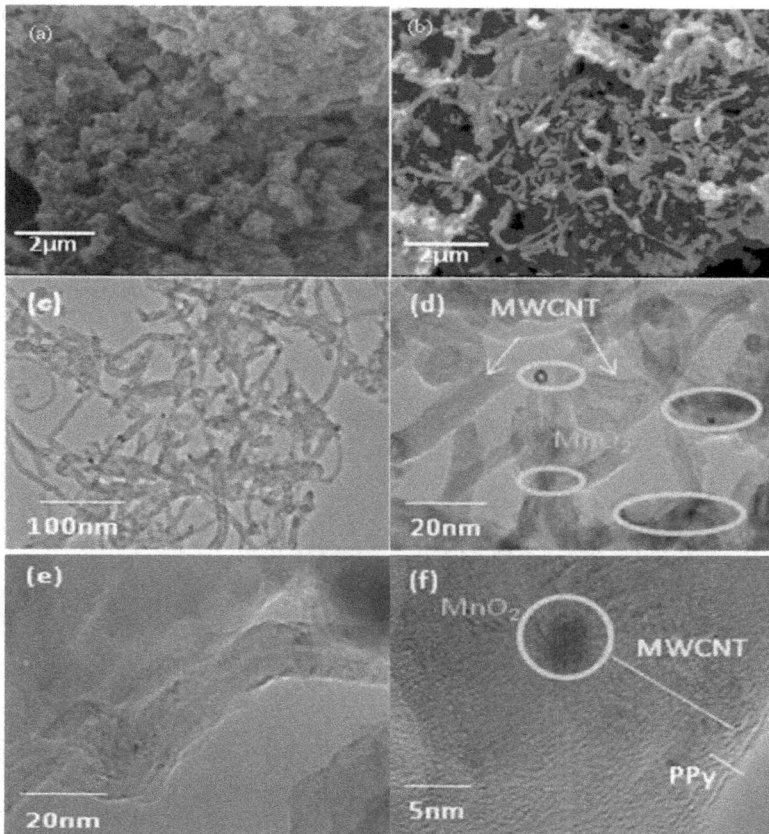

FIGURE 11.3 Scanning electron microscopy and high-resolution transmission electron microscopy (HRTEM) images of synthesized MWCNTs/MnO$_2$ and MWCNTs/MnO$_2$/Ppy composite using an electrochemical method, Reprinted with permission from [85]. Copyright (2016) International Journal of Hydrogen Energy.

faster with a controlled thickness. The MO material can be amorphous or crystalline in the form of NPs, NRs, or nanobeads. Several in situ synthesis processes, such as electrochemical [77], chemical reduction and oxidation [78], electrodeposition [79], sol-gel process [80], hydrothermal and aerosol techniques [81], and gas-phase deposition (thermal evaporation, sputtering, PLD, CVD, ALD) [82–84] are used for synthesizing composites and thin films of MO and CNTs. In the coming section, wet-chemical synthesis processes are presented.

Mishra and Jain synthesized MWCNTs and MnO_2/Ppy composites using electrochemical deposition [85]. MWCNTs were ultrasonicated for 30 minutes in deionized (DI) water (10 mL) and 0.3 mL of PTSA. MWCNTs were covered by PTSA, which stabilizes the nanotubes by static charge repulsion in water. About 50 mL of 3-mM manganese acetate solution was prepared and added to the NT solution, and stirred for 14 hours. After that, the solution was filtered and washed several times using DI water. The resulting MWCNTs/Mn^{2+} was oxidized in 5mM $kMnO_4$ to obtain MWCNT/MnO_2. A solution of 0.2 M pyrrole was prepared in 20 mL of DI water, and 0.5 mg MWCNT/MnO_2 was added and sonicated for 30 minutes. The solution was washed and dried at room temperature for 4 hours, resulting in the formation of an MWCNT/MnO_2/Ppy nanocomposite. Figure 11.3 shows the scanning electron microscopy (SEM) and transmission electron microscopy (TEM) images of synthesized MWCNT/MnO_2/Ppy and MWCNTs/MnO_2 composites [85]. Spencer et al. carried out electrodepositing of CNT foam with molybdenum trioxide (MoO3) [86], and the freestanding CNTs on the foam acted as working electrodes and were covered with MoO_3 MO, as shown in Figure 11.4.

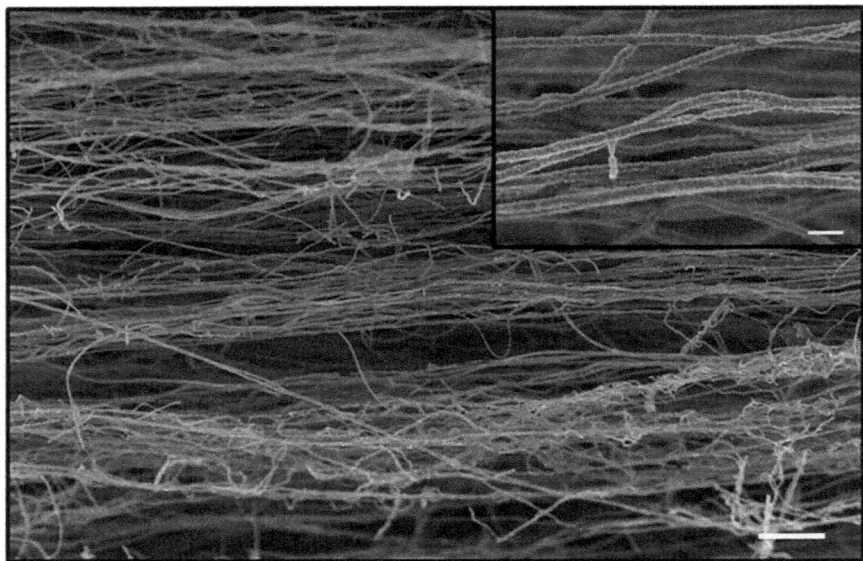

FIGURE 11.4 Scanning electron microscopy micrograph of an MoO_3/CNTs composite, Reprinted with permission from [86]. Copyright (2021) Energy & Fuels, American Chemical Society.

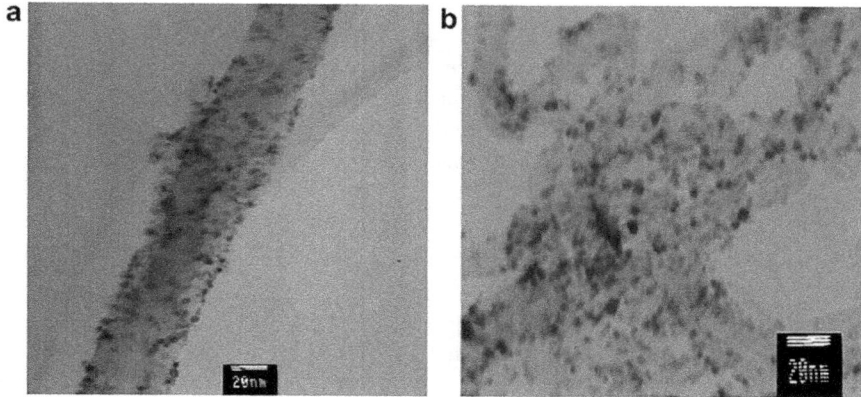

FIGURE 11.5 Transmission electron microscopy images of TiO$_2$/MWCNTs supported by Pt NPs, Reprinted with permission from [87]. Copyright (2021) Electrochemistry Communications, Elsevier.

Song et al. prepared Pt-TiO$_2$/MWCNTs composite using tetrabutyl titanate dissolved in ethanol. The solution was mixed with acetic acid to obtain TiO$_2$ sol, which was diluted using 5% of ethanol. Finally, MWCNTs were added to a solution, agitated for 20 min, dried at 80°C, and annealed at 600°C for 2 hours. Pt doping was carried out by a reduction of chloroplatinic acid by ethylene glycol. Figure 11.5 shows the TiO$_2$/MWCNT supported with Pt particles [87]. Byrappa et al. prepared two separate ZnO/CNT and TiO$_2$/CNT composites using a hydrothermal method. Precursors of the ZnO/CNTs and TiO$_2$/CNTs were prepared and transferred into Teflon-lined autoclaves. The reaction time took 40 hours at 150°C and 240°C, respectively. The XRD pattern confirmed the crystalline growth of materials [81].

11.6 FUTURE CHALLENGES

CNTs are one the highly researched materials in gas sensing because of their large surface area, hollow interiors, outside sensing walls, and prominently temperature-dependent resistivity. Furthermore, linking CNTs with MO material enhances sensing parameters like sensitivity, selectivity, low detection limits, and multiple analyte detection. TiO$_2$/CNTs are the superior candidates because of their ability to provide chemical stability, nonhazardous nature, and photo-oxidation. Also, SnO$_2$, ZrO$_2$, and MnO$_2$ have phenomenal characteristics that make them ideal for gas-sensing applications. An MO/CNT-based gas sensor's real-time response can be limited by the surrounding humidity. It has positive and negative impacts on the sensor, which depend on the analyte concentration and operating temperature. Although CNTs are highly sensitive, efforts are required to increase their reproducibility, uniformity, recovery, and stability of functional groups and make them defect-free. Also, a low-cost and more controlled synthesis and handling system is immensely needed for large-scale production of CNTs.

11.7 SUMMARY

CNTs and MO composite-based nanosensors have seen tremendous progress due to their unique morphological and electrical properties. These composites are adopted for sensor fabrication to detect toxic and explosive gases. In this chapter, various aspects of sensing mechanisms and materials for nanosensors were summarized. The synthesis methodology of MO/CNTs composites and their functionalization were presented. The behavior of MOs/CNTs highly depends on the synthesis method, functionalization, and coating quality of the MO.

LIST OF ABBREVIATIONS IN THIS CHAPTER

ALD	atomic layer deposition
CNTs	carbon nanotubes
MOs	metal oxides
MOS	metal oxide semiconductor
MWCNTs	multiwalled carbon nanotubes
NFs	nanofibers
NPs	nanoparticles
NRs	nanorods
NTs	nanotubes
NWs	nanowires
PLD	pulsed laser deposition
PTSA	p-toluene sulfonic acid
SWCNTs	single-walled carbon nanotubes
XRD	X-ray diffraction

REFERENCES

1. C. Sarafoleanu, C. Mella, M. Georgescu, and C. Perederco, "The importance of the olfactory sense in the human behavior and evolution," *J. Med. Life*, vol. 2, no. 2, p. 196, Apr. 2009.
2. A. Morim and G. T. Guldner, *Chlorine Gas Toxicity*. StatPearls Publishing, 2020.
3. A. Biswas, I. S. Bayer, A. S. Biris, T. Wang, E. Dervishi, and F. Faupel, "Advances in top–down and bottom–up surface nanofabrication: Techniques, applications & future prospects," *Adv. Colloid Interface Sci.*, vol. 170, no. 1–2, pp. 2–27, Jan. 2012, doi: 10.1016/J.CIS.2011.11.001.
4. K. Sasaki, Z. Y. Shi, R. Kopelman, and H. Masuhara, "Three-dimensional pH microprobing with an optically-manipulated fluorescent particle," *Chem. Lett.*, no. 2, pp. 141–142, Mar. 2006, doi: 10.1246/CL.1996.141.
5. J. Geng, M. D. R. Thomas, D. S. Shephard, and B. F. G. Johnson, "Suppressed electron hopping in a Au nanoparticle/H2S system: Development towards a H2S nanosensor," *Chem. Commun.*, no. 14, pp. 1895–1897, Mar. 2005, doi: 10.1039/B418559E.
6. K. Wu *et al.*, "Magnetic-nanosensor-based virus and pathogen detection strategies before and during covid-19," *ACS Appl. Nano Mater.*, vol. 3, no. 10, pp. 9560–9580, Oct. 2020.
7. E. A. Wachter and T. Thundat, "Micromechanical sensors for chemical and physical measurements," *Rev. Sci. Instrum.*, vol. 66, no. 6, p. 3662, Sep. 1998, doi: 10.1063/1.1145484.

8. W. Guan, N. Tang, K. He, X. Hu, M. Li, and K. Li, "Gas-sensing performances of metal oxide nanostructures for detecting dissolved gases: A mini review," *Front. Chem.*, vol. 8, p. 76, Feb. 2020, doi: 10.3389/FCHEM.2020.00076/BIBTEX.

9. L. Xu *et al.*, "Micro/nano gas sensors: A new strategy towards in-situ wafer-level fabrication of high-performance gas sensing chips," *Sci. Rep.*, vol. 5, no. 1, pp. 1–12, May 2015, doi: 10.1038/srep10507.

10. C. Wang, L. Yin, L. Zhang, D. Xiang, and R. Gao, "Metal oxide gas sensors: Sensitivity and influencing factors," *Sensors (Basel).*, vol. 10, no. 3, p. 2088, Mar. 2010, doi: 10.3390/S100302088.

11. P. K. Clifford and D. T. Tuma, "Characteristics of semiconductor gas sensors I: Steady state gas response," *Sensors and Actuators*, vol. 3, no. C, pp. 233–254, Jan. 1982, doi: 10.1016/0250-6874(82)80026-7.

12. H. Bai and G. Shi, "Gas sensors based on conducting polymers," *Sensors*, vol. 7, no. 3, pp. 267–307, Mar. 2007, doi: 10.3390/S7030267.

13. L. Zhang, K. Khan, J. Zou, H. Zhang, and Y. Li, "Recent advances in emerging 2D material-based gas sensors: Potential in disease diagnosis," *Adv. Mater. Interfaces*, vol. 6, no. 22, p. 1901329, Nov. 2019, doi: 10.1002/ADMI.201901329.

14. S. Majumdar, P. Nag, and P. S. Devi, "Enhanced performance of CNT/SnO2 thick film gas sensors towards hydrogen," *Mater. Chem. Phys.*, vol. 147, no. 1–2, pp. 79–85, Sep. 2014, doi: 10.1016/J.MATCHEMPHYS.2014.04.009.

15. S. Mao, G. Lu, and J. Chen, "Nanocarbon-based gas sensors: Progress and challenges," *J. Mater. Chem. A*, vol. 2, no. 16, pp. 5573–5579, Mar. 2014, doi: 10.1039/C3TA13823B.

16. K. Arshak, E. Moore, G. M. Lyons, J. Harris, and S. Clifford, "A review of gas sensors employed in electronic nose applications," *Sens. Rev.*, vol. 24, no. 2, pp. 181–198, 2004, doi: 10.1108/02602280410525977/FULL/PDF.

17. R. A. Zargar, "ZnCdO thick film: A material for energy conversion devices," *Mater. Res. Express*, vol. 6, pp. 095909, Jul. 2019, doi: 10.1088/2053-1591/ab2fb6.

18. S. D. Han *et al.*, "Self-doped nanocolumnar vanadium oxides thin films for highly selective NO2 gas sensing at low temperature," *Sensors Actuators B Chem.*, vol. 241, pp. 40–47, Mar. 2017, doi: 10.1016/J.SNB.2016.10.029.

19. J. H. Kim, P. Wu, H. W. Kim, and S. S. Kim, "Highly selective sensing of CO, C6H6, and C7H8 gases by catalytic functionalization with metal nanoparticles," *ACS Appl. Mater. Interfaces*, vol. 8, no. 11, pp. 7173–7183, Mar. 2016, doi: 10.1021/ACSAMI.6B01116.

20. R. A. Zargar *et al.*, "Structural, optical, photoluminescence, and EPR behaviour of novel Zn0·80Cd0·20O thick films: An effect of different sintering temperatures," *J. Luminescence*, vol. 245, p. 118769, Jan. 2022, doi: 10.1016/j.jlumin.2022.118769.

21. M. V. Nikolic *et al.*, "Humidity sensing properties of nanocrystalline pseudobrookite (Fe2TiO5) based thick films," *Sensors Actuators B Chem.*, vol. 277, pp. 654–664, Dec. 2018, doi: 10.1016/J.SNB.2018.09.063.

22. S. Capone, P. Siciliano, F. Quaranta, R. Rella, M. Epifani, and L. Vasanelli, "Analysis of vapours and foods by means of an electronic nose based on a sol–gel metal oxide sensors array," *Sensors Actuators B Chem.*, vol. 69, no. 3, pp. 230–235, Oct. 2000, doi: 10.1016/S0925-4005(00)00496-2.

23. H. W. Ryu *et al.*, "ZnO sol–gel derived porous film for CO gas sensing," *Sensors Actuators B Chem.*, vol. 96, no. 3, pp. 717–722, Dec. 2003, doi: 10.1016/J.SNB.2003.07.010.

24. H. C. Chiu and C. S. Yeh, "Hydrothermal synthesis of SnO2 nanoparticles and their gas-sensing of alcohol," *J. Phys. Chem. C*, vol. 111, no. 20, pp. 7256–7259, May 2007, doi: 10.1021/JP0688355.

25. L. Zhu, Y. Li, and W. Zeng, "Hydrothermal synthesis of hierarchical flower-like ZnO nanostructure and its enhanced ethanol gas-sensing properties," *Appl. Surf. Sci.*, vol. 427, pp. 281–287, Jan. 2018, doi: 10.1016/J.APSUSC.2017.08.229.

26. C. Yang, X. Su, J. Wang, X. Cao, S. Wang, and L. Zhang, "Facile microwave-assisted hydrothermal synthesis of varied-shaped CuO nanoparticles and their gas sensing properties," *Sensors Actuators B Chem.*, vol. 185, pp. 159–165, Aug. 2013, doi: 10.1016/J. SNB.2013.04.100.

27. L. A. Mercante, R. S. Andre, L. H. C. Mattoso, and D. S. Correa, "Electrospun ceramic nanofibers and hybrid-nanofiber composites for gas sensing," *ACS Appl. Nano Mater.*, vol. 2, no. 7, pp. 4026–4042, Jul. 2019, doi: 10.1021/ACSANM.9B01176/ASSET/ IMAGES/ACSANM.9B01176.SOCIAL.JPEG_V03.

28. N. Nafarizal, "Precise Control of Metal Oxide Thin Films Deposition in Magnetron Sputtering Plasmas for High Performance Sensing Devices Fabrication," *Procedia Chem.*, vol. 20, pp. 93–97, Jan. 2016, doi: 10.1016/J.PROCHE.2016.07.016.

29. D. Mutschall, K. Holzner, and E. Obermeier, "Sputtered molybdenum oxide thin films for NH3 detection," *Sensors Actuators B Chem.*, vol. 36, no. 1–3, pp. 320–324, Oct. 1996, doi: 10.1016/S0925-4005(97)80089-5.

30. J. Huotari, V. Kekkonen, J. Puustinen, J. Liimatainen, and J. Lappalainen, "Pulsed laser deposition for improved metal-oxide gas sensing layers," *Procedia Eng.*, vol. 168, pp. 1066–1069, Jan. 2016, doi: 10.1016/J.PROENG.2016.11.341.

31. J. Xie, X. Liu, S. Jing, C. Pang, Q. Liu, and J. Zhang, "Chemical and electronic modulation via atomic layer deposition of NiO on porous In2O3Films to boost NO2Detection," *ACS Appl. Mater. Interfaces*, vol. 13, no. 33, pp. 39621–39632, Aug. 2021, doi: 10.1021/ ACSAMI.1C11262/ASSET/IMAGES/ACSAMI.1C11262.SOCIAL.JPEG_V03.

32. A. Mirzaei, B. Hashemi, and K. Janghorban, "α-Fe2O3 based nanomaterials as gas sensors," *J. Mater. Sci. Mater. Electron.*, vol. 27, no. 4, pp. 3109–3144, Dec. 2015, doi: 10.1007/S10854-015-4200-Z.

33. Z. Li *et al.*, "A fast response & recovery H2S gas sensor based on α-Fe2O3 nanoparticles with ppb level detection limit," *J. Hazard. Mater.*, vol. 300, pp. 167–174, Dec. 2015, doi: 10.1016/J.JHAZMAT.2015.07.003.

34. Z. Liu, T. Yamazaki, Y. Shen, T. Kikuta, N. Nakatani, and Y. Li, "O2 and CO sensing of Ga2O3 multiple nanowire gas sensors," *Sensors Actuators B Chem.*, vol. 129, no. 2, pp. 666–670, Feb. 2008, doi: 10.1016/J.SNB.2007.09.055.

35. S. Park, S. An, H. Ko, C. Jin, and C. Lee, "Synthesis of nanograined ZnO nanowires and their enhanced gas sensing properties," *ACS Appl. Mater. Interfaces*, vol. 4, no. 7, pp. 3650–3656, Jul. 2012, doi: 10.1021/AM300741R/SUPPL_FILE/AM300741R_ SI_001.PDF.

36. C. H. Lin, S. J. Chang, and T. J. Hsueh, "A low-temperature ZnO nanowire ethanol gas sensor prepared on plastic substrate," *Mater. Res. Express*, vol. 3, no. 9, p. 095002, Sep. 2016, doi: 10.1088/2053-1591/3/9/095002.

37. Y. T. Lim, J. Y. Son, and J. S. Rhee, "Vertical ZnO nanorod array as an effective hydrogen gas sensor," *Ceram. Int.*, vol. 39, no. 1, pp. 887–890, Jan. 2013, doi: 10.1016/J. CERAMINT.2012.06.035.

38. W. Zheng *et al.*, "A highly sensitive and fast-responding sensor based on electrospun In2O3 nanofibers," *Sensors Actuators B Chem.*, vol. 142, no. 1, pp. 61–65, Oct. 2009, doi: 10.1016/J.SNB.2009.07.031.

39. X. Jia, H. Fan, and W. Yang, "Hydrothermal synthesis and primary gas sensing properties of CuO nanosheets," *Journal of Dispersion Science and Technology*, vol. 31, no. 7, pp. 866–869, Jun. 2010, doi: 10.1080/01932690903223641.

40. L. Song *et al.*, "Facile synthesis of hierarchical tin oxide nanoflowers with ultra-high methanol gas sensing at low working temperature," *Nanoscale Res. Lett.*, vol. 14, no. 1, pp. 1–11, Mar. 2019, doi: 10.1186/S11671-019-2911-4/FIGURES/12.

41. T. Akamatsu, T. Itoh, N. Izu, and W. Shin, "Effect of noble metal addition on Co 3 O 4-based gas sensors for selective NO detection," *Sensors Mater.*, vol. 28, no. 11, pp. 1191–1201, 2016.

42. J. Ma *et al.*, "Pt nanoparticles sensitized ordered mesoporous WO3 semiconductor: Gas sensing performance and mechanism study," *Adv. Funct. Mater.*, vol. 28, no. 6, p. 1705268, Feb. 2018, doi: 10.1002/ADFM.201705268.

43. Y. Tan and Y. Lei, "Atomic layer deposition of Rh nanoparticles on WO3 thin film for CH4 gas sensing with enhanced detection characteristics," *Ceram. Int.*, vol. 46, no. 7, pp. 9936–9942, May 2020, doi: 10.1016/J.CERAMINT.2019.12.094.

44. L. Y. Zhu *et al.*, "Fabrication of heterostructured p-CuO/n-SnO2 core-shell nanowires for enhanced sensitive and selective formaldehyde detection," *Sensors Actuators B Chem.*, vol. 290, pp. 233–241, Jul. 2019, doi: 10.1016/J.SNB.2019.03.092.

45. M. H. Raza, N. Kaur, E. Comini, and N. Pinna, "Toward optimized radial modulation of the space-charge region in one-dimensional SnO2-NiO core-shell nanowires for hydrogen sensing," *ACS Appl. Mater. Interfaces*, vol. 12, no. 4, pp. 4594–4606, Jan. 2020, doi: 10.1021/ACSAMI.9B19442/SUPPL_FILE/AM9B19442_SI_001.PDF.

46. J. H. Yang *et al.*, "Facile synthesis of α-Fe2O3/ZnO core-shell nanowires for enhanced H2S sensing," *Sensors Actuators B Chem.*, vol. 307, p. 127617, Mar. 2020, doi: 10.1016/J.SNB.2019.127617.

47. K. P. Yuan *et al.*, "Precise preparation of WO3@SnO2 core shell nanosheets for efficient NH3 gas sensing," *J. Colloid Interface Sci.*, vol. 568, pp. 81–88, May 2020, doi: 10.1016/J.JCIS.2020.02.042.

48. S. Iijima, "Helical microtubules of graphitic carbon," *Nature*, vol. 354, no. 6348, pp. 56–58, 1991, doi: 10.1038/354056a0.

49. S. Iijima and T. Ichihashi, "Single-shell carbon nanotubes of 1-nm diameter," *Nature*, vol. 363, no. 6430, pp. 603–605, 1993, doi: 10.1038/363603a0.

50. M. Michálek, J. Sedláček, M. Parchoviansky, M. Michálková, and D. Galusek, "Mechanical properties and electrical conductivity of alumina/MWCNT and alumina/zirconia/MWCNT composites," *Ceram. Int.*, vol. 40, no. 1, pp. 1289–1295, Jan. 2014, doi: 10.1016/J.CERAMINT.2013.07.008.

51. J. Wu, H. Zhang, Y. Zhang, and X. Wang, "Mechanical and thermal properties of carbon nanotube/aluminum composites consolidated by spark plasma sintering," *Mater. Des.*, vol. 41, pp. 344–348, Oct. 2012, doi: 10.1016/J.MATDES.2012.05.014.

52. R. A. Zarga, M. M. Hassan, K. Kumar, V. Nagal, A. Bashir, B. Alshahrani, T. Alshahrani, and M. Shkir, "Development and characterization of $(ZnO)_{0.90}(CNT)_{0.10}$ thick film for photovoltaic application," *Optik*, vol. 248, p. 167975, Sep. 2021, doi: 10.1016/j.ijleo.2021.167975.

53. R. A. Zarga, M. Arora, T. Alshahrani, and M. Shkir, "Screen printed novel ZnO/MWCNTs nano composite thick films," *Ceram. Int.*, vol. 47, pp. 6084–6093, Oct. 2021, doi: 10.1016/j.ceramint. 2020.10.185.

54. A. Boyd, I. Dube, G. Fedorov, M. Paranjape, and P. Barbara, "Gas sensing mechanism of carbon nanotubes: From single tubes to high-density networks," *Carbon N. Y.*, vol. 69, pp. 417–423, Apr. 2014, doi: 10.1016/J.CARBON.2013.12.044.

55. Z. Hou, D. Xu, and B. Cai, "Ionization gas sensing in a microelectrode system with carbon nanotubes," *Appl. Phys. Lett.*, vol. 89, no. 21, p. 213502, Nov. 2006, doi: 10.1063/1.2392994.

56. J. T. W. Yeow, and J. P. M. She, "Carbon nanotube-enhanced capillary condensation for a capacitive humidity sensor," *Nanotechnol.*, vol. 17, no. 21, p. 5441, Oct. 2006, doi: 10.1088/0957-4484/17/21/026.

57. E. S. Snow, F. K. Perkins, E. J. Houser, S. C. Badescu, and T. L. Reinecke, "Chemical detection with a single-walled carbon nanotube capacitor," *Science*, vol. 307, no. 5717, pp. 1942–1945, Mar. 2005, doi: 10.1126/SCIENCE.1109128/SUPPL_FILE/SNOW.SOM.PDF.

58. S. Chopra, K. McGuire, N. Gothard, A. M. Rao, and A. Pham, "Selective gas detection using a carbon nanotube sensor," *Appl. Phys. Lett.*, vol. 83, no. 11, p. 2280, Sep. 2003, doi: 10.1063/1.1610251.

59. S. Chopra, A. Pham, J. Gaillard, A. Parker, and A. M. Rao, "Carbon-nanotube-based resonant-circuit sensor for ammonia," *Appl. Phys. Lett.*, vol. 80, no. 24, p. 4632, Jun. 2002, doi: 10.1063/1.1486481.

60. J. Li, Y. Lu, Q. Ye, M. Cinke, J. Han, and M. Meyyappan, "Carbon nanotube sensors for gas and organic vapor detection," *Nano Lett.*, vol. 3, no. 7, pp. 929–933, Jul. 2003, doi: 10.1021/NL034220X/ASSET/IMAGES/NL034220X.SOCIAL.JPEG_V03.

61. J. Zhang, A. Boyd, A. Tselev, M. Paranjape, and P. Barbara, "Mechanism of NO2 detection in carbon nanotube field effect transistor chemical sensors," *Appl. Phys. Lett.*, vol. 88, no. 12, p. 123112, Mar. 2006, doi: 10.1063/1.2187510.

62. C. Piloto *et al.*, "Room temperature gas sensing properties of ultrathin carbon nanotube films by surfactant-free dip coating," *Sensors Actuators B Chem.*, vol. 227, pp. 128–134, May 2016, doi: 10.1016/J.SNB.2015.12.051.

63. Y. L. Liu, H. F. Yang, Y. Yang, Z. M. Liu, G. L. Shen, and R. Q. Yu, "Gas sensing properties of tin dioxide coated onto multi-walled carbon nanotubes," *Thin Solid Films*, vol. 497, no. 1–2, pp. 355–360, Feb. 2006, doi: 10.1016/J.TSF.2005.11.018.

64. N. Van Hieu, L. T. B. Thuy, and N. D. Chien, "Highly sensitive thin film NH3 gas sensor operating at room temperature based on SnO2/MWCNTs composite," *Sensors Actuators B Chem.*, vol. 129, no. 2, pp. 888–895, Feb. 2008, doi: 10.1016/J. SNB.2007.09.088.

65. C. Bittencourt *et al.*, "WO3 films modified with functionalised multi-wall carbon nanotubes: Morphological, compositional and gas response studies," *Sensors Actuators B Chem.*, vol. 115, no. 1, pp. 33–41, May 2006, doi: 10.1016/J.SNB.2005.07.067.

66. Y. J. Kwon *et al.*, "Synthesis, characterization and gas sensing properties of ZnO-decorated MWCNTs," *Appl. Surf. Sci.*, vol. 413, pp. 242–252, Aug. 2017, doi: 10.1016/J. APSUSC.2017.03.290.

67. C. Hua *et al.*, "A flexible gas sensor based on single-walled carbon nanotube-Fe2O3 composite film," *Appl. Surf. Sci.*, vol. 405, pp. 405–411, May 2017, doi: 10.1016/J. APSUSC.2017.01.301.

68. F. Özütok, I. K. Er, S. Acar, and S. Demiri, "Enhancing the Co gas sensing properties of ZnO thin films with the decoration of MWCNTs," *J. Mater. Sci. Mater. Electron.*, vol. 30, no. 1, pp. 259–265, Jan. 2019, doi: 10.1007/S10854-018-0288-2/TABLES/1.

69. G. Chimowa *et al.*, "Improving methane gas sensing properties of multi-walled carbon nanotubes by vanadium oxide filling," *Sensors Actuators B Chem.*, vol. 247, pp. 11–18, Aug. 2017, doi: 10.1016/J.SNB.2017.02.167.

70. S. Navazani, M. Hassanisadi, M. M. Eskandari, and Z. Talaei, "Design and evaluation of SnO2-Pt/MWCNTs hybrid system as room temperature-methane sensor," *Synth. Met.*, vol. 260, p. 116267, Feb. 2020, doi: 10.1016/J.SYNTHMET.2019.116267.

71. X. V. Le, T. L. A. Luu, H. L. Nguyen, and C. T. Nguyen, "Synergistic enhancement of ammonia gas-sensing properties at low temperature by compositing carbon nanotubes with tungsten oxide nanobricks," *Vacuum*, vol. 168, p. 108861, Oct. 2019, doi: 10.1016/J. VACUUM.2019.108861.

72. J. Fan, P. Liu, X. Chen, H. Zhou, S. Fu, and W. Wu, "Carbon nanotubes-CuO/SnO2 based gas sensor for detecting H2S in low concentration," *Nanotechnology*, vol. 30, no. 47, p. 475501, Sep. 2019, doi: 10.1088/1361-6528/AB3CB3.

73. X. Jia, C. Cheng, S. Yu, J. Yang, Y. Li, and H. Song, "Preparation and enhanced acetone sensing properties of flower-like α-Fe2O3/multi-walled carbon nanotube nanocomposites," *Sensors Actuators B Chem.*, vol. 300, p. 127012, Dec. 2019, doi: 10.1016/J. SNB.2019.127012.

74. R. Guo *et al.*, "The enhanced ethanol sensing properties of CNT@ZnSnO3 hollow boxes derived from Zn-MOF(ZIF-8)," *Ceram. Int.*, vol. 46, no. 6, pp. 7065–7073, Apr. 2020, doi: 10.1016/J.CERAMINT.2019.11.198.

75. G. X. Wang, B. L. Zhang, Z. L. Yu, and M. Z. Qu, "Manganese oxide/MWNTs composite electrodes for supercapacitors," *Solid State Ionics*, vol. 176, no. 11–12, pp. 1169–1174, Mar. 2005, doi: 10.1016/J.SSI.2005.02.005.

76. J. Zhou *et al.*, "Interaction between Pt nanoparticles and carbon nanotubes—An X-ray absorption near edge structures (XANES) study," *Chem. Phys. Lett.*, vol. 437, no. 4–6, pp. 229–232, Apr. 2007, doi: 10.1016/J.CPLETT.2007.02.026.

77. D. J. Guo, and H. L. Li, "High dispersion and electrocatalytic properties of Pt nanoparticles on SWNT bundles," *J. Electroanal. Chem.*, vol. 573, no. 1, pp. 197–202, Nov. 2004, doi: 10.1016/J.JELECHEM.2004.07.006.

78. Y. Lin, X. Cui, C. Yen, and C. M. Wai, "Platinum/carbon nanotube nanocomposite synthesized in supercritical fluid as electrocatalysts for low-temperature fuel cells," *J. Phys. Chem. B.*, vol. 109, no. 30, pp. 14410–14415, Aug. 2005, doi: 10.1021/JP0514675.

79. I.-H. Kim, J.-H. Kim, Y.-H. Lee, and K.-B. Kim, "Synthesis and characterization of electrochemically prepared ruthenium oxide on carbon nanotube film substrate for supercapacitor applications," *J. Electrochem. Soc.*, vol. 152, no. 11, p. A2170, Sep. 2005, doi: 10.1149/1.2041147/XML.

80. W. Q. Han, and A. Zettl, "Coating single-walled carbon nanotubes with tin oxide," *Nano Lett.*, vol. 3, no. 5, pp. 681–683, May 2003, doi: 10.1021/NL034142D/ASSET/IMAGES/NL034142D.SOCIAL.JPEG_V03.

81. R. A. Zargar, M. Arora, R. A. Bhat, "Study of nanosized copper doped ZnO dilute magnetic semiconductor thick films for spintronic device applications". *J. Appl. Phys–A*, vol. 124, p. 36. Dec. 2018, doi: 10.1007/s00339-017-1457-5.

82. Y. Zhu *et al.*, "Multiwalled carbon nanotubes beaded with ZnO nanoparticles for ultra-fast nonlinear optical switching," *Adv. Mater.*, vol. 18, no. 5, pp. 587–592, Mar. 2006, doi: 10.1002/ADMA.200501918.

83. T. Ikuno *et al.*, "Insulator-coated carbon nanotubes synthesized by pulsed laser deposition," *Japanese J. Appl. Physics, Part 2 Lett.*, vol. 42, no. 11 B, p. L1356, Nov. 2003, doi: 10.1143/JJAP.42.L1356/XML.

84. Q. Kuang *et al.*, "Controllable fabrication of SnO2-coated multiwalled carbon nanotubes by chemical vapor deposition," *Carbon N. Y.*, vol. 44, no. 7, pp. 1166–1172, Jun. 2006, doi: 10.1016/J.CARBON.2005.11.001.

85. P. Mishra, and R. Jain, "Electrochemical deposition of MWCNT-MnO2/PPy nanocomposite application for microbial fuel cells," *Int. J. Hydrogen Energy*, vol. 41, no. 47, pp. 22394–22405, Dec. 2016, doi: 10.1016/J.IJHYDENE.2016.09.020.

86. M. A. Spencer, O. Yildiz, I. Kamboj, P. D. Bradford, and V. Augustyn, "Toward deterministic 3D energy storage electrode architectures via electrodeposition of molybdenum oxide onto CNT foams," *Energy and Fuels*, vol. 35, no. 19, pp. 16183–16193, Oct. 2021, doi: 10.1021/ACS.ENERGYFUELS.1C02352/SUPPL_FILE/EF1C02352_SI_002.MP4.

87. H. Song, X. Qiu, F. Li, W. Zhu, and L. Chen, "Ethanol electro-oxidation on catalysts with TiO2 coated carbon nanotubes as support," *Electrochem. Commun.*, vol. 9, no. 6, pp. 1416–1421, Jun. 2007, doi: 10.1016/J.ELECOM.2007.01.048.

12 Metal Oxide– Based Carbon Nanocomposites for Pollutant Nanosensing

Krishna Kumari Swain, Rodrigo A. Abarza Munoz, Sapna Sudan, Bijay Kumar Behera, and Rayees Ahmad Zargar

CONTENTS

12.1 INTRODUCTION

Population growth and aging, as well as the growing industries in the agrochemical, chemical, cosmetic, and pharmaceutical fields, led to the synthesis and increased manufacturing of several chemicals. These chemicals are now consumed and discarded on a daily basis by millions of people all over the world as shown in Figure 12.1. The presence of emerging contaminants, such as pharmaceuticals and personal care products, endocrine-disrupting compounds, heavy metals, flame retardants, pesticides, and artificial sweeteners in soil, aquatic environments, sediments, and atmosphere, poses a significant threat to the environment and human health [1]. As a result, there has been a higher demand for real-time management and monitoring of all aspects of our environment's health and safety.

DOI: 10.1201/9781003323464-12

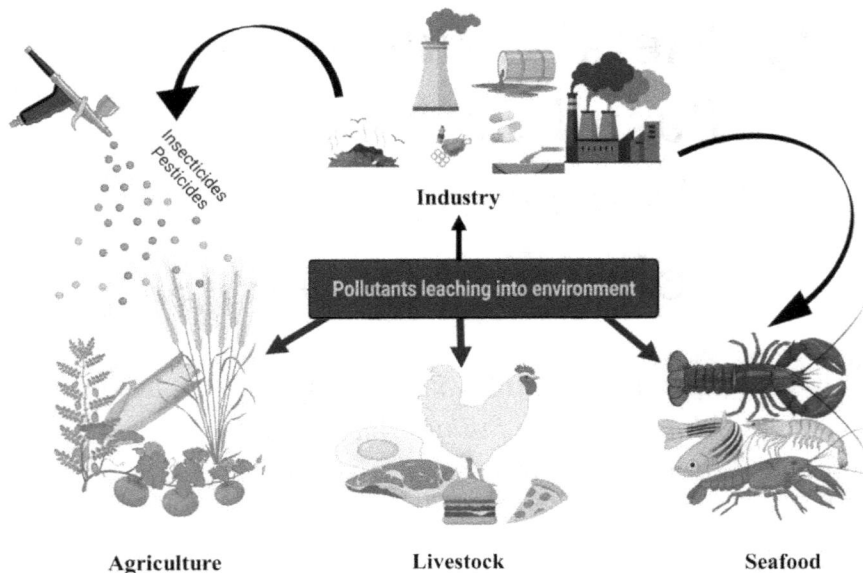

FIGURE 12.1 Schematic representation of environmental pollutants leaching into the environment.

There are numerous conventional analytical techniques reported for the analysis of contaminants in the environment such as atomic absorption spectrometry (AAS), inductively coupled plasma mass spectroscopy (ICP-MS), inductively coupled plasma optical emission spectroscopy (ICP-OES), cold vapor atomic fluorescence spectrometry (CV-AFS), reverse-phase high-performance liquid chromatography (RP-HPLC), and electrothermal vaporization-inductively coupled plasma mass spectrometry (ETV-ICP-MS) [2–7]. Despite providing highly precise, sensitive, stable, and accurate results, these instrumental approaches require expensive and heavy instrumentation, need specially trained operators, and tedious and complicated sample preparation and are not suited for on-site analysis. They also utilize a large volume of samples for analysis. Hence, there is a critical need for the design and development of a low-cost, simple, high-throughput, yet sensitive method for monitoring contaminants and pollutants in the environment, particularly for human beings.

Nanotechnology is one of the most important technologies in the 21st century. Nanotechnology is being widely used in a variety of sectors, including biological and chemical analysis. Nanomaterials, owing to their nanometer size, increase in the surface/volume ratio with decreasing size, and spatial confinement effects, make a remarkable difference in their physicochemical properties from their bulk-sized counterparts [8]. Nanomaterial-enabled sensors are an exciting technology that can detect environmental contaminants at nanomolar to sub-picomolar levels. The potential for simple, in-field contaminant detection without the need for expensive lab equipment has piqued interest in these sensors.

Among all nanomaterials, carbon nanotubes (CNTs) have emerged as promising candidates because of their unique features. Different CNTs like multiwalled carbon nanotubes (MWCNTs) and single-walled carbon nanotubes (SWCNTs) have become some of the most attractive nanomaterials in the nanotechnology revolution. CNTs possess exceptional catalytic properties like high electron transfer capacity, high interfacial adsorption properties, and thermal conductivity. In addition to that, the special biocompatibility nature and enhanced electrocatalytic activity make CNTs better electrochemical sensors for the sensing of analytes. Moreover, due to their large surface area, CNTs act as a better support for binding nanoparticles. Nanoparticle-decorated CNTs are significantly used for different applications like electromagnetic wave–absorbing processes, the development of highly sensitive and selective sensors, electrochemical energy storage devices, and water oxidation processes. In particular, metal oxide nanoparticle (MON)–decorated CNTs (MONP-CNTs) have been highlighted because of their conjoining properties, such as optical, thermal, electrical, and magnetic properties; surface-enhanced Raman scattering; plasmonic resonance energy transfer; and magneto-optical effect, which improves the performance of sensors for detection of analytes in environmental matrices. In addition to that, the surface of MONP-CNTs provide suitable and enough space for the immobilization of a large number of biomolecules for the development of biosensor. Therefore, recently researchers across the world have been showing great interest in the production, characterization, and application of MONP-CNTs sensors for the sensing of gas, environmental contaminants, and clinical diagnostics.

Due to these extraordinary features, nanomaterials have been one of the most popular alternatives for meeting the desired requirements for the building of exceptionally sensitive sensors.

12.2 CARBON NANOTUBES

Out of all the materialistic perspectives, carbon nanostructured materials such as carbon nanotubes (CNTs) have piqued researchers' interest as an exciting platform for contamination and pollutant analysis. Because of their unique biocompatibility, CNTs can be used as carriers in various research fields. The chemical and electrical properties and high scalability of CNTs make them as excellent electrocatalysts. Due to their infinite functionalization and possibilities with an array of inorganic nanomaterials and biomolecules, CNTs have captivated particular interest in biomedical science [9].

In 1991, the first CNT was invented by S. Injima in Japan. One of the fascinating things about nanotubes is their unique physiochemical properties and shape. CNTs are elongated cylindrical structures made up of one or more hexagonal graphite planes rolled in tubes with diameters ranging from 1to several dozens of nanometers and lengths ranging from 1to several microns. Their surface is made up of hexagonal carbon cycles in a regular pattern called hexagons. CNTs are carbonaceous material and have unique characteristics such as crystal lattice, allowing modification of the cell parameters and electrochemical properties that control the bandgap by altering their surface properties, conductivity, and chemical reactivity, which made them highly potent to be used as sensing elements [10]. CNTs are one of the most promising heterogeneous photocatalytic possibilities for organic compound degradation, heavy metal reduction, and selective oxidative reactions. Depending on the synthesis circumstances of CNTs, one- or multilayered tubulenes with open or closed terminations may arise. Tubulenes are commonly described as having endless cylinder surfaces that accommodate carbon atoms integrated into a single network with hexagonal cells (sp^2 network). CNTs can be used either as single-walled (SWCNT), multiwalled (MWCNT), or functionalized nano constructs. Based on their geometry, CNTs are subdivided into two categories, that is, achiral and chiral. Chiral tubulenes in particular exhibit a screw symmetry, whereas achiral tubulenes have a cylindrical symmetry [11].

12.3 SYNTHESIS OF CNTS

CNTs have transformed the state-of-the-art into nanotechnology due to their exceptional mechanical, electrical, thermal, optical, and chemical capabilities. There are three types of CNTs, that is, SWNTs, MWNTs, and surface-functionalized CNTs. There are a variety of ways to make CNTs, but following the three are the most critical and widely utilized.

12.3.1 CHEMICAL VAPOR DEPOSITION METHOD

It is critical to note that high-quality, high-purity CNTs necessitate very reliable synthesis procedures. What we require is an understanding of the influencing elements and control conditions of specialized carbon nanotube synthesis. The chemical vapor deposition (CVD) process involves breaking a carbon atom–containing gas that is continuously flowing past the catalyst nanoparticle to form carbon atoms, which are subsequently deposited as CNTs on the catalyst's or substrate's surface. CVD is distinct from the other CNT production techniques. The CVD approach to producing CNTs offers the benefits of a large yield of nanotubes and a reduced temperature requirement (550–1000°C), making it both cheaper and more accessible for lab use. The CVD process also enables control over the shape and structure of the CNTs generated, as well as the development of aligned nanotubes in a specified direction. The preparation of MWCNTs is now well developed, and industrial manufacturing has been achieved using CVD. SWCNTs are still relatively expensive to produce, and macroscopic arrays of variously oriented SWCNTs have yet to be accomplished. The nanotubes created by the CVD process, however, are more structurally faulty than those formed by laser evaporation or arc discharge. Arc discharge and laser vaporization are both operated under high temperatures (above 3000 K) with a short span of reaction time (micro- to milliseconds), whereas catalytic CVD can be operated under intermediate temperatures (700–1473 K) with a lengthy time (minutes to hours) [12, 13].

12.3.2 ELECTRIC ARC DISCHARGE METHOD

Various synthesis techniques have been developed and used during the last few years in an effort to produce high-quality CNTs in bulk. Arc discharge is one of the oldest and most effective methods for making high-quality CNTs. Despite the fact that this synthesis approach has been studied for a long time, the nanotube growth mechanism is still unknown, and the growth conditions have nothing to do with the final result [14]. A voltage of 20–25 V is supplied across pure graphite electrodes maintained by 1millimeter and kept at 500 torr pressure of circulating helium gas within the quartz chamber in this approach. An electric arc is created when the electrodes are made to hit each other under these conditions. The arc's energy is transmitted to the anode, which ionizes the carbon atoms in the anode's pure graphite and creates C+ ions, forming plasma (plasma is atoms or molecules in a vapor state at high temperature). The length of the anode reduces as the CNTs grow longer, but the electrodes are adjusted to maintain a 1-millimeter gap between the two electrodes. If the electrodes are properly cooled, a homogeneous deposition of CNTs is generated on the cathode, which is done by using an inert gas at the right pressure. MWCNTs are produced using this approach, while SWCNTs are created using catalyst nanoparticles of iron, cobalt, and nickel that are kept inside the center of the positive electrode. Furthermore, these obtained CNTs are refined to get pure CNTs [15].

12.3.3 Laser Ablation Method

The laser ablation technique (LA) involves the physical vapor deposition process in which a laser source is used to vaporize a graphite target. The LA procedure consists of graphite (as a target) being put in the center of a chamber (made of quartz), filled with argon gas, and kept at 1200°C. A continuous laser source, like a CO_2 laser, and a pulsed laser source, like a Nd-doped yttrium aluminum garnet laser, can be used to vaporize the target material into target vapor atoms. The flow of argon gas sweeps the evaporated target atoms (carbon) toward the cooled copper collector. On a cooled copper collector, carbon atoms are deposited and grown as CNTs. This approach produces MWCNTs, while catalyst nanoparticles of Fe, Co, and Ni are utilized to produce SWCNTs. Thereafter, the pure CNTs are acquired by refining the obtained CNTs. Also, there is another technique called pulsed laser ablation in liquid (PLAL) is an appealing and promising approach that has several benefits over other techniques, including simplicity, low cost, no requirement for a catalyst, no vacuum, high-purity product, and good control over product size and shape. The PLAL nanomaterial characteristics are highly influenced by laser parameters such as laser fluence, pulse width, repetition rate, and wavelength. PLAL nanoparticles have been employed in a variety of applications, including gas photocatalysts, antibacterial agents, and optoelectronic devices [16, 17].

12.4 FUNCTIONALIZATION OF CNTS

CNTs, which are one-dimensional allotropes of carbon, have piqued researchers' attention since their discovery in 1991, owing to their huge aspect ratio, low mass density, and unique chemical, physical, and electrical characteristics that provide fascinating nanoscale applications. However, while working with this type of nanomaterial, two key difficulties should be considered: their strong agglomerating propensity, since they are generally present as bundles or ropes of nanotubes, and the metallic impurities and carbonaceous pieces that come with them. The uniform dispersion of CNTs and the creation of a strong chemical contact with the polymeric matrix are required for their effective use in a wide range of applications, particularly in the field of polymer composites [18]. The chemical modification and solubilization of CNTs area new frontier in nanotube-based materials research. Direct attachment of functional groups to the graphitic surface and the usage of nanotube-bound carboxylic acids are the two broad types of these processes. For SWCNTs and MWCNTs, several research groups have reported various effective functionalization processes such as metal oxide nanoparticle decoration on the surface of CNTs [19].

Because CNTs' surfaces are inert, they require a chemical treatment to make them more active in chemical reactions. From the literature survey, it has been reported that there are numerous chemical techniques for attaching gold nanoparticles (NPs) onto MWCNTs in AuNPs chemically linked CNTs nanocomposites [20]. Ultraviolet (UV) irradiation was used by Zhang et al. to generate AuNPs on the surface of MWCNTs, resulting in Au-CTN nanocomposites. Other functionalization procedures have been effectively exploited for this purpose, such as the two-step synthesis of gold/iron-oxide magnetic NP–decorated CNTs (Au/MNP-CNTs). AuNPs have also been self-assembled using oxidized CNTs covered with poly-(diallyl dimethyl ammonium) chloride [21].

Mustafa K A Mohammed et al. reported that the surface of SWCNTs and MWCNTs are modified with sulfuric acid and nitric acid (3:1 v/v) by chemical precipitation treatment. The process was used to adorn SWCNTs and MWCNTs with Ag-doped ZnO and Au-doped ZnO nanoparticles. Because of the mutual effect between the large surface area and powerful adsorption property of CNTs (SWCNTs, MWCNTs) and the high bactericidal and catalytic activities of metal/ZnO nanoparticles, the synergistic effect of metal/ZnO NPs and CNTs (SWCNTs, MWCNTs) has shown a high potential for biomedicine applications [22].

A highly sensitive laser-assisted electrochemical sensor for the detection of hazardous pollutant 4-NP was first developed by Kuang-Yow Lian et al. The technique uses the ZnO NPs@fMWCNTs nanocomposite for sensing the analyte. The synthesis of nanocomposites was done from three-dimensional flower-like ZnO NPs by the ultra-sonication method [23]. Another study by N.L. Mary et al. reported an enhanced electrochemical sensor based on acid-modified CNTs decorated with ZnO NPs. The synthesis was assisted by a microwave technique. A high value was achieved for CNT-COOH/ZnO shows a super-specific capacitance for the electrochemical sensor [24].

12.5   METAL OXIDE–DECORATED CNT GAS/HUMIDITY SENSORS

The unique features of CNTs have also been used to construct gas sensors with various transduction principles, such as changes in electrical properties, changes in mass, or changes in optical properties, among others. When exposed to the target gas or humidity analytes, the electrical characteristics of CNTs change substantially. Outside of the CNTs, a rich-electron conjugation forms, making them electrochemically active and susceptible to charge transfer and chemical doping effects by diverse compounds. When the groups like NO_2 or O_2 (electron-withdrawing) or NH_3 (electron-donating) interact with CNTs, which is basically a p-type semiconductor, then the density of holes in the nanotube, as well as conductivity, changes. SWCNTs have a low adsorption rate as well as low affinity and a long recovery time. As a result, they can't be employed as the foundation for a gas sensor. These drawbacks can be overcome by the surface modification (or decorations) of CNTs by some functionalizing materials like metal oxide NPs. Metal oxide nanoclusters or nanoparticles can also be used to adorn the sidewalls of CNTs as shown in Figure 12.2. Owing to their high catalytic activity, efficient charge transfer properties, and advanced physicochemical features, such as adsorption capacity and metal oxide nanoclusters, they exhibit a wide variety of reaction when comes in contact with various gases [25].

The first semiconductor metal oxide gas sensor was developed in 1962, which is widely utilized for gas detection. Metal oxides and CNTs have lately been employed as materials for semiconductor gas sensors, lithium-ion batteries, and catalysts. Gas sensors can be made from a variety of materials such as SnO_2-decorated CNTs, TiO_2-decorated CNTs, Fe_2O_3-decorated CNTs, WO_3-decorated CNTs, and Co_3O_4-decorated CNT composites. SnO_2-decorated CNTs can sense NO_2, NH_3, and xylene gases at 220°C. In 2010, R. Leghrib et al. reported the sensing of NO_2 in the ppb range using SnO_2-decorated CNTs. The MWCNT-SnO_2 nanoclusters were prepared by the precipitation method wherebyaSnO$_2$ colloidal suspension was dispersed in the presence of CNTs in a water-free acetic acid [27]. Similarly Lee H. et al. have reported tin oxide–modified CNT gas sensor system for the detection of NH_3 gas [28]. MWCNTs decorated with iridium oxide nanoparticles for

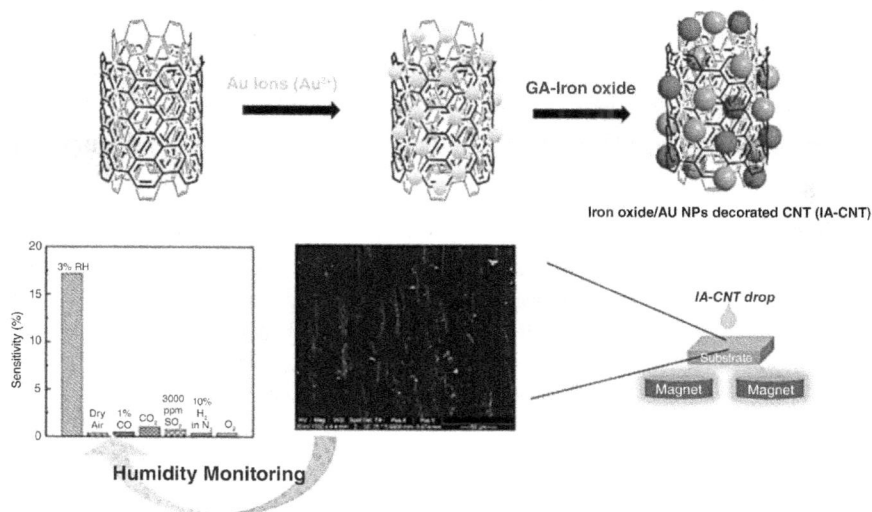

FIGURE 12.2 Schematic diagram of iron oxide/gold nanoparticle–decorated CNT for humidity sensing. (Reprinted with permission from Jaewook Lee, Suresh Mulmi, Venkataraman Thangadurai, et al. [26] copyright 2015 American Chemical Society.)

the analysis of NH_3 and NO_2 gas were reported by J.C-Cháfer et al. [29]. Another group, Jaewook Lee et al., have reported on a mixed metal oxide–decorated CNT (i.e., magnetically aligned Fe_2O_3/Au-decorated CNT)–based humidity sensor [30]. Acetone gas sensing was reported by S. J. Young et al., who used MWCNT-decorated silver nanoparticles [31]. M. S. Choi et al. have reported and discussed dual sensitization gas sensors for selective detection of H_2S and C_2H_5OH gas using ZnO/CuO-decorated MWCNTs at working temperatures of 100°C and 200°C, respectively [32]. In a similar way, Jung and Lee's group demonstrated a humidity sensor using MnO_2-coated CNT yarn. The result shows that the MnO_2-coated sensor gave better sensitivity than the uncoated sensor [33]. Yong Jung Kwon and co-workers have reported that the decoration of ZnO nanoparticle on MWCNTs greatly enhance the gas sensing properties as compared to bare MWCNTs [34]. A toluene gas sensor based on TiO_2-decorated 3D-grapheneCNTs was reported by Yotsarayuth Seekaew and co-workers. They suggested that the decoration of TiO_2 on CNTs greatly enhances the sensing of toluene more than 7 times more than the bare CNTs [35]. George Chimowa and co-workers have reported that vanadium oxide–decorated MWCNTs improve methane gas sensing from 0.5% to 1.5% at room temperature. The group said that the enhanced response is due to the result of the encapsulated metal oxide in unfilled CNTs, which increases the density of states around the Fermi level of the composite material [36]. In the similar way, a hybrid of metal oxide (SnO_2/CuO)–decorated graphene was used by Dongzhi Zhang and co-workers for the sensing a formaldehyde and NH_3gas mixture [37]. From the previously reported literature, it can be concluded that the metal oxide–decorated CNTs show a better performance for gas sensing as compared to bare CNTs. A summary of various metal oxide–decorated CNTs for gas and humidity sensing is reported in Table 12.1.

TABLE 12.1

A Summary of Various Metal Oxide–Decorated CNTs for Gas and Humidity Sensing

Metal Oxide–Decorated CNTs	Analyte	Limit of Detection (LOD)	Ref.
IrOx	NO_2, NH_3	1 ppb	[38]
Ag np	Acetone	50 ppm	[39]
ZnO	NO_2	1 ppm	[40]
TiO_2	Toluene gas	50–500 ppm	[41]
TiO_2	NO_2 and O_2	5 ppm	[42]
Pt nanoparticle	Toluene	1 and 5 ppm	[43]
SnO_2	Ethanol, methanol, and H_2S	30, 30, and 9 ppm	[44]
SnO_2 and WO_3	NO_2, NH_3, acetone, and EtOH	200 ppb, 8 ppb	[45]
Cu np	H_2S	5 ppm	[46]
WO_3 NPs	NO_2	1 ppm	[47]
Au and TiO_2 np	NO_2	1 ppm	[48]
Fe_2O_3	H_2S	20 ppm	[49]

12.6 METAL OXIDE–DECORATED CNT-BASED BIO- AND ELECTROCHEMICAL SENSORS

In recent years, screening of a variety of hazardous elements and compounds utilizing chemical/biosensing techniques has sparked a lot of attention. Biosensors are capable of replacing these traditional approaches and overcoming these constraints since they allow real-time analysis with minimal or no sample preprocessing shown in Figure 12.3. A biosensor is a combination of the physical and chemical sensing technique. According to the International Union of Pure and Applied Chemistry (IUPAC) *Gold Book* (Compendium of Chemical Terminology, 2nd edition), "[a]biosensor is a device that uses specific biochemical reactions mediated by isolated enzymes, immuno systems, tissues, organelles or whole cells to detect chemical compounds usually by electrical, thermal or optical signals." The transducer transformed the biochemical signal into a measurable electronic signal, which is directly or indirectly to the concentration of a group of analytes or a specific single analyte present in the matrix.

Due to the characteristics of high speed as well as high sensitivity, excellent specificity, and real-time remote monitoring capability in a biosensor device, it is developing as one of the most significant diagnostic techniques for food, clinical, and environmental monitoring. The most common types of transducers are electrochemical, bioluminescent, piezoelectric, calorimetric, and optical transducers. Biosensors use biomolecules for signal generation, whereas electrochemical sensors generate signals using an electric field around the biomolecule. Electrochemical sensors are further subdivided into potentiometric, amperometric, and impedance sensors. In order to improve sensor performance, the surface of the transducer can be changed using a variety of functional materials. Among these functional materials, nanomaterial-based techniques are practically convenient and sensitive analyzers that sense chemical and biological agents. There are numerous reports devoted to the functionalization of nanomaterials on the surface of CNTs for developing electrochemical and biochemical sensors. Ghazala Ashraf et al. reported the in vitro electrochemical detection of serotonin in biological matrices by using Cu_2O decorated with Pt on the surface of CNTs [50]. Using the CNT-Cu_2O-CuO@Pt nanostructure-modified electrode, a lower detection limit of 3 nM was reported for serotonin. S. S. Jyothirmayee Aravind and S. Ramaprabhu have reported SnO_2 and

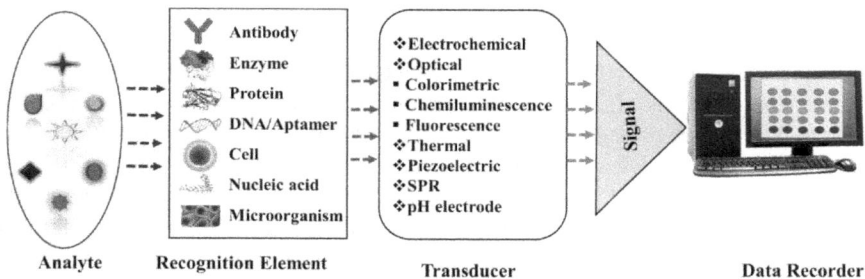

FIGURE 12.3   Schematic representation of biosensors.

ZnO nanoparticle–decorated MWCNTs for the sensing of dopamine. The synthesis of MWCNTs and the surface modification of MWCNTs were carried out by using a chemical deposition method and a chemical treatment method. The sensor showed a best limit of detection of 1 µM of dopamine using the cyclic voltammetry (CV) technique [51].

In 2012, Zhiyu Yang and co-workers developed a CuO nanoleaf–decorated MWCNTs for the detection of glucose. The hybrid nanocomposites were prepared by a simple chemical precipitation method. The amperometric signal of CuO nanoleaf–decorated MWCNT electrode showed high sensitivity with an excellent detection limit of 5.7 µM of glucose compared to other nonenzymaticglucose-sensing techniques. These excellent enhanced properties of CuO nanoleaf–decorated MWCNT sensors may be due to the unique structure and large surface area of the CuO nanoleaves present on the CNT surface [52]. Jingjing Li et al. fabricated a glucose biosensor by using Pt nanoparticle–decorated Fe_yO_x-MWCNT composite. The formation of Pt@ Fe_yO_x/MWCNT hybrid composites was done by the electrodeposition method of Pt on anFe_yO_x-MWCNT-modified glassy carbon electrode (GCE). The suggested biosensor showed a high electrocatalytic activity and a superior glucose response performance. The Pt@ Fe_yO_x/MWCNTs gave a response with a detection limit of 2.0×10^{-6} M of glucose [53]. Jaewook Lee and co-workers produced a multifunctional Au/iron-oxide-decorated CNT hybrid material for detecting virus DNA. The developed sensor detects a different category of virus DNA, such as influenza and norovirus with a detection limit of 8.4 and 8.8 pM. The detection potential of Au/MNP-CNT-based DNA sensing system was outstanding [54]. In 2021, Nandini Nataraj et al. developed a highly sensitive electrochemical sensor based on Ni-ZrO_2 nanoparticle–decorated MWCNT nanocomposite for the determination of 5-amino salicylic acid (anti-inflammatory drug). A simple coprecipitation method was involved to produce Ni-ZrO_2 nanoparticles, which further were deposited onto the surface of MWCNTs via an ultrasonication technique. The developed sensor Ni-ZrO_2@MWCNTs not only improves catalytic capabilities, but it also increases surface area, electrical conductivity, and the electron transfer process significantly. Using CV and differential pulse voltammetry (DPV), the electrochemical activity of the produced Ni-ZrO_2@ MWCNT was comprehensively examined and found to have a lower detection limit of 0.0029 µM of 5-amino salicylic acid. Furthermore, in the presence of several interfering species, the biosensor displayed outstanding repeatability, reproducibility, stability, and high specificity for 5-ASA detection [55]. Hassan Karimi-Maleh et al. in 2021 reported on a guanine-based DNA biosensor for the detection of daunorubicin (an anticancer drug). The experiment is investigated by a electrochemical technique whereby ds-DNA was fabricated on GCE amplified with Pt/SWCNT composites. The synthesis of Pt/SWCNT composites was performed by using a polyol method. The sensing result of sensor ds-DNA/Pt/SWCNT showed that the sensor has a high surface area, high conductivity, and high oxidation signal as compared to the other guanine-based DNA sensor and a favorable shift after interacting with the anticancer medication daunorubicin. A lower detection limit of 1.0 nM for determining daunorubicin was obtained for a ds-DNA/Pt/SWCNT sensor [56]. A summary of various metal oxide–decorated CNT-based bio- and electrochemical sensors is reported in Table 12.2.

TABLE 12.2

A Summary of Reported Various Metal Oxide–Decorated CNT-Based Bio-and Electrochemical Sensors

Receptor	Response Signal	Analyte	LOD	Ref.
Ferrocene-Au@ CNTs	Electrochemical	Serotonin	17 nM	[57]
Ferrocene-grated CNTs	Electrochemical	H_2O_2	0.49 μM	[58]
PQQ-decorated CNTs	Electrochemical	Tris(2-carboxyethyl) phosphine	5 pg ml^{-1}	[59]
Pt-MWCNTs nanocomposite	Electrochemical	Cell-secreted dopamine	2 nM	[60]
Nd_2O_3-SiO_2-decorated carboxylated SWCNTs	Electrochemical	L-DOPA	0.70μmol L^{-1}	[61]
SWCNT-mesoporous silicon nanocomposite	Electrochemical	Glucose	9.6 μM	[62]
3DAu@Pt-decorated GO/ MWCNT composite	Electrochemical	Glucose	4.2×10^{-8} M	[63]
Au/ZnO/MWCNTs	Electrochemical	Cholesterol	0.1 μM	[64]
Ti_3C_2/graphitized-MWCNT/ ZnO nanocomposites	Electrochemical	Dopamine	3.3 nM	[65]
Lac/Fe_3O_4/MWCNTs	Electrochemical	Guaiacol	34.3 nM	[66]
CTAB-functionalized ZnO-decorated CNTs on silver films	Electrochemical	Catechol	0.1 μM	[67]
Peptide-decorated Au np/CNT	Electrochemical	Matrix metalloproteinase-7	6 pg mL^{-1}	[68]
CdO-decorated CNT	Electrochemical	Glutathione	30.0 pM	[69]
ZnO-decorated CNT	Electrochemical	Glucose	5 μM	[70]
rGO/MWCNT-decorated with Fe_3O_4 np	Electrochemical	Hydrazine	0.75 μM	[71]
Cu-BTA@ MWCNTs	Electrochemical	H_2O_2	2.21×10^{-7} M	[72]
Co_3O_4-rGO/CNTs	Electrochemical	Nitrite	0.016 μM	[73]
AuNPs/MWCILE	Impedimetric immunosensor	HER2	7.4 ng mL^{-1}	[74]
NiO·CNT NCs	Electrochemical	4-aminophenol	15.0± 0.1 pM	[75]
Cr_2O_3-CNT NC	Electrochemical	4-methoxyphenol	0.06428 ± 0.0002 nM	[76]

* rGO—reduced graphene oxide, Lac—laccase, Np—nanoparticle, Au—gold, Cu-BTA—copper (II)/1Hbenzotriazole, CdO—cadmium oxide, Nd_2O_3—neodymium oxide, HER2—human epidermal growth factor receptor 2, AuNP/MWCILE—gold nanoparticle–decorated multiwall carbon nanotube–ionic liquid electrode, NiO·CNT NCs—nickel oxide nanoparticle–decorated carbon nanotube nanocomposites, Cr_2O_3–CNT NC—chromium (III) oxide nanomaterial–decorated carbon nanotubes.

For a variety of heavy metals, nano-enabled sensors have been successfully developed, and we look at mercury, lead, cadmium, and chromium detection in this section. To detect these environmental contaminants, a variety of transducers and nanoparticles are used, all with the goal of developing sensitive and selective sensors. Lead (Pb) is a heavy metal contaminant that has been linked to an increased risk of cancer as well as subtle cognitive and neurological deficits. For sensitive Pb(II) detection, both labeled and label-free nanosensors have been reported. The recognition element 8–17 DNA zyme, a catalytic nucleic acid, as well as a class of oligonucleotides that form G-quadruplexes in the presence of lead, has been used for label-based detection. Although there is less research on nano-enabled sensors for cadmium (Cd) detection than for mercury and lead, detection limits on the order of nanomolar have been made. Quantum dot (QD) SWCNTs and antimony nanoparticles are just a few of the nanomaterials that have been studied.

12.7 CONCLUSION AND OUTLOOK

Nanosensor development for environmental contaminants is accelerating, and nanomaterials and recognition agents are being combined in novel and creative ways, as described throughout this review. Sensor design advances in recent years have aimed to address issues like nonspecific binding, particle size variation, nanoparticle aggregation, and nanoparticle stability that plagued first-generation sensors. Because of their unique properties and structure to hold the metal oxide nanoparticle within, CNTs can be utilized as active elements of sensors to detect a variety of molecules, including gases and organic chemicals. Because of their high electric catalytic activity and quick electron transfer, as well as the exceptional stability of nanotube complexes with redox polymers, CNTs can be used as electrochemical biosensors. The addition of functional groups to the surface of CNTs, such as metal oxides dramatically increase the selectivity of the sensor's detector. Metal oxide–decorated CNTs for gas, bio- and electrochemical sensors are still a long way off, and substantial advancements is still required. The fabrication of metal oxide nanoparticles on the surface of CNTs is one of the most challenging problems, and additional research is needed to create cost-effective, scalable production processes that preserve the materials' fundamental features. Evidence that the biopersistence and pro-inflammatory action of metal oxide–coated CNTs in vivo are hindering the progress of application is another major challenge. If individuals are to be empowered to analyze their environment, field deployable sensors must be robust. As a result, more research is needed to learn as well as create awareness about how to safely handle, operate, and recycle metal oxide–decorated CNTs in gas, bio-, and chemical sensors.

ACKNOWLEDGEMENT

KKS would like to thank all the co-authors for their collaboration and helpful discussions. All the authors would like to thank the editor of the book for giving such opportunity to publish the chapter in this reputed CRC press Tylor and Francis publishing group.

CONFLICTS OF INTEREST

The authors declare no financial and commercial conflict of interest.

REFERENCES

1. Chaturvedi, M., Mishra, A., Sharma, K., Sharma, G., Saxena, G., and Singh, A. K. (2021) Emerging contaminants in wastewater: Sources of contamination, toxicity, and removal approaches. In *Emerging Treatment Technologies for Waste Management.* Springer, pp. 103–132.
2. Adu, J. K., Fafanyo, D., Orman, E., Ayensu, I., Amengor, C. D., and Kwofie, S. (2020) Assessing metal contaminants in milled maize products available on the Ghanaian market with atomic absorption spectrometry and instrumental neutron activation analyser techniques. *Food Control* 109, 106912.
3. Lai, Y., Dong, L., Li, Q., Li, P., Hao, Z., Yu, S., and Liu, J. (2021) Counting nanoplastics in environmental waters by single particle inductively coupled plasma mass spectroscopy after cloud-point extraction and in situ labeling of gold nanoparticles. *Environmental Science & Technology* 55, 4783–4791.
4. Ebrahimi-Najafabadi, H., Pasdaran, A., Bezenjani, R. R., and Bozorgzadeh, E. (2019) Determination of toxic heavy metals in rice samples using ultrasound assisted emulsification microextraction combined with inductively coupled plasma optical emission spectroscopy. *Food Chemistry* 289, 26–32.
5. Astolfi, M. L., Conti, M. E., Ristorini, M., Frezzini, M. A., Papi, M., Massimi, L., and Canepari, S. (2021) An analytical method for the biomonitoring of mercury in bees and beehive products by cold vapor atomic fluorescence spectrometry. *Molecules* 26, 4878.
6. De Pra, M., Greco, G., Krajewski, M. P., Martin, M. M., George, E., Bartsch, N., and Steiner, F. (2020) Effects of titanium contamination caused by iron-free high-performance liquid chromatography systems on peak shape and retention of drugs with chelating properties. *Journal of Chromatography A* 1611, 460619.
7. Wu, C.-H., Jiang, S.-J., and Sahayam, A. (2018) Using electrothermal vaporization inductively coupled plasma mass spectrometry to determine S, As, Cd, Hg, and Pb in fuels. *Spectrochimica Acta Part B: Atomic Spectroscopy* 147, 115–120.
8. Zargar, R. A., Kumar, K., Arora, M., Shkir, M., Somaily, H. H., Algarni, H., and Al Faify, S. (2022) Structural, optical, photoluminescence, and EPR behaviour of novel $Zn_{0.80}Cd_{0.20}O$ thick films: An effect of different sintering temperatures. *Journal of Luminescence* 245, 118769.
9. Soni, S. K., Thomas, B., and Kar, V. R. (2020) A comprehensive review on CNTs and CNT-reinforced composites: Syntheses, characteristics and applications. *Materials Today Communications* 25, 101546.
10. Zargar, R. A., Hassan, M. M., Kumar, K., Nagal, V., Bashir, A., Alshahrani, B., Alshahrani, T., and Shkir, M. (2021) Development and characterization of (ZnO)0.90(CNT)0.10 thick film for photovoltaic application. *Optik* 248, 167975.
11. Eatemadi, A., Daraee, H., Karimkhanloo, H., Kouhi, M., Zarghami, N., Akbarzadeh, A., Abasi, M., Hanifehpour, Y., and Joo, S. W. (2014) Carbon nanotubes: Properties, synthesis, purification, and medical applications. *Nanoscale Research Letters* 9, 393.
12. Manawi, Y. M., Samara, A., Al-Ansari, T., and Atieh, M. A. (2018) A review of carbon nanomaterials' synthesis via the chemical vapor deposition (CVD) method. *Materials* 11, 822.
13. Wang, X., Vinodgopal, K., and Dai, G. (2019) Synthesis of carbon nanotubes by catalytic chemical vapor deposition. In H. E. Saleh, and S. M. M. El-Sheikh (eds.), *Perspective of Carbon Nanotubes.* IntechOpen.

14. Arora, N., and Sharma, N. N. (2014) Arc discharge synthesis of carbon nanotubes: Comprehensive review. *Diamond and Related Materials* 50, 135–150, ISSN 0925-9635.

15. Jagadeesan, A. K., Thangavelu, K., and Dhananjeyan, V. (2020) Carbon nanotubes: Synthesis, properties and applications. In P. Pham, P. Goel, S. Kumar, and K. Yadav (eds.), *21st Century Surface Science—A Handbook*. IntechOpen.

16. Ismail, R. A., Mohsin, M. H., Ali, A. K., Hassoon, K. I., and Erten-Ela, S. (2020) Preparation and characterization of carbon nanotubes by pulsed laser ablation in water for optoelectronic application. *Physica E: Low-dimensional Systems and Nanostructures* 119, 113997.

17. Lu, Z., Raad, R., Safaei, F., Xi, J., Liu, Z., and Foroughi, J. (2019) Carbon nanotube based fiber supercapacitor as wearable energy storage. *Frontiers in Materials*, 6, 138.

18. Díez-Pascual, A. M. (2021) Chemical functionalization of carbon nanotubes with polymers: A brief overview. *Macromol* 1(2), 64–83.

19. Liu, X., Marangon, I., Melinte, G., Wilhelm, C., Menard-Moyon, C., Pichon, B. P., Ersen, O., Aubertin, K., Baaziz, W., Pham-Huu, C., and Begin-Colin, S. (2014) Design of covalently functionalized carbon nanotubes filled with metal oxide nanoparticles for imaging, therapy, and magnetic manipulation. *ACS Nano*, 8(11), 11290–11304.

20. Tobias, G., Mendoza, E., and Ballesteros, B. (2016) Functionalization of carbon nanotubes. In B. Bhushan (ed.), *Encyclopedia of Nanotechnology*. Springer.

21. Duc Chinh, V., Speranza, G., Migliaresi, C., Van Chuc, N., Minh Tan, V., and Phuong, N. T. (2019) Synthesis of gold nanoparticles decorated with multiwalled carbon nanotubes (Au-MWCNTs) via cysteaminium chloride functionalization. *Scientific Reports* 9, 5667. https://doi.org/10.1038/s41598-019-42055-7.

22. Zargar, R. A., Arora, M., and Bhat, R. A. (2018) Study of nanosized copper-doped ZnO dilute magnetic semiconductor thick films for spintronic device applications. *Applied Physics A* 124, 1–9.

23. Balram, D., Lian, K. Y., and Sebastian, N. (2020) Ultrasound-assisted synthesis of 3D flower-like zinc oxide decorated fMWCNTs for sensitive detection of toxic environmental pollutant 4-nitrophenol. *Ultrasonicssonochemistry* 60, 104798.

24. Chakraborty, S., Simon, R., Antonia Trisha Zac, R., Anoop, V., and Mary, N. L. (2022) Microwave-assisted synthesis of ZnO decorated acid functionalized carbon nanotubes with improved specific capacitance. *Journal of Applied Electrochemistry* 52, 1–12.

25. Jung, D., Kim, J., and Lee, G. S. (2015) Enhanced humidity-sensing response of metal oxide coated carbon nanotube. *Sensors and Actuators A: Physical*, 223, 11–17.

26. Lee, J., Mulmi, S., Thangadurai, V., and Park, S. S. (2015) Magnetically aligned iron oxide/gold nanoparticle-decorated carbon nanotube hybrid structure as a humidity sensor. *ACS Applied Materials & Interfaces* 7(28), 15506–15513.

27. Leghrib, R., Pavelko, R., Felten, A., Vasiliev, A., Cané, C., Gràcia, I., Pireaux, J. J., and Llobet, E. (2010) Gas sensors based on multiwall carbon nanotubes decorated with tin oxide nanoclusters. *Sensors and Actuators B: Chemical* 145(1), 411–416.

28. Lee, H., Lee, S., Kim, D. H., Perello, D., Park, Y. J., Hong, S. H., Yun, M., and Kim, S. (2012) Integrating metal-oxide-decorated CNT networks with a CMOS readout in a gas sensor. *Sensors* 12(3), 2582–2597.

29. Casanova-Cháfer, J., Navarrete, E., Noirfalise, X., Umek, P., Bittencourt, C., and Llobet, E. (2018) Gas sensing with iridium oxide nanoparticle decorated carbon nanotubes. *Sensors* 19(1), 113.

30. Lee, J., Morita, M., Takemura, K., and Park, E. Y. (2018) A multi-functional gold/iron-oxide nanoparticle-CNT hybrid nanomaterial as virus DNA sensing platform. *Biosensors and Bioelectronics* 102, 425–431.

31. Seekaew, Y., Wisitsoraat, A., Phokharatkul, D., and Wongchoosuk, C. (2019) Room temperature toluene gas sensor based on TiO2 nanoparticles decorated 3D graphene-carbon nanotube nanostructures. *Sensors and Actuators B: Chemical*, 279, 69–78.

32. Choi, M. S., Bang, J. H., Mirzaei, A., Na, H. G., Kwon, Y. J., Kang, S. Y., Choi, S. W., Kim, S. S., and Kim, H. W. (2018) Dual sensitization of MWCNTs by co-decoration with p-and n-type metal oxide nanoparticles. *Sensors and Actuators B: Chemical* 264, 150–163.

33. Jung, D., Kim, J., and Lee, G. S. (2015) Enhanced humidity-sensing response of metal oxide coated carbon nanotube. *Sensors and Actuators A: Physical* 223, 11–17.

34. Zargar, R. A., Arora, M., Alshahrani, T., and Shkir, M. (2021) Screen printed novel ZnO/MWCNTs nanocomposite thick films. *Ceramics International* 47(5), 6084–6093.

35. Seekaew, Y., Wisitsoraat, A., Phokharatkul, D., and Wongchoosuk, C. (2019) Room temperature toluene gas sensor based on TiO2 nanoparticles decorated 3D graphene-carbon nanotube nanostructures. *Sensors and Actuators B: Chemical* 279, 69–78.

36. Chimowa, G., Tshabalala, Z. P., Akande, A. A., Bepete, G., Mwakikunga, B., Ray, S. S., and Benecha, E. M. (2017) Improving methane gas sensing properties of multi-walled carbon nanotubes by vanadium oxide filling. *Sensors and Actuators B: Chemical* 247, 11–18.

37. Zhang, D., Liu, J., Jiang, C., Liu, A., and Xia, B. (2017) Quantitative detection of formaldehyde and ammonia gas via metal oxide-modified graphene-based sensor array combining with neural network model. *Sensors and Actuators B: Chemical* 240, 55–65.

38. Casanova-Cháfer, J., Navarrete, E., Noirfalise, X., Umek, P., Bittencourt, C., and Llobet, E. (2018) Gas sensing with iridium oxide nanoparticle decorated carbon nanotubes. *Sensors* 19(1), 113.

39. Young, S. J., Liu, Y. H., Lin, Z. D., Ahmed, K., Shiblee, M. N. I., Romanuik, S., Sekhar, P. K., Thundat, T., Nagahara, L., Arya, S., and Ahmed, R. (2020) Multi-walled carbon nanotubes decorated with silver nanoparticles for acetone gas sensing at room temperature. *Journal of The Electrochemical Society*, 167(16), 167519.

40. Kwon, Y. J., Mirzaei, A., Kang, S. Y., Choi, M. S., Bang, J. H., Kim, S. S., and Kim, H. W. (2017) Synthesis, characterization and gas sensing properties of ZnO-decorated MWCNTs. *Applied Surface Science* 413, 242–252.

41. Seekaew, Y., Wisitsoraat, A., Phokharatkul, D., and Wongchoosuk, C. (2019) Room temperature toluene gas sensor based on TiO2 nanoparticles decorated 3D graphene-carbon nanotube nanostructures. *Sensors and Actuators B: Chemical* 279, 69–78.

42. Marichy, C., Donato, N., Latino, M., Willinger, M. G., Tessonnier, J. P., Neri, G., and Pinna, N. (2014) Gas sensing properties and p-type response of ALD TiO2 coated carbon nanotubes. *Nanotechnology* 26(2), 024004.

43. Kwon, Y. J., Na, H. G., Kang, S. Y., Choi, S. W., Kim, S. S., and Kim, H. W. (2016) Selective detection of low concentration toluene gas using Pt-decorated carbon nanotubes sensors. *Sensors and Actuators B: Chemical* 227, 157–168.

44. Mendoza, F., Hernández, D. M., Makarov, V., Febus, E., Weiner, B. R., and Morell, G. (2014) Room temperature gas sensor based on tin dioxide-carbon nanotubes composite films. *Sensors and Actuators B: Chemical* 190, 227–233.

45. Evans, G. P., Buckley, D. J., Skipper, N. T., and Parkin, I. P. (2014) Single-walled carbon nanotube composite inks for printed gas sensors: Enhanced detection of NO 2, NH 3, EtOH and acetone. *RSC Advances* 4(93), 51395–51403.

46. Asad, M., Sheikhi, M. H., Pourfath, M., and Moradi, M. (2015) High sensitive and selective flexible H2S gas sensors based on Cu nanoparticle decorated SWCNTs. *Sensors and Actuators B: Chemical* 210, 1–8.

47. Yaqoob, U., Uddin, A. I., and Chung, G. S. (2016) A high-performance flexible NO2 sensor based on WO3 NPs decorated on MWCNTs and RGO hybrids on PI/PET substrates. *Sensors and Actuators B: Chemical* 224, 738–746.

48. Fort, A., Panzardi, E., Al-Hamry, A., Vignoli, V., Mugnaini, M., Addabbo, T., and Kanoun, O. (2019) Highly sensitive detection of NO2 by Au and TiO2 nanoparticles decorated SWCNTs sensors. *Sensors* 20(1), 12.

49. Hua, C., Shang, Y., Wang, Y., Xu, J., Zhang, Y., Li, X., and Cao, A. (2017) A flexible gas sensor based on single-walled carbon nanotube-Fe2O3 composite film. *Applied Surface Science* 405, 405–411.

50. Ashraf, G., Asif, M., Aziz, A., Iftikhar, T., and Liu, H. (2021) Rice-spikelet-like copper oxide decorated with platinum stranded in the CNT network for electrochemical in vitro detection of serotonin. *ACS Applied Materials & Interfaces* 13(5), 6023–6033.

51. Aravind, S. J., and Ramaprabhu, S. J. N. M. (2012) Dopamine biosensor with metal oxide nanoparticles decorated multi-walled carbon nanotubes. *Nanoscience Methods* 1(1), 102–114.

52. Yang, Z., Feng, J., Qiao, J., Yan, Y., Yu, Q., and Sun, K. (2012) Copper oxide nanoleaves decorated multi-walled carbon nanotube as platform for glucose sensing. *Analytical Methods*, 4(7), 1924–1926.

53. Li, J., Yuan, R., Chai, Y., and Che, X. (2010). Fabrication of a novel glucose biosensor based on Pt nanoparticles-decorated iron oxide-multiwall carbon nanotubes magnetic composite. *Journal of Molecular Catalysis B: Enzymatic*, 66(1-2), pp.8-14.

54. Lee, J., Morita, M., Takemura, K., and Park, E. Y. (2018). A multi-functional gold/iron-oxide nanoparticle-CNT hybrid nanomaterial as virus DNA sensing platform. *Biosensors and Bioelectronics*, 102, 425–431.

55. Nataraj, N., Krishnan, S. K., Chen, T. W., Chen, S. M., and Lou, B. S. (2021) Ni-doped ZrO₂ nanoparticles decorated MW-CNT nanocomposite for the highly sensitive electrochemical detection of 5-amino salicylic acid. *Analyst* 146(2), 664–673.

56. Karimi-Maleh, H., Alizadeh, M., Orooji, Y., Karimi, F., Baghayeri, M., Rouhi, J., Tajik, S., Beitollahi, H., Agarwal, S., Gupta, V. K., Rajendran, S., Rostamnia, S., Fu, L., Saberi-Movahed, F., and Malekmohammadi, S. (2021) A multi-functional gold/iron-oxide nanoparticle-CNT hybrid nanomaterial as virus DNA sensing platform. *Journal of Theoretical and Applied Physics* 14, 339–348.

57. Wu, B., Yeasmin, S., Liu, Y. and Cheng, L.J., (2022) Sensitive and selective electrochemical sensor for serotonin detection based on ferrocene-gold nanoparticles decorated multiwall carbon nanotubes. *Sensors and Actuators B: Chemical*, 354, p. 131216.

58. Wu, B., Yeasmin, S., Liu, Y. and Cheng, L.J., (2022) Ferrocene-grafted carbon nanotubes for sensitive non-enzymatic electrochemical detection of hydrogen peroxide. *Journal of Electroanalytical Chemistry*, 908, p. 116101.

59. Ma, X., Deng, D., Xia, N., Hao, Y. and Liu, L., (2021) Electrochemical immunosensors with PQQ-decorated carbon nanotubes as signal labels for electrocatalytic oxidation of tris (2-carboxyethyl) phosphine. *Nanomaterials*, 11(7), 1757.

60. Li, J., Huang, X., Shi, W., Jiang, M., Tian, L., Su, M., Wu, J., Liu, Q., Yu, C. and Gu, H., (2021) Pt nanoparticle decorated carbon nanotubes nanocomposite based sensing platform for the monitoring of cell-secreted dopamine. *Sensors and Actuators B: Chemical*, 330, p. 129311.

61. Đurđić, S., Stanković, V., Vlahović, F., Ognjanović, M., Kalcher, K., Manojlović, D., Mutić, J. and Stanković, D.M., (2021) Carboxylated single-wall carbon nanotubes decorated with SiO2 coated-Nd2O3 nanoparticles as an electrochemical sensor for L-DOPA detection. *Microchemical Journal*, 168, p. 106416.

62. Ahmed, J., Rashed, M. A., Faisal, M., Harraz, F. A., Jalalah, M., and Alsareii, S. A. (2021) Novel SWCNTs-mesoporous silicon nanocomposite as efficient non-enzymatic glucose biosensor. *Applied Surface Science* 552, 149477.

63. Wang, R., Liu, X., Zhao, Y., Qin, J., Xu, H., Dong, L., Gao, S., and Zhong, L. (2022) Novel electrochemical non-enzymatic glucose sensor based on 3D Au@Pt core–shell nanoparticles decorated graphene oxide/multi-walled carbon nanotubes composite. *Microchemical Journal* 174, 107061.

64. Ghanei Agh Kaariz, D., Darabi, E., and Elahi, S. M. (2020) Fabrication of Au/ZnO/MWCNTs electrode and its characterization for electrochemical cholesterol biosensor. *Journal of Theoretical and Applied Physics* 14, 339–348.

65. Ni, M., Chen, J., Wang, C., Wang, Y., Huang, L., Xiong, W., Zhao, P., Xie, Y., and Fei, J. (2022) A high-sensitive dopamine electrochemical sensor based on multi-layer Ti3C2 MXene, graphitized multi-walled carbon nanotubes and ZnO nanospheres. *Microchemical Journal*, 178, 107410.

66. Uc-Cayetano, E., Villanueva-Mena, I., Estrella-Gutiérrez, M., Ordóñez, L., Aké-Uh, O., and Sánchez-González, M. (2020) Study of amperometric response of guaiacol biosensor using multiwalled carbon nanotubes with laccase immobilized. *ECS Journal of Solid State Science and Technology* 9, 115009.

67. Pathak, A., and Gupta, B. D. (2020) Fiber-optic plasmonic sensor utilizing CTAB-functionalized ZnO nanoparticle-decorated carbon nanotubes on silver films for the detection of catechol in wastewater ACS. *Applied Nano Materials* 3, 2582–2593.

68. Palomar, Q., Xu, X., Selegård, R., Aili, D., and Zhang, Z. (2020) Peptide decorated gold nanoparticle/carbon nanotube electrochemical sensor for ultrasensitive detection of matrix metalloproteinase-7. *Sensors and Actuators B: Chemical* 325, 128789.

69. Rahman, M. M., Hussain, M. M., and Asiri, A. M. (2016) A glutathione biosensor based on a glassy carbon electrode modified with CdO nanoparticle-decorated carbon nanotubes in a nafion matrix. *Microchimica Acta* 183, 3255–3263.

70. Ibrahim, A. A., Umar, A., Ahmad, R., Kumar, R., and Baskoutas, S. (2016) Fabrication and characterization of highly sensitive and selective glucose biosensor based on ZnO decorated carbon nanotubes. *Nanoscience and Nanotechnology Letters* 8, 853–858.

71. Nehru, S., Sakthinathan, S., Tamizhdurai, P., Chiu, T.-W., and Shanthi, K. (2020) Reduced graphene oxide/multiwalled carbon nanotube composite decorated with Fe3O4 magnetic nanoparticles for electrochemical determination of hydrazine in environmental water. *Journal of Nanoscience and Nanotechnology* 20, 3148–3156.

72. Yang, Y. J., and Li, W. (2017) Multiwalled carbon nanotubes decorated with copper(II)/1H-benzotriazole complex via π–π interaction for enhanced electroreduction of hydrogen peroxide. *Sensor Letters* 15, 639–646.

73. Zhao, Z., Zhang, J., Wang, W., Sun, Y., Li, P., Hu, J., Chen, L., and Gong, W. (2019) Synthesis and electrochemical properties of Co3O4-rGO/CNTs composites towards highly sensitive nitrite detection. *Applied Surface Science* 485, 274–282.

74. Arkan, E., Saber, R., Karimi, Z., and Shamsipur, M. (2015) A novel antibody–antigen based impedimetric immunosensor for low level detection of HER2 in serum samples of breast cancer patients via modification of a gold nanoparticles decorated multiwall carbon nanotube-ionic liquid electrode. *Analytica Chimica Acta* 874, 66–74.

75. Hussain, M. M., Rahman, M. M., and Asiri, A. M. (2017) Ultrasensitive and selective 4-aminophenol chemical sensor development based on nickel oxide nanoparticles decorated carbon nanotube nanocomposites for green environment. *Journal of Environmental Sciences* 53, 27–38.

76. Rahman, M. M., Balkhoyor, H. B., and Asiri, A. M. (2017) Phenolic sensor development based on chromium oxide-decorated carbon nanotubes for environmental safety. *Journal of Environmental Management* 188, 228–237.

Index

For Product Safety Concerns and Information please contact our EU
representative GPSR@taylorandfrancis.com
Taylor & Francis Verlag GmbH, Kaufingerstraße 24, 80331 München, Germany